*Introduction
to Classical and Modern*
OPTICS

4TH EDITION

Introduction
to Classical and Modern
OPTICS

JURGEN R. MEYER-ARENDT, M.D.
Pacific University

Prentice Hall, Englewood Cliffs, New Jersey 07632

Library of Congress Cataloging-in-Publication Data

Meyer-Arendt, Jurgen R.
  Introduction to classical and modern optics/Jurgen R. Meyer
-Arendt.—4th ed.
    p.   cm.
  Includes bibliographical references and index.
  ISBN 0-13-124356-X
  1. Optics.   I. Title
QC355.2.M49   1995   94-30562
535—dc20   CIP

Acquisitions editor: *Ray Henderson*
Editorial/production supervision
  and interior design: *Kathleen M. Lafferty*
Proofreader: *Bruce D. Colegrove*
Cover designer: *Karen Salzbach*
Manufacturing buyer: *Trudy Pisciotti*

Published by Prentice-Hall, Inc.
A Simon & Schuster Company
Englewood Cliffs, New Jersey 07632

ISBN 0-13-124356-X

90000

9 780131 243569

Printed in the United States of America

10 9 8 7 6 5 4 3 2

ISBN 0-13-124356-X

Prentice-Hall International (UK) Limited, *London*
Prentice-Hall of Australia Pty. Limited, *Sydney*
Prentice Hall Canada Inc., *Toronto*
Prentice-Hall Hispanoamericana, S.A., *Mexico*
Prentice-Hall of India Private Limited, *New Delhi*
Prentice-Hall of Japan, Inc., *Tokyo*
Simon & Schuster Asia Pte. Ltd., *Singapore*
Editora Prentice-Hall do Brasil, Ltda., *Rio de Janeiro*

Contents

PART 2
WAVE OPTICS

PART 3
QUANTUM OPTICS

Preface

The purpose and emphasis in this fourth edition remain the same as before: to provide a concise, and still readable, *Introduction to Classical and Modern Optics,* written for the advanced undergraduate and for a course spanning two semesters or the equivalent.

Compared with the earlier editions, I have again made major changes. I continue, however, using the rational *Cartesian sign convention,* long familiar from ray tracing and from ophthalmic optics but sometimes slighted in physics. This convention is essential for the concept of *vergence* and mandatory in any *computer-aided applications.* This dual connection identifies the two groups of readers to which this *Introduction to Classical and Modern Optics* is addressed in particular: those interested in the *scientific and engineering applications of optics* and those preparing for the *ophthalmic professions.*

The field of optics is often divided into geometrical optics and physical optics. Actually, there is no dichotomy. Optics is an indivisible whole. Consider only what seems to be a simple example: image formation. Image formation can be discussed from a purely geometrical point of view, but that is the old approach dating back hundreds of years. Today we discuss image formation in terms of diffraction and Fourier transformation; that is the modern approach.

Part 1, Geometrical Optics, and the first several chapters open with the same combination of three lenses, a *triplet.* Each time, the light progresses a little further: from refraction at a single surface to thin lenses to a combination of lenses to the use of matrix algebra, aberrations, stops and pupils, gradient-index optics, and lens design. Lens design in particular shows how optics is handled today, using a detailed example and computer spreadsheets.

Part 2, Wave Optics, begins with "Interference," which covers double-slit interference, wave phenomena in general, and other types of interference. "Thin Films" and "Diffraction" follow. Maxwell's equations have become part of "Light Scattering" and Fourier transforms part of "Optical Data Processing." "Holography" discusses a multitude of examples, from point-of-sale scanners to pattern recognition.

"Light Sources and Detectors" opens Part 3, Quantum Optics, followed by "Photometry" and "Absorption." "Lasers," of course, take a prominent place, with emphasis on their theory, their types, their applications, and the necessary safety precautions. "Relativistic Optics" points to the future.

As before, ample space has been allocated to classical topics such as thin lenses, optical systems, and polarization. But modern subjects such as gradient-index optics, antireflection coatings, polarization-based modulators, and light-emitting diodes are also presented in detail.

As a prerequisite, all that is needed is a background in physics and a working knowledge of how to use a hand-held calculator.

Numerous worked-out examples, from optical path length to relativistic reflection, are presented in detail throughout the text. None of the examples or problems is intended as "busy work"; some of them, in fact, were drawn from consulting work I do from time to time. A summary of equations, problems, and suggestions for further reading are found at the end of each chapter. The problems are arranged in the order in which they occur in the text and in terms of increasing difficulty. Helpful hints accompany some of them.

Many of the examples and problems are new. All have been use-tested extensively and modified where needed. (Answers to the odd-numbered problems are found in the back of the book.) Lecture demonstrations are referred to on occasion. A great many historical footnotes have been included to show the human side of optics' great masters, adding color to the description of their accomplishments.

I have enjoyed writing this new edition. It took hundreds of hours of lecturing. It also took the patience of a great many students, questioning, challenging, attentive, and at times not so attentive, for me to discover—often by trial and error—how to get a concept across. How do I best present the idea of principal planes, the use of Cornu's spiral, or the essence of holography? How do I present the wide field of optics at a reasonable level and within a reasonable length of time? There is no final answer to that, just steps of successive improvement.

ACKNOWLEDGMENTS

It is a pleasure to acknowledge the assistance I have received from others. Many of my colleagues have helped, in ways large and small, with the preparation and revision of this new edition. Much appreciation also goes to my students. But comments, both laudatory and critical, have also come from readers I have never met. In particular, I wish to thank Joel H. Blatt, Willard B. Bleything, James K. Boger, Harold A. Daw, Eugene D. Farley, Yukap Hahn, Jerald R. Izatt, Michael S. Mason, Daniel J. McLaughlin, Sergio L. Monteiro, Brian J. Thompson, Patricia R. Wakeling, and Lesley Walls. I also express my gratitude to the editorial and production staff of Prentice-Hall, Inc., especially Ray Henderson, acquisitions editor, and Kathleen Lafferty, production editor; because of them my manuscript became the book it now is.

Jurgen R. Meyer-Arendt

Introduction

I begin this introduction to optics by presenting right away a practical problem. Look at Figure 1. This is a combination of lenses, a lens *system*, containing a positive lens in front (on the left), a negative lens behind it, and another positive lens in the back (on the right). These lenses have certain surface characteristics, they have certain thicknesses, and they are certain distances apart. Probably, they are made out of different types of glass.

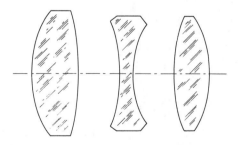

Figure 1 Combination of three lenses, used for introducing various aspects of optics.

How does the system work? Why are the lenses of the type shown preferable to other lenses? Why is this system superior to other systems? These are questions that we are not yet ready to answer. We will use this system as an example and a guide to introduce many of the concepts of optics.

Some phenomena in optics are easy to see. Others are very subtle. Look at a street lamp through the fabric of an open umbrella. You will see light *fans*, extending in various directions. These are due to diffraction. Or look at a fairly bright star (without the umbrella). You may see light fans or "points" that seem to emanate from the star. In reality, there are no such points. They are merely due to occasional straight segments in the otherwise round pupil of the eye caused by a hardening of the smallest arteries, arteriolosclerosis.

And so, in the art of all ages, stars have traditionally been represented as objects with a multitude of points. However, a star with an odd number of points, as shown in Figure 2, is wave-optically impossible because the light fans must, by necessity, always occur in pairs.

Figure 2 Artist's conception of a star. Stained-glass window in St. John's Church, Herford, Germany, fourteenth century.

Optics is a field of science that is particularly lucid, logical, challenging, and beautiful. Most of our appreciation of the outside world—nature, art—comes to us through light. Let us now discuss its many aspects.

Reflection
and Refraction

Look at sunlight that, on a misty morning, breaks through the dense foliage of a tree. The light, made visible by the moisture in the air, travels along straight lines called *rays*. Rays are characterized by *rectilinear propagation*. But, why are the rays in Figure 1-1 diverging? That is merely an illusion; in reality, the rays are parallel, just as railroad tracks are parallel even if they seem to converge toward the horizon.

PROPAGATION OF LIGHT

Indeed, light can often be considered to consist of *rays*. At other times we need to think of light as a sequence of *waves*. In that case, we describe the properties of light using terms that also pertain to waves in general.

Consider the waves that move across a lake. These waves may have different *heights*. The height of a wave can be expressed in terms of *amplitude*, which is the height of the wave above the average level, as in Figure 1-2.

Another property is the *length* of the wave, the distance from the crest of one wave to the crest of the next. These lengths are called *wavelengths*, λ. Wavelengths are measured in the same units as length in general. The basic unit of length in the Système International d'Unités, the International System of Units,

Figure 1-1   Sunlight passing through the foliage of a tree.

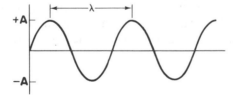

Figure 1-2   Amplitude, **A**, and wavelength, λ, of a sinusoidal wave.

SI, is the meter, m. But since the wavelength of light is rather small, fractional units of a meter are needed. These are:

1 millimeter, mm, 10^{-3} m. 1 mm = 1000 μm.

1 mi'crometer, μm, 10^{-6} m, 1/1000 of a millimeter. The prefix *micron*, μ, alone should not be used although it often is. (Do not confuse with micro'-meter, accent on the *o*, which is an instrument for measuring small distances.)

1 nanometer, nm, 10^{-9} m, 1/1000 of a micrometer. This is the *preferred unit of wavelength of light* (in the visible part of the spectrum).

1 Ångström, Å, = 0.1 nm.

Frequency, ν, is the number of oscillations per second. The unit of frequency is the *hertz*, Hz.* Think again of a sequence of water waves that pass a given point, making perhaps two complete (up and down) oscillations per second. Their frequency, therefore, is

$$\nu = \frac{2 \text{ osc}}{1 \text{ sec}} = 2 \text{ Hz}$$

At which *velocity* will these waves move? Assume that the waves, from crest to crest, measure 75 cm, $\lambda = 0.75$ m. If two such waves per second pass the point, their velocity of propagation is

$$(0.75 \text{ m})(2 \text{ Hz}) = 1.5 \text{ m/s}$$

which shows that the velocity of a wave is the product of wavelength and frequency,

$$\boxed{v = \lambda \nu}$$ [1-1]

This is an equation that holds for any wave.

The Electromagnetic Spectrum

Electromagnetic waves, in general, extend across a large range of frequencies. They form a continuous *spectrum*. Light is part of this spectrum. The wavelength of visible light extends from about 380 nm for violet-blue to about 750 nm for deep red. "Pure" blue has a peak wavelength of about 475 nm, green of about 520 nm, yellow 580 nm, and red 630 nm. Below 380 nm there is the *ultraviolet*, UV, above 750 nm the *infrared*, IR (Figure 1-3).

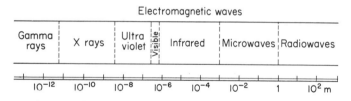
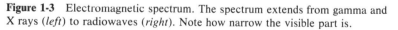

Figure 1-3 Electromagnetic spectrum. The spectrum extends from gamma and X rays (*left*) to radiowaves (*right*). Note how narrow the visible part is.

X rays have wavelengths from as short as 10^{-12} m to about 50 nm. Ultraviolet of less than 200 nm is absorbed by air and therefore is called *vacuum ultraviolet*. *Microwaves* extend from 1 mm to 30 cm and *radio waves* from there up to perhaps 30 km wavelength.

---

* Named after Heinrich Rudolf Hertz. When in 1924 the German Physical Society proposed the name "hertz" for the unit cycles per second, Walter Hermann Nernst (1864–1941), German physical chemist, objected, saying: "I do not see the need for introducing a new name; by the same reasoning one could as well call one liter per second a 'falstaff'."

Velocity of Light

In general, waves can have different velocities, from very slow to very fast. Light represents a type of wave that travels very fast; indeed, there is nothing in our daily experience to suggest that the propagation of light is not instantaneous or its speed infinite. Still, it seems that Galileo was the first to suggest that it may take light a finite time to travel from one point to another. In his book *Discorsi*, he relates his thoughts using two fictitious characters, Sagredo and a somewhat foolish man, Simplicio. Here is part of what they say:

> SIMPLICIO: Everyday experience shows that the propagation of light is instantaneous; for when we see a piece of artillery fired, at a great distance, the flash reaches our eyes without delay, but the sound reaches the ear only after a noticeable interval.
> SAGREDO: Well, Simplicio, the only thing I can infer from this is that sound, in reaching our ears, travels more slowly than light; it does not inform me whether light travels instantaneous or whether, although very fast, it still takes time.

Since then, over several centuries, the technology of determining the speed of light has gradually progressed, using at first astronomical and then, as more precise time-of-flight measurements became possible, direct terrestrial methods.

Today we know that the velocity of light in a vacuum is the same at all wavelengths: *The velocity of light is a fundamental constant of nature*. In fact, as we will see later in more detail, the special theory of relativity requires the velocity of light to be a universal constant. The velocity of light, therefore, has become a *defined* quantity; its magnitude, adopted by international agreement, is

$$c = 299\ 792\ 458 \text{ m/s} \qquad\qquad [1\text{-}2]$$

Conversely, the meter has become a *derived* quantity; it is defined as "the length of the path traveled by light in a vacuum during a time interval of 1/299 792 458 of a second."* This means that there is no need anymore to rely on a laboriously calibrated meter bar; the standard meter can be reproduced using the speed of light, and any future refinement would change, not the velocity of light, but the length of the meter.

---

* Definition adopted by the 17th General Conference on Weights and Measures (Conférence Générale des Poids et Mesures, CGPM), Paris, France, 1983.

All these limits are didactic and for tabulation only. They do not represent division lines. The properties of one category merge with those of the next, and the methods of production and detection overlap.

Shadows

One of the consequences of the rectilinear propagation of light is the formation of *shadows*. With a single point source, an opaque object will cast a uniformly dark shadow. Outside the shadow the screen is fully illuminated [Figure 1-4 (top)]. With two point sources, there is a central dark shadow or *umbra*, and next to it

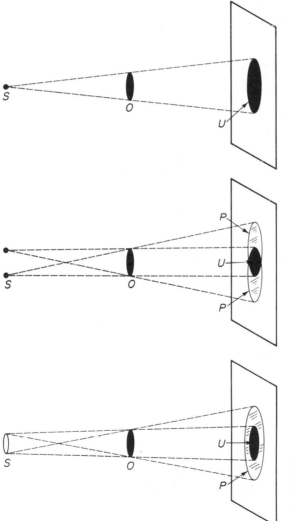

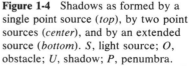

Figure 1-4 Shadows as formed by a single point source (*top*), by two point sources (*center*), and by an extended source (*bottom*). *S*, light source; *O*, obstacle; *U*, shadow; *P*, penumbra.

two partial shadows or *penumbras*, each of the penumbras receiving additional light from the source not casting the shadow (center). With an extended source, the penumbra assumes the shape of a ring, surrounding the umbra (bottom).

Optical Path Length

Assume that light travels from one point, *A*, to another point, *B*. The distance as such has nothing to do with the medium that may be present between *A* and *B*. For example, if the path is in air and the distance from *A* to *B* is *L*, then, if the path were in water, the distance would still be *L*.

Now let *light* travel from *A* to *B*. In air it takes the light a certain time to go from *A* to *B*. If the path is in water, however, it takes the light *longer*. The reason is that the light interacts with, and is hindered by, the molecules of the water. Therefore, another "length" (in place of *L*) is needed to account for the delay. This length is the *optical path length*, *S*, the product of length (distance) and *refractive index*, *n*,

$$\boxed{S = Ln} \qquad [1\text{-}3]$$

The term *refractive index* is of the greatest significance. It is closely related to the velocity of the light. In free space the velocity of light is highest. In matter, such as in water or glass, it is less. The ratio of the two velocities, the velocity in free space, *c*, to the velocity of light in matter, *v*, is the refractive index, *n*, of the particular matter,

$$n = \frac{c}{v} \qquad [1\text{-}4]$$

Because of the higher refractive index (of matter), light passing through matter seems to go through a *longer optical path*. This is shown schematically in Figure 1-5 (top).

But then look at the situation from the opposite point of view. An observer may view an object *A* through a medium of index *n* (bottom). In reality the object is in plane *A*, but to the observer the object *appears* to be in plane *A'*. These two distances, the optical path length in air, Ln_0, and the optical path length in matter, $L'n$, must be the same:

$$Ln_0 = L'n$$

and therefore, since the index of air $n_0 \approx 1$,

$$\boxed{L' = \frac{L}{n}} \qquad [1\text{-}5]$$

where *L'* is called the *reduced distance*: the higher the index, the shorter the reduced distance.

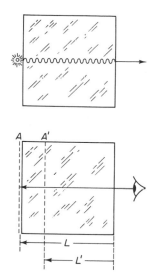

Figure 1-5 While light passing through a block of glass encounters a *longer* optical path (*top*), to an observer it appears as if *A* has moved to *A'* and hence is *closer* to the eye (*bottom*).

Refractive index is a dimensionless quantity. It has no units. Consequently, the actual distance, the optical path length, and the reduced distance are all measured in the same units as length in general, preferably in millimeters or centimeters.

Example

If two points, *A* and *B*, are 10 cm apart, their *actual distance* is $L = 10$ cm. But if the two points and the path between them are immersed in a liquid of refractive index 1.6:

(a) What is the *optical path length*?
(b) How much closer will point *A* appear to an observer at *B*?

Solution: (a) Whereas the actual distance from *A* to *B* is 10 cm, with a medium of index 1.6 present, it will take the light 1.6 times as long to go from *A* to *B*, and thus

$$S = (10)(1.6) = \boxed{16 \text{ cm}}$$

(b) To an observer at *B*, point *A* will appear at a distance of

$$L' = \frac{10}{1.6} = 6.25 \text{ cm}$$

and, hence, it will appear

$$10 - 6.25 = \boxed{3.75 \text{ cm}}$$

closer to *B*.

Sometimes it may seem difficult to decide when to use optical path length and when reduced distance. From the "point of view" of the light, use optical path length. From the point of view of the observer, use reduced distance.

REFLECTION AND REFRACTION

Next we consider the *direction* of the light and any changes that may occur. Assume that the light, as shown in Figure 1-6, is incident on the surface between two media. Part of the light is returned to the medium from which it came. That is called *reflection*. Another part of the light is entering the second medium; this part also changes direction. That is called *refraction*.

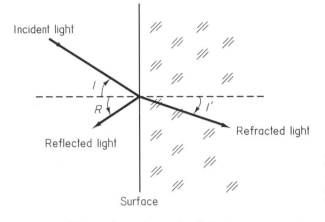

Figure 1-6 Reflection and refraction of light at the surface between two media (air to the left of the surface, glass to the right).

The point where the light intersects the surface is called the *point of incidence*. A line constructed at this point, perpendicular (*normal*) to the surface, is the *surface normal*. The angle subtended by the surface normal and the incident ray is the *angle of incidence*, I. The angle subtended by the surface normal and the reflected ray is the *angle of reflection*, R, and the angle subtended by the surface normal and the refracted ray is the *angle of refraction*, I'. The incident ray, the reflected ray, the transmitted ray, and the surface normal all lie in the same plane; they are *coplanar*. The various angles are measured *from* the surface normal, *toward* the ray.

Now consider two points, A and B, located on the same side of the surface (Figure 1-7). While we could draw an infinite number of paths, each coming from A and after reflection going to B, there is only one path actually taken by the light. This path, according to *Fermat's principle*, is shorter than any other path: it is *the path of least time*.*

---

    * This theorem was first formulated, in 1657, by Pierre de Fermat (1601–1665), French jurist and mathematician, counsel to the Toulouse parliament. Fermat used to write his thoughts about analytic geometry, calculus, probability, and number theory into the margins of other books, pursuing his avocation mostly for his own enjoyment. Five years after his death his notes were published by his son. Fermat justified his principle of least time on the grounds that nature is "economical." Indeed, the time spent by light traveling from one point to another is most often a minimum.

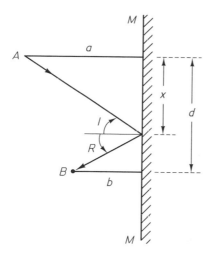

Figure 1-7 Reflection of light at a plane surface.

The distance from *A* to the surface *M-M* is called *a* and the distance from *M-M* to *B* is called *b*. From the construction and from Pythagoras' theorem it follows that the total path, from *A* to the surface *M-M* and from there to *B*, has a length

$$L = \sqrt{a^2 + x^2} + \sqrt{b^2 + (d - x)^2}$$

and the time needed for the light to go through the path is

$$t = \frac{L}{v}$$

Now we differentiate, setting the derivative of *t* with respect to *x* equal to zero,

$$\frac{dt}{dx} = 0$$

from which

$$\frac{x}{\sqrt{a^2 + x^2}} = \frac{d - x}{\sqrt{b^2 + (d - x)^2}}$$

In other words, the two ratios are equal. They are the sine of the angle of incidence and the sine of the angle of reflection,

$$\sin I = \sin R$$

The two angles, therefore, are *numerically equal* as well (although their signs are opposite).

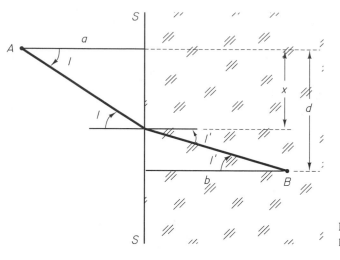

Figure 1-8   Refraction of light at a plane surface.

Next we turn to *Fermat's principle for refraction* (Figure 1-8). Now we have a surface, *S-S*, that separates two media but again the light goes from *A* to *B*. If the refractive index on either side of the surface were the same, no matter what its magnitude, the path from *A* to *B* would be a straight line. But if the two media have different indices, the path is no longer straight. Instead, the light is *refracted*.

Consider the velocities of the light, *v* and *v'*, on both sides of the surface. Since $v = L/t$, the time it takes the light to travel from *A* to the point of incidence and from there to *B* is

$$t = \frac{\sqrt{a^2 + x^2}}{v} + \frac{\sqrt{b^2 + (d - x)^2}}{v'}$$

Again we differentiate,

$$\frac{dt}{dx} = \frac{x}{v\sqrt{a^2 + x^2}} - \frac{d - x}{v'\sqrt{b^2 + (d - x)^2}} = 0$$

from which

$$\frac{\sin I}{v} - \frac{\sin I'}{v'} = 0$$

and

$$\frac{\sin I}{\sin I'} = \frac{v}{v'}$$

When we substitute $v = c/n$ and $v' = c/n'$, we find that

$$\frac{\sin I}{\sin I'} = \frac{c/n}{c/n'} = \frac{n'}{n}$$

and therefore

$$n \sin I = n' \sin I'$$

[1-6]

which is *Snell's law of refraction.** Table 1-1 lists some typical figures.

Table 1-1 Indices of refraction
(for helium d light, 587.6 nm)

Vacuum	1.000000
Air	1.0003
Water	4/3
Spectacle crown, C-1	1.5230
Extra dense flint, EDF-3	1.7200
Diamond	2.42

Critical Angle

There are two limiting cases that follow from Figure 1-8. Let the light, as shown, pass from a rarer medium (of lower refractive index) to a denser medium (of higher index). At first assume that the light is incident normal, with the angle of incidence zero, $I = 0$. At that angle the light is going through without deviation [Figure 1-9 (left), ray 1].

But, as the angle of incidence is gradually made larger, the angle of refraction becomes larger too, although not as fast as I. The light, therefore, is *bent toward the surface normal.* As I reaches 90° and the ray is tangent to the surface (ray 3), $\sin I$ becomes unity and Snell's law reduces to

$$n = n' \sin I'$$

where I' is the *limiting angle of refraction.*

Then let the light come from the side of the *higher* index [Figure 1-9 (right)]. As angle I is now gradually made larger, the light is *bent away* from the normal, at a rate *faster* than the increase of I. The angle of refraction, in fact, increases until a critical angle of incidence is reached for which the angle of refraction is $I' = 90°$ and $\sin I' = 1$. Snell's law then becomes

$$n \sin I = n'$$

---

* Named after Willebrord Snel van Royen (1591–1626), Dutch astronomer and mathematician. At age 21, Snell succeeded his father as professor of mathematics at the University of Leiden. At 26, he determined the size of Earth from measurements of its curvature between Alkmaar and Bergen-op-Zoom in The Netherlands. Snell's original statement, based on empirical observation, was that the ratio of the cosecants of the angles of incidence and refraction is constant. The law in its present form is due to René Descartes (Renatus Cartesius) (1596–1650), French philosopher and mathematician, who published it in his *La Dioptrique*, 1637.

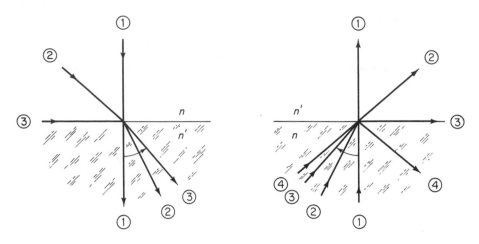

Figure 1-9 Limiting angle of refraction (*left*) and minimum angle of total internal reflection (*right*).

where *I* is now the critical *minimum angle of total internal reflection*. At angles larger than that, the light is returned to the first medium. The limiting angle (of refraction) and the minimum angle (of total internal reflection) are numerically equal; they both lie in the higher-index medium.

PRISMS

Reflecting Prisms

We distinguish two major groups of prisms, *reflecting prisms* and *refracting prisms*. Reflecting prisms, unless they have a reflective coating, make use of total internal reflection. In Figure 1-10 we show several examples. In a *right-angle prism* the light is reflected at the hypotenuse. If the same prism is oriented so that

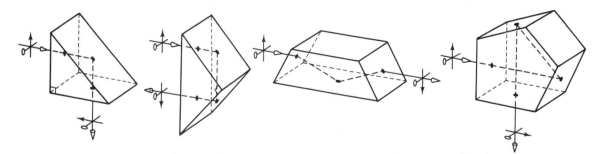

Figure 1-10 Some reflecting prisms, based on total internal reflection are (*from left to right*) right-angle prism, roof prism, Dove prism, and penta prism.

the light is passing (twice) through the hypotenuse, we have a *roof* or *Porro prism*. Still the same prism, with or without its apex cut off, is called a *Dove prism*. As the Dove prism is rotated about the line of sight, the image rotates through twice the angle. A *pentagonal prism* deflects the light through a constant 90°, without changing the "handedness" of the image.

Refracting Prisms

A refracting prism has two plane surfaces that subtend a certain angle A, the *apex angle*. The face opposite the apex is called the *base*. The total angle by which the light changes direction is the *angle of deviation*, D.

When the light passes through the prism *symmetrically* (with equal angles of incidence and emergence), then from the construction in Figure 1-11 it follows that

$$I_1 = I_2'$$

$$I_1' = I_2 = \frac{A}{2}$$

$$D_1 = D_2 = \frac{D}{2} = I_2' - I_2 = I_2' - \frac{A}{2}$$

so that

$$I_2' = \frac{A}{2} + \frac{D}{2}$$

Now we apply Snell's law, $n \sin I = n' \sin I'$, to the second (right-hand) surface:

$$n_{\text{prism}} \sin I_2 = n_0 \sin I_2'$$

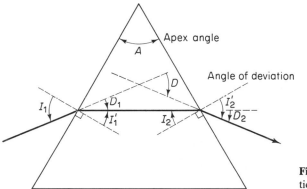

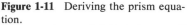

Figure 1-11 Deriving the prism equation.

where n_0 is the index outside the prism. Solving for $\sin I_2'$ gives

$$\sin I_2' = \left(\frac{n_{\text{prism}}}{n_0}\right) \sin I_2$$

and thus

$$\sin\left(\frac{A + D}{2}\right) = \left(\frac{n_{\text{prism}}}{n_0}\right) \sin\left(\frac{A}{2}\right)$$

$$\boxed{\frac{n_{\text{prism}}}{n_0} = \frac{\sin \tfrac{1}{2}(A + D)}{\sin \tfrac{1}{2}A}} \qquad [1\text{-}7]$$

which is the *prism equation*. It determines the *angle of minimum deviation*. For light not passing through symmetrically, the deviation is larger.

Example

Look at the moon, best on a cold evening and with the sky slightly overcast. If you see a *halo*, it is due to the refraction of light in ice crystals floating in the upper atmosphere. The crystals have the hexagonal shape shown in Figure 1-12 but, in essence, they are 60° prisms with the corners cut off. If we assume that the refractive index of ice is 1.31, somewhat less than the index of water (why?), what is the angle of minimum deviation?

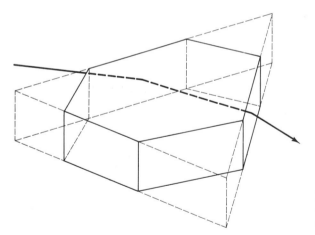

Figure 1-12   Hexagonal ice crystal (*solid lines*) and the triangular 60° prism equivalent to it (*dashed lines*).

Solution: From the prism equation, setting $n = 1.31$ and $A = 60°$, we find that

$$\frac{1.31}{1.00} = \frac{\sin \tfrac{1}{2}(60° + D)}{\sin \tfrac{1}{2}(60°)}$$

We solve $\sin \tfrac{1}{2}(60°) = \sin 30° = 0.5$ and multiply with the ratio on the left:

$$0.655 = \sin \tfrac{1}{2}(60° + D)$$

Taking the arcsine on both sides gives

$$40.92° = \tfrac{1}{2}(60° + D) = 30° + \frac{D}{2}$$

and thus

$$10.92° = \frac{D}{2}$$

$$D = \boxed{21.84°}$$

That means that the halo subtends with the axis of observation an angle of approximately 22°, the *22° halo*. Because of additional *dispersion* (to be discussed shortly), the halo is faintly red inside, blue outside.

Thin Prisms

Thin prisms have apex angles and angles of deviation much smaller than those of other prisms. The sines of these angles may be set equal to the angles themselves, measured in *radians*.* If the prism is in air ($n_0 \approx 1$), Equation [1-7] reduces to

$$D = A(n - 1) \qquad\qquad [1\text{-}8]$$

Seen through a thin prism, a target always appears displaced toward the apex. As the prism is rotated about the line of sight, the displacement changes with the angle of rotation (Figure 1-13). The angle of displacement is best given in units of *centrad*, $^\triangledown$, defined as $\frac{1}{100}$ of a radian, 1 cm at a distance of 1 meter.

Figure 1-13   Displacement of a target line as seen through a thin prism.

DISPERSION

When a bundle of light passes through a prism, each wavelength of the light will be refracted according to the index of the prism for that wavelength. For different wavelengths these indices are different: at *longer* wavelengths the refractive index

---

* There are two systems of units for measuring angles, based on *degrees* and on *radians*. One radian is the size of the angle at the center of a circle subtended by an arc equal to the radius of the circle: 1 rad $\approx$ 57.296°. To convert an angle in degrees, D, to an angle in radians, R, multiply D by $\pi/180$: $R = D \times \pi/180$.

is *less*. That is called *normal dispersion*. It holds for most cases (except near regions of absorption). A beam of light that contains the colors red and blue, for example, will spread apart as shown in Figure 1-14. The angle subtended by the two colors is called *angular dispersion*.

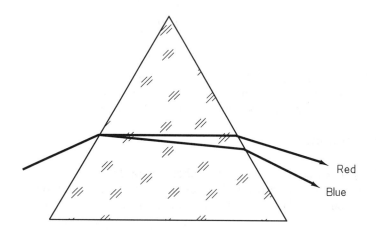

Figure 1-14   Dispersion of light by a prism.

Example

A hollow 60° prism is filled with carbon disulfide, whose index of refraction for blue light is 1.652, for red light 1.618. What is the angular dispersion?

Solution: Using Equation [1-7], we determine first the deviation for blue:

$$\frac{1.652}{1.000} = \frac{\sin \frac{1}{2}(60° + D)}{\sin \frac{1}{2}(60°)}$$

$$0.826 = \sin \tfrac{1}{2}(60° + D)$$

$$111.38° = 60° + D$$

$$D = 51.38°$$

For red,

$$0.809 = \sin \tfrac{1}{2}(60° + D')$$

$$108° = 60° + D'$$

$$D' = 48.0°$$

The difference between the two angles is the angular dispersion between blue and red,

$$51.38° - 48.00° = \boxed{3.38°}$$

But refraction and dispersion bear no simple relationship to one another: some glasses have a high index of refraction and little dispersion, others have just the opposite (Figure 1-15). Clearly, to characterize a glass we need more than one refractive index.

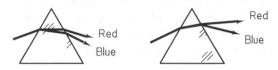

Figure 1-15   (*Left*) Prism of high refraction and low dispersion (note how much the light is deviated but how close the two colors are). (*Right*) Prism of low refraction and high dispersion (little deviation but the two colors are spread apart much farther).

In practice, three refractive indices are chosen, one for each of three wavelengths: the green mercury *e* line, wavelength 546.1 nm, defining the *main index* n_e; the blue cadmium *F'* line, 480.0 nm, defining $n_{F'}$; and the red cadmium *C'* line, 643.8 nm, defining $n_{C'}$.* The ratio

$$\frac{n_e - 1}{n_{F'} - n_{C'}} = V \qquad\qquad [1\text{-}9]$$

is the *V-value*, also called *Abbe's number*.

Glasses of low dispersion (where the difference $n_{F'} - n_{C'}$ is small and the *V*-value high, above 55) are customarily called *crowns*. Glasses of high dispersion (where *V* is below 50) are called *flints*. Some typical examples are shown in Table 1-2.

Table 1-2   Refractive indices of crown and flint at different wavelengths

Fraunhofer Line	Color	Wavelength (nm)	Spectacle Crown, C-1	Extra Dense Flint, EDF-3
F	Blue	486.1	1.5293	1.7378
d	Yellow	587.6	1.5230	1.7200
C	Red	656.3	1.5204	1.7130
			Abbe's number	
			58.8	29.0

---

* The letter designations and subscripts *e*, *F*, *C*, and others go back to Joseph Fraunhofer who used them to identify the absorption lines that he saw in the spectrum of the sun.

The lines mentioned here replace the wavelengths used before: the blue hydrogen *F* line, 486.1 nm; the yellow sodium doublet, 589.3 nm, or helium *d*, 587.6 nm; and the red hydrogen *C*, 656.3 nm. Measuring the new lines using a Hg-Cd-lamp is easier than using a H- and a He-lamp. Sodium *D* light anyway is not precise because it is a doublet.

Applications

The principal use of dispersing prisms is in spectroscopy. *Spectroscopy* refers to the study and analysis of spectra. More specifically, a spectro*scope* is used for viewing a spectrum. A spectro*meter* is used for measuring wavelengths. A spectro*graph* is built for photographing a spectrum (Figure 1-16). In a spectro*radiometer* a photodetector takes the place of the photographic film. A spectro*gram* is the result of spectrography or spectroradiometry.

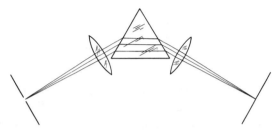

Figure 1-16   The two lenses of a *prism spectrograph* (collimating lens left, focusing lens right) let parallel light pass through the prism oriented for minimum deviation.

SUMMARY OF EQUATIONS

Velocity, wavelength, and frequency:

$$v = \lambda \nu \qquad\qquad [1\text{-}1]$$

Optical path length:

$$S = Ln \qquad\qquad [1\text{-}3]$$

Snell's law:

$$n \sin I = n' \sin I' \qquad\qquad [1\text{-}6]$$

Prism equation:

$$\frac{n_{\text{prism}}}{n_0} = \frac{\sin \frac{1}{2}(A + D)}{\sin \frac{1}{2}A} \qquad\qquad [1\text{-}7]$$

Thin prism:

$$D = A(n - 1) \qquad\qquad [1\text{-}8]$$

Dispersion, Abbe's number:

$$V = \frac{n_e - 1}{n_{F'} - n_{C'}} \qquad\qquad [1\text{-}9]$$

PROBLEMS

1-1. A 4-m-long rope is set in oscillatory motion so that $2\frac{1}{2}$ waves are present on the rope at any one time. If the waves travel at a velocity of 10 m/s, what is the frequency of oscillation?

1-2. A 60-cm-long string is oscillating at a frequency of 4 Hz, with three waves present on the string at any one time. How fast do the waves move forward?

1-3. What is the frequency of red light of 625 nm wavelength?

1-4. The blue F' cadmium line has a wavelength of 480 nm.
    (a) How many waves are there in 1 mm in air?
    (b) What is its frequency?

1-5. A small hole is made in the shutter of a dark room, and a screen is placed at a distance of 1.5 m from the shutter. If a tree outside is 30 m away from the shutter and if the tree casts on the screen an image 20 cm high, how tall is the tree?

1-6. If in a *pinhole camera* (Figure 1-17) the distance between an object, left, and its image, right, is 1.5 m, how long must the camera be in order to make the image one-fifth the size of the object?

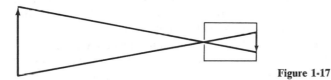

Figure 1-17

1-7. A pinhole camera produces a 5-cm-high image of a telephone pole. Moving the camera 5 m farther away from the pole reduces the image size to 4 cm. To make the image 5 cm high again, the camera has to be made 4 cm longer. How tall is the telephone pole?

1-8. An extended light source is 10 cm in diameter. A solid round object is placed 2 m to the right of the source and a screen 1.5 m to the right of the object. How wide is the penumbra?

1-9. A sheet of newsprint is viewed through a plate of glass that is placed on the newsprint and known to have a refractive index of $n = 1.8$. When looking through the plate, the newsprint appears to be 5 mm closer to the observer than without it. How thick is the plate?

1-10. A cylindrical tube is closed at both ends by glass plates 10 mm thick. If the *inside* length of the tube is 7.5 cm, the refractive index of the glass is $n = 1.5$, and the tube is filled with water ($n = \frac{4}{3}$), what is the optical path length between the *outer* surfaces?

1-11. How far from true vertical does the setting sun appear to an observer under water ($n = \frac{4}{3}$)?

1-12. What is the minimum angle of total internal reflection for light passing from glass of $n = 1.5396$ to water ($n = \frac{4}{3}$)?

1-13. An open cylindrical container, resting on its circular base, is 20 cm in diameter and 11.2 cm deep. An observer is looking into the empty container from such a direction that he or she can just see the opposite bottom corner (Figure 1-18). When the container is filled with a liquid and the observer keeps looking from the same direction, he or she can see the *center* of the bottom. What is the refractive index of the liquid?

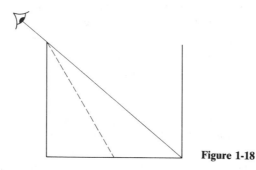

Figure 1-18

1-14. (a) Assume that light is reflected at the hypotenuse of a *right-angle prism* (see Figure 1-10). How will the image be *oriented*, that is, when looking through the prism at the letter R, how will it change? **(b)** Now let the light be reflected at the two smaller faces, as in a *Porro prism*. How will the image change?

1-15. A prism with a 60° apex angle shows minimum deviation for a particular wavelength at an angle of 56°. What is the refractive index of the prism for that wavelength?

1-16. A hollow (and empty) 60° prism is immersed in a liquid of index 1.74. What is the angle of minimum deviation?

1-17. A prism, made of glass of $n = 1.62$, shows an angle of minimum deviation of 48.2°. What is the apex angle?

1-18. Look through a thin prism at a meter stick 2 m away. If you see the marks on the stick displaced by 3.5 cm, **(a)** what is the angle of deviation, in centrads? **(b)** If $n = 1.625$, what is the apex angle?

1-19. An object 1.5 m away is viewed through a 2^∇ prism. By how much, and in which direction, will the object appear to be displaced?

1-20. The light forming a *rainbow* is refracted when it enters each drop of water and refracted again when it leaves. In a *primary* rainbow the light in addition is reflected *once*, at the rear face of the drop [Figure 1-19 (left)]. Show how a blue ray and a red ray are proceeding.

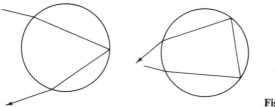

Figure 1-19

1-21. In a *secondary* rainbow the light is reflected *twice* [Figure 1-19 (right)]. Again, follow a blue ray and a red ray.

1-22. If white light is incident at an angle of 30° on a slab of glass that for blue light has a refractive index of 1.7 and for red light of 1.6, what is the angular dispersion between blue and red inside the glass?

1-23. Consider a direct-vision prism (which has no deviation for yellow).

    (a) If the crown component ($n = 1.523$) has an apex angle of 11°, what is the apex angle of the flint component ($n = 1.72$)?

    (b) What is the angular dispersion for blue and red?

SUGGESTIONS FOR FURTHER READING

GIACOMO, P. "The New Definition of the Meter." *Am. J. Phys.* **52** (1984), 607–13.

GREENLER, R. *Rainbows, Halos, and Glories*. New York: Cambridge University Press, 1980.

SCHOLZE, H. *Glass: Nature, Structure, and Properties*. New York: Springer-Verlag, 1991.

YOUNG, M. "Pinhole Imagery." *Am. J. Phys.* **40** (1972), 715–20.

2

Thin Lenses

Now we come to the first lens of an optical system. As an example, look at Figure 2-1. Note that a lens has two surfaces that enclose a medium of a refractive index different from the index outside the lens. Hence, before we discuss lenses as such, we ask what happens to the light as it passes through a single curved *surface*.

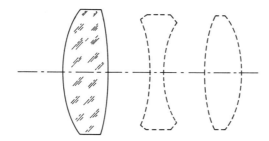

Figure 2-1 Front lens (*solid outline*) of a three-lens system.

SINGLE REFRACTING SURFACE

Vergence

As the light is incident on the first surface, it has a certain *vergence*. Vergence is a term whose significance to optics can hardly be overemphasized. Light that spreads out is called *divergent*. Such light may come from a point source on the

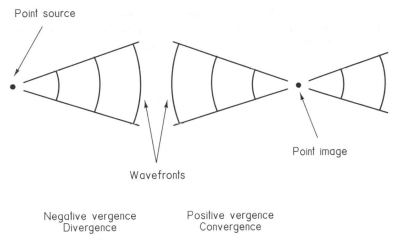

Point source

Wavefronts

Point image

Negative vergence
Divergence

Positive vergence
Convergence

Figure 2-2 Vergence of light coming from a point source (*left*) and proceeding toward a point image (*right*).

left, causing the light's wavefronts to be curved concave to the left (Figure 2-2). The centers of curvature of the wavefronts are then on the left and their radii of curvature are measured from the wavefronts to the left.

Light that comes together is called *convergent*. Its wavefronts are curved concave to the right, their centers of curvature are on the right, and their radii of curvature are measured to the right.

Quantitatively, vergence, both divergence and convergence, is given in units of reciprocal meter, m^{-1}, commonly called *diopter*, *D*. The diopter, in other words, is the reciprocal of the distance in meters.* Light that diverges from a point 2 m away, for example, has a vergence of $-0.5\ D$. Light that converges to a point 4 meters away has a vergence of $+0.25\ D$. But the light may travel in a medium other than air. A medium of index *n* will delay the light by a factor of *n*; hence we define the *reduced vergence*, **V**, as

$$\mathbf{V} = \frac{n}{R} \qquad\qquad [2\text{-}1]$$

Now the light is reaching the surface. Any refractive surface has a certain *refractive power*. Refractive power is the reciprocal of the focal length, in meters. As with vergence, a surface that is *concave to the right* is *positive*. If the power is

---

*The unit of diopter was introduced in 1872 by Ferdinand Monoyer (1836–1912), French ophthalmologist. The son of an army physician, Monoyer in 1870 succeeded his stepfather as chairman of the Department of Ophthalmology at Strassburg, in 1872 became professor of ophthalmology in Nancy, and in 1876 professor of medical physics in Lyon. It was in Lyon where he did most of his optics work, introducing the metric system into ophthalmic measurements. The unit of diopter occurs first on page 111 of his contribution "Sur l'introduction du système métrique dans le numérotage des verres de lunette, et sur le choix d'une unité de réfraction," *Ann. Ocul.* **68** (1872), 105–117.

positive and *higher* than the (negative) vergence (of the light), some excess (positive) vergence remains and the light after refraction becomes *convergent* [Figure 2-3 (top)]. Such light will form a *real image*.

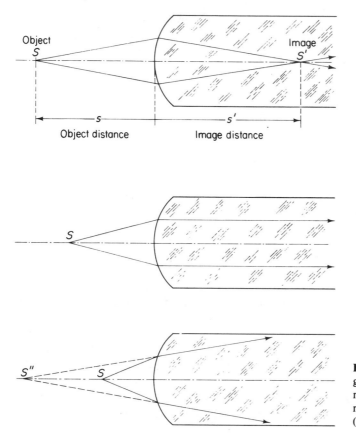

Figure 2-3 When an object, *S*, is gradually brought closer to a positive refracting surface, the image is first real, *S'* (*top*); projected out to infinity (*center*); and then virtual, *S"* (*bottom*).

If the object is moved closer to the surface, the radius of curvature of a wavefront just entering the surface is shorter. The (negative) vergence of the light becomes more negative, and may just be counterbalanced by the (positive) power of the surface. The resultant vergence then is zero, which means that the light emerges parallel and the image is projected out to infinity (center).

If the object is brought even closer, the vergence of the light becomes still more negative, the surface does not have enough power to compensate for it, and the emergent light *remains divergent* (bottom). Now the rays will *not* come together on the right-hand side. Instead, the light seems to come from a point, *S"*, to the left of the surface. Such a point is part of a *virtual image*.*

---

*A good example of a virtual image is what you see in a mirror. The object and the observer are both in front of the mirror, but the image is behind it. No light is actually going to where the image

Next, we consider the *focal points* of a surface. If *parallel light* is incident on a positive surface, as shown in Figure 2-4, top left, the light comes together at the *second focal point*, F_2. If the light is *made parallel* by the surface, the light comes from the *first focal point*, F_1 (top right). The same convention applies to a negative surface: parallel light that is incident relates to F_2 (bottom left), and parallel light that emerges relates to F_1 (bottom right).

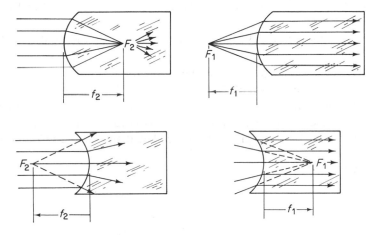

Figure 2-4   Focal points, F, and focal lengths, f.

The line connecting the two focal points is the *optical axis*. The point where the optical axis intersects the surface is the *vertex* (of the surface). The distance from the vertex to the object is the *object distance*, s. The distance from the vertex to the image is the *image distance*, s'. Note that these distances are measured *from* the vertex, *toward* the object or the image, respectively. But the image may be either real or virtual, and therefore may lie either to the right or the left of the surface; the *image space*, in other words, extends from infinity on one side to infinity on the other. The same holds for the *object space*; both spaces completely overlap. Whether a given point is in the object space or the image space depends on whether it is part of a ray *before* or *after* refraction.

Objects, and images, are composed of a great many tiny "*picture cells*," called *pixels*. Object pixels and image pixels that correspond (belong) to each other are called *conjugate*. Distances and other parameters that correspond to each other are called conjugate also.

---

appears to be: the image is *virtual*. In a way, the terms "real" and "virtual" are misleading. The word "real" seems to imply that "virtual" has something "unreal" to it. That is not so; a virtual image can be seen, and be photographed, just as well as a real image. The only difference is that (without an additional lens) it cannot be received on a screen.

Sign Convention

There are two major sign conventions in optics. In the *empirical system*, the distance of a real object is taken as positive. Intuitively, that seems to be a good choice. However, there are many advantages, from the definition of vergence to computer-aided lens design, that show why it is better to consider the distance of a real object *negative* and to use the *rational, Cartesian sign convention,** as we are doing it here. In the Cartesian sign convention, illustrated in Figure 2-5:

1. All figures and ray diagrams are drawn with the light traveling from left to right.
2. *Distances* measured to the left (of a surface or lens), in a direction opposite to the propagation of the light, are considered *negative*. Distances measured to the right are *positive*. Therefore, the distance of a typical (real) object, because it is measured to the left, is negative. The distance of a real image, measured to the right, is positive. The distance of a virtual image is negative.

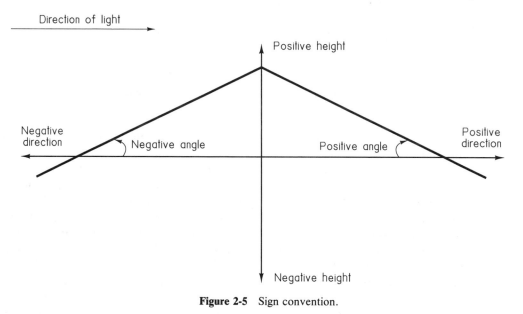

Figure 2-5   Sign convention.

---

*Named after René Descartes (1596–1650), French mathematician and philosopher. Educated at a Jesuit college, Descartes graduated with a law degree, then joined the army to work on problems of ballistics and navigation. Shortly after the discovery of Snell's law, and most likely independent of it, he found the same law and published it in his *La Dioptrique* (Leiden: 1637). He also showed how every point in space can be represented by a system of three numbers, the *Cartesian system of coordinates* (from the latinized version of the name he used in his writings, Renatus Cartesius). Descartes often started out with a central principle and moved outward to find the phenomena of nature which follow it, an approach opposite to that promulgated by Newton.

3. *Radii of curvature* are measured *from* the surface, *toward* the center of curvature. Radii that extend to the left (from the surface) are negative; radii that extend to the right are positive. For example, when we look (in the same direction in which the light is going) at a *convex* surface, that surface has a *positive* radius. A concave surface has a negative radius.

4. By the same reasoning, *divergent* light, because the radii of curvature of the wavefronts are measured to the left, has negative vergence; *convergent* light has positive vergence. Parallel light has zero vergence.

5. The *refractive power* of a surface, or lens, that makes light more convergent, or less divergent, is positive. The *second focal length* of such a surface, or lens, is also positive. The power and the (second) focal length of a diverging lens are negative.

6. *Heights* above the optical axis are positive, heights below the axis are negative.

7. *Angles* measured *clockwise* from the optical axis, or from a surface normal, are *positive*. Angles measured counterclockwise are negative.

Gauss' Formula and the Surface Power Equation

Consider the single surface shown in Figure 2-6. A ray originates at an axial point object, S. At S the ray subtends with the axis an angle $+U$. At the image, S', the ray subtends an angle $-U'$. From the construction it follows that, to the left of the surface,

$$I = U + \phi$$

and to the right

$$I' = \phi + U'$$

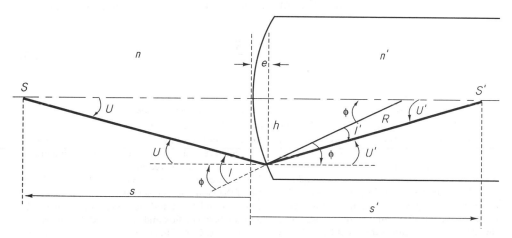

Figure 2-6 Refraction at a single spherical surface.

If the ray is close to the optical axis, the angles involved are small and their sines and tangents are nearly equal to the angles themselves, measured in radians. Such rays are called *paraxial rays*.* Snell's law,

$$n \sin I = n' \sin I'$$

then becomes

$$nI \approx n'I'$$

Substituting for I and I' the terms just derived, we obtain

$$nU + n\phi = n'\phi + n'U' \tag{2-2}$$

For paraxial rays the distance e, compared to s and s', is small and may be neglected and thus, following the notation in Figure 2-6,

$$+U = \frac{-h}{-s}, \quad -U' = \frac{-h}{+s'}, \quad \text{and} \quad +\phi = \frac{-h}{+R}$$

Substituting these terms in Equation [2-2], canceling h, and rearranging then leads to

$$\boxed{\frac{n}{s} + \frac{n' - n}{R} = \frac{n'}{s'}} \tag{2-3}$$

which is *Gauss' formula for refraction at a single surface*,† written in a form that is easy to remember: Just think of the light propagating from object (s) to surface (R) to image (s'). Although I have derived Gauss' formula for a convex surface, and for a real object and image, the formula holds as well for all other conditions.

If the object moves to infinity, its distance becomes $-\infty$, the first term in Equation [2-3] drops out, and the image moves to the second focal point so that $s' \rightarrow f_2$:

$$f_2 = \frac{n'R}{n' - n} \tag{2-4}$$

---

*From the Greek $\pi\alpha\rho\grave{\alpha}$ = nearby, closely adjacent; paraxial = close to the optical axis.

†Named after Johann Carl Friedrich Gauss (1777–1855), German mathematician. The son of a bricklayer, Gauss, at the age of three, found an error in his father's payroll. While still in his teens, he discovered the method of least squares, used for fitting a curve to a series of points minimizing human error. Soon after, he calculated the (elliptical) orbits of asteroids, showed how to construct polygons using straightedge and compass only, and made significant contributions to number theory including complex numbers. In 1817 he was appointed director of the Göttingen Observatory. He became adept at using telescopes, derived the surface power equation and the thin-lens equation, designed the eyepiece named after him, and discovered the relationship between flux and charge and the divergence theorem, important forerunners of Clerk Maxwell's equations. One of his books, *Dioptrische Untersuchungen*, was published in 1843.

Since by definition n'/f_2 is the refractive power, P, of the surface,

$$\boxed{P = \frac{n' - n}{R}}$$ [2-5]

which is the *surface power equation*.

Recall that vergence is inversely proportional to distance. That means, if we also include the refractive index of the medium to the left of the surface, that the first term in Gauss' formula,

$$\frac{n}{s} + \frac{n' - n}{R} = \frac{n'}{s'}$$

is simply the vergence of the light as it enters the surface, the *entrance vergence*, **V**. The last term is the vergence of the light as it leaves the surface, the *exit vergence*, **V'**. Substituting the surface power equation for the second term, we conclude that

$$\boxed{\mathbf{V} + P = \mathbf{V'}}$$ [2-6]

This is a beautifully simple equation. It shows how the power of the lens acts on the incoming wavefronts, with their entrance vergence, and changes them to the outgoing wavefronts, with their exit vergence. The units of vergence and power are the same, diopters.

Example 1

A point light source is placed 50 cm to the left of a single surface of +6.00 diopters power, ground on a rod of glass of index 1.6 (Figure 2-7). What is the image distance (inside the rod)?

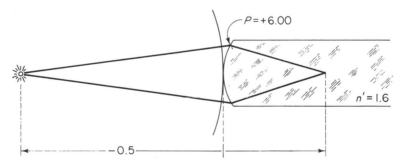

Figure 2-7 Image formation by a positive surface. Distances in meters.

Solution: The *entrance vergence*, as the light enters the surface, is

$$\mathbf{V} = \frac{1}{-0.5 \text{ m}} = -2 \text{ diopters}$$

The *exit vergence*, from Equation [2-6], is

$$\mathbf{V'} = \mathbf{V} + P = (-2) + (+6) = +4 \text{ diopters}$$

Note the reversal of the signs of the two vergences. If the exit vergence is positive, as here, the image is inverted and real.

The light now proceeds in the medium of higher index, $n' = 1.6$. The image distance, therefore, is

$$s' = \frac{n'}{V'} = \frac{1.6}{+4} = \boxed{+40 \text{ cm}}$$

If the surface had only $+1.50$ diopters power, instead of $+6.00$, how would the image change? In that case

$$V' = (-2) + (+1.5) = -0.5 \text{ diopters}$$

which means that the light remains divergent, and the image is virtual and *to the left* of the surface [as in Figure 2-3 (bottom)]. The image distance in that case would be

$$s' = \frac{1.6}{-0.5} = -3.2 \text{ m}$$

This is an important case. While the medium to the left of the surface is air ($n \approx 1.0$), the *index related to the image space is 1.6*. In other words, with any image, real or virtual, *always use the image space index*. Do not use the index of the medium where the image merely *seems* to be.

Example 2

A *reduced eye* is a much simplified model of the human eye. It has only *one* refracting surface, assumed to have a radius of curvature of 5.7 mm (Figure 2-8). The refractive index to the left of the surface is 1.00, the index to the right, 1.34. Find the focal length(s) and power(s) of the surface.

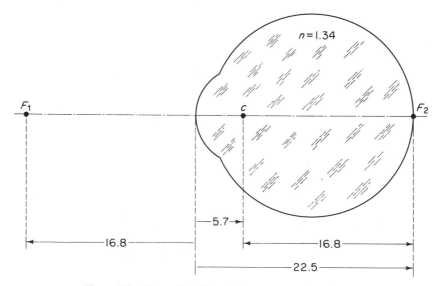

Figure 2-8 The reduced eye. Dimensions in millimeters.

Solution: On the left (outside the eye, in the air) the rays are parallel to each other. Such rays come together on the right, inside the eye, forming the second focal length, f_2. We use Gauss' formula, Equation [2-3], noting that for the rays on the left to be parallel, the object distance must be infinite, $s \rightarrow \infty$. Inserting that in Gauss' formula gives

$$\frac{n}{\infty} + \frac{n' - n}{R} = \frac{n'}{s'}$$

and therefore, as in Equation 2-4,

$$f_2 = \frac{n'R}{n' - n}$$

Inserting the actual values gives

$$f_2 = \frac{(1.34)(5.7)}{1.34 - 1.00} = \boxed{+22.5 \text{ mm}}$$

Now consider the *first* focal length, f_1. We let the light come from the *right*, passing through the cornea and coming together on the left at F_1. This time $s \rightarrow f_1$ and $s' \rightarrow \infty$. Gauss' formula then becomes

$$\frac{n}{f_1} + \frac{n' - n}{R} = \frac{n'}{\infty}$$

which is equal to

$$\frac{n}{f_1} = -\left(\frac{n' - n}{R}\right) = \frac{n - n'}{R}$$

and, solving for f_1,

$$f_1 = \frac{nR}{n - n'}$$

Inserting the actual values gives

$$f_1 = \frac{(1.00)(5.7)}{1.00 - 1.34} = \boxed{-16.8 \text{ mm}}$$

In short, we have *two focal lengths*. The reason is that on one side, but not on the other, we have a medium of index 1.34,

$$\frac{22.5 \text{ mm}}{1.34} = 16.8 \text{ mm}$$

But we have only *one power*,

$$P = \frac{n' - n}{R} = \frac{1.34 - 1.00}{0.0057} = \boxed{+60 \text{ diopters}}$$

THIN LENSES

A *thin lens* is defined as a lens whose thickness is small relative to its focal length, to the two radii of curvature, and to the object and image distance. A lens has two centers of curvature, one for each surface. The line connecting the two centers is the *optical axis of the lens*, not to be confused with the optical axis of the system as a whole. The points where the optical axis of the lens intersects the two surfaces of the lens are the *front vertex* and the *back vertex*, respectively.

If parallel light is incident on the lens, it forms the *second focal point*, no matter whether the lens is positive or negative. A positive, or converging, lens makes parallel light convergent. Such a lens is thicker in the center than at the periphery. A negative, or diverging, lens makes parallel light divergent; such a lens is thinner in the center than at the periphery (Figure 2-9).

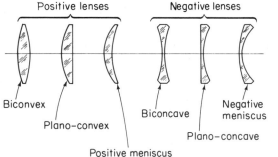

Figure 2-9   Types of lenses.

With a positive lens, the second focal point, F_2, lies to the right of the lens, and for a negative lens it lies to the left. That is the same as for a single surface.

When the light is *made parallel* by the lens, the light comes from the *first focal point*, F_1. Customarily, whenever reference is made to "*the* focal length," that always means the *second focal length*. As the object is brought closer to the lens, or moved away from it, the image distance changes accordingly (Figure 2-10).

Graphical Construction

In recent years the art of graphical ray tracing has lost some of its earlier prominence. This is due to the growing sophistication of computer-aided ray tracing, which we discuss later, in Chapter 8.

In order to locate the image by graphical ray tracing, we need at least two rays, out of the three shown in Figure 2-11:

1. The *parallel ray*, 1 in Figure 2-11, is at first parallel to the axis and then, after refraction, passes through F_2.
2. The *focal ray*, 2, passes through F_1 and then, after refraction, is parallel to the axis.

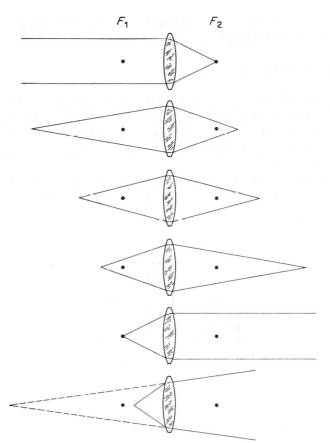

F_1 F_2

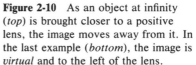

Figure 2-10 As an object at infinity (*top*) is brought closer to a positive lens, the image moves away from it. In the last example (*bottom*), the image is *virtual* and to the left of the lens.

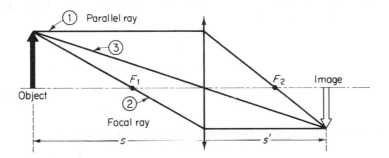

Figure 2-11 Designation of rays used for image construction. The object is *outside* the focal length of a converging lens.

3. Ray 3 goes through the center of the lens, without deviation and, since the lens is thin, with only negligible displacement.

The resulting image is real and inverted.

If the object is *inside* the focal length, we follow the same procedure, using the same rays 1-2-3 and their definitions. After refraction, though, these rays are still divergent and do not intersect on the right. Instead they seem to come from a point on the left. That point is part of a *virtual image*, an image that, as we see, is upright and magnified (Figure 2-12).

A virtual *object* is usually a virtual image formed by a preceding lens. Finally, with a *diverging* lens, the image, no matter where the object is placed, is always virtual, and reduced in size.

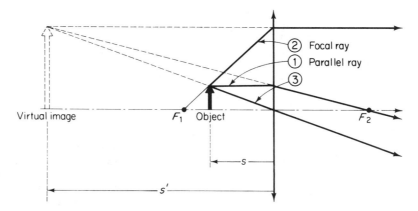

Figure 2-12   The object is *inside* the focal length of the lens.

Lens Equations: The Lens-Makers Formula

With a *thin* lens, the refractive power of the lens is simply the algebraic sum of the powers of the two surfaces:

$$P_{\text{lens}} = P_1 + P_2 \qquad [2\text{-}7]$$

We replace both P_1 and P_2 by the surface power equation,

$$P_1 = \frac{n_{\text{lens}} - n_0}{R_1} \quad \text{and} \quad P_2 = \frac{n_0 - n_{\text{lens}}}{R_2} = \frac{n_{\text{lens}} - n_0}{-R_2}$$

and substitute these terms in Equation [2-7]. That gives

$$\boxed{P_{\text{lens}} = (n_{\text{lens}} - n_0)\left(\frac{1}{R_1} - \frac{1}{R_2}\right)} \qquad [2\text{-}8]$$

which is the *lens-makers formula*. It tells us what the lens-maker needs to know to grind a lens of a given power: the refractive index of the glass and the two radii of curvature.

Example

A biconvex lens of 50 mm focal length is made of glass of $n = 1.52$. If the radius of curvature of the back surface is twice as long as the radius of the front surface, what are the two radii?

Solution: First we convert the focal length of the lens into power:

$$P = \frac{1}{f} = \frac{1}{50 \text{ mm}} = \frac{1}{0.05 \text{ m}} = 20 \text{ diopters}$$

The second radius, R_2, is twice as long as the first, R_1. But the first surface is convex to the left and the second surface convex to the right; hence the signs are opposite:

$$R_2 = -2R_1$$

Then from the lens-makers formula:

$$20 = (1.52 - 1.00)\left(\frac{1}{R_1} - \frac{1}{R_2}\right)$$

$$= (0.52)\left(\frac{1}{R_1} - \frac{1}{-2R_1}\right)$$

$$= 0.78 \, \frac{1}{R_1}$$

$$R_1 = \frac{0.78}{20} = 0.039 = \boxed{+39 \text{ mm}}$$

$$R_2 = -2R_1 = (-2)(39) = \boxed{-78 \text{ mm}}$$

Newton's Lens Equation

Now we derive two "lens-users equations"; they connect the focal length of the lens with the object and image distances.

From Figure 2-13 we see that the object size, y, and the image size, y', are connected through similar triangles. To the left of the lens we have

$$\frac{y}{N} = \frac{y'}{f_1}$$

and to the right of the lens

$$\frac{y}{f_2} = \frac{y'}{N'}$$

Combining the two equations by eliminating y/y' gives

$$\frac{N}{f_1} = \frac{f_2}{N'}$$

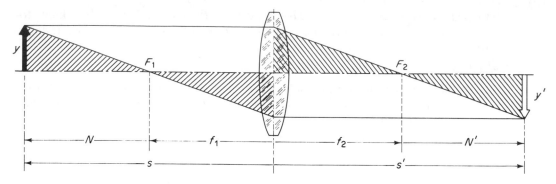

Figure 2-13 Newton's lens equation.

and thus

$$NN' = f_1 f_2$$ [2-9]

which is *Newton's lens equation*. Note that N and N' are measured from the focal points of the lens, rather than from the lens as such; they are *extrafocal* distances. Consequently, Newton's equation can be used with both "thin" lenses and "thick" lenses.

Thin-Lens Equation

Again we use Gauss' formula,

$$\frac{n}{s} + \frac{n' - n}{R} = \frac{n'}{s'}$$

We substitute the surface power equation for the center term,

$$\frac{n' - n}{R} = P$$

But P is the reciprocal of the focal length and thus if the lens is in air,

$$\frac{1}{s} + \frac{1}{f_2} = \frac{1}{s'}$$ [2-10]

which is the *thin-lens equation*. As before, we proceed from object (s) to lens (f) to image (s'). But remember that the object distance, s, is measured to the left and therefore is negative, while both f_2 and s' are measured to the right and are positive. Solving Equation [2-10] for the individual terms gives for the

Object distance: $s = \dfrac{s'f}{f - s'}$ [2-11a]

Image distance: $s' = \dfrac{sf}{s + f}$ [2-11b]

$$\text{Focal length:} \qquad f = \frac{ss'}{s - s'} \qquad\qquad \text{[2-11c]}$$

Example 1

When focusing at a target 24.5 cm from the object-side focal plane of a camera lens, the film in the camera must be moved 5 mm away from the position used for objects at infinity. What is the focal length of the lens?

Solution: For the second of the two positions, and from Figure 2-14 and *Newton's equation*,

$$NN' = f_1 f_2$$

Since $-f_1 = f_2$,

$$-NN' = f_2^2$$

and thus,

$$f_2 = \sqrt{-NN'} = \sqrt{-(-245 \text{ mm})(5 \text{ mm})}$$

$$= \sqrt{1225} = \boxed{+35 \text{ mm}}$$

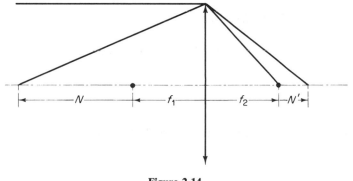

Figure 2-14

Example 2

When an object is placed 9 cm in front of a converging lens, its image is three times as far away from the lens as when the object is at infinity. What is the focal length of the lens?

Solution: While for the object at infinity,

$$s' = f$$

for the close-by object,

$$s'' = 3f$$

Then from the *thin-lens equation* solved for f,

$$f = \frac{ss''}{s - s''} = \frac{(s)(3f)}{(s) - (3f)} = \frac{(-9)(3f)}{(-9) - (3f)}$$

$$f(-9 - 3f) = (-9)(3f)$$

$$-9f - 3f^2 = -27f$$

$$9 + 3f = 27$$

$$3f = 18$$

$$f = \frac{18}{3} = \boxed{+6 \text{ cm}}$$

MAGNIFICATION

The term *magnification* means many things. Whatever its type, though, magnification always refers to a *comparison*, comparing the size of the image to the size of the object. We discuss three types of magnification: *transverse magnification*, *axial magnification*, and *angular magnification*.

Transverse Magnification

Transverse, or *lateral*, magnification, M_T, is defined as the ratio of image size, y', to object size, y:

$$\boxed{M_T = \frac{y'}{y}} \qquad [2\text{-}12]$$

From the similar triangles in Figure 2-15 we see that

$$\frac{y}{s} = \frac{y'}{s'}$$

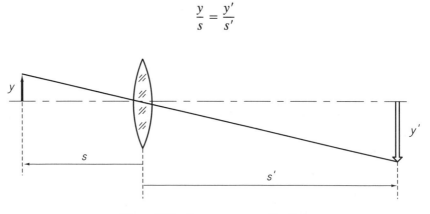

Figure 2-15   Transverse magnification.

and therefore, transverse magnification can also be defined as the ratio of image *distance*, *s'*, to object *distance*, *s*:

$$M_T = \frac{s'}{s} \qquad [2\text{-}13]$$

But this holds only if the refractive indices on both sides are the same. With a *single surface*, by necessity, they are different. Consider such a single surface and draw a ray from the point of the arrow to the vertex (Figure 2-16). The ray will be refracted at the surface according to Snell's law, which for paraxial rays simplifies to

$$nI = n'I' \qquad [2\text{-}14]$$

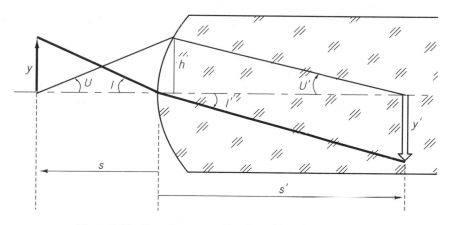

Figure 2-16   Transverse magnification with a single surface.

Again, for rays close to the optical axis the angles of incidence and refraction are

$$I = \frac{y}{s} \quad \text{and} \quad I' = \frac{y'}{s'}$$

Substituting in Equation [2-14] gives

$$n\frac{y}{s} = n'\frac{y'}{s'} \qquad [2\text{-}15]$$

But the ratio n/s is the entrance vergence of the light, **V**, and the ratio n'/s' is the exit vergence, **V'**; thus,

$$\mathbf{V}y = \mathbf{V}'y'$$

Rearranging and substituting in Equation [2-12] then shows that transverse magnification can also be defined as

$$\boxed{M_T = \frac{\mathbf{V}}{\mathbf{V}'}} \qquad [2\text{-}16]$$

From Figure 2-16 we also see that

$$U = \frac{h}{s} \quad \text{and} \quad U' = \frac{h}{s'}$$

Solving for s and s', substituting in Equation [2-15], and eliminating h results in

$$\boxed{nyU = n'y'U'} \qquad [2\text{-}17]$$

which is the *Smith–Helmholtz relationship*, also known as *Lagrange's theorem*.* The product nyU is called the *optical invariant*.

I don't like the term "demagnification." Does it mean that a preceeding process of magnification has been cancelled or nullified? Probably not. I dislike "minification" as well. If I want to say that an image has the same size as the object, $y' = y$, I say that the magnification is unity, $M_T = 1$. That happens when the object distance is equal to the image distance and either is equal to twice the focal length, $s = s' = 2f$. Whenever the object distance is less than $2f$, the image is larger than the object and the magnification is larger than 1. And, whenever the object distance is larger than $2f$, the image is smaller than the object and the magnification is less than 1. Finally, following Equation [2-12], if an image is inverted, its height, and therefore M_T, are negative. If the image is upright, its height, and M_T, are positive.

---

*Robert Smith (1689–1768), British physicist, Plumian professor of physics at Trinity College. Smith seems to have been the first to find this relationship. He wrote two books, *A Compleat System of Opticks in Four Books, viz. A Popular, a Mathematical, a Mechanical, and a Philosophical Treatise* (Cambridge, 1738), a textbook widely used in the eighteenth century, and *Harmonics, or the Philosophy of Musical Sounds* (Cambridge, 1749).

The modern form, Equation [2-17], is due to Hermann Ludwig Ferdinand von Helmholtz (1821–1894). German physician and physicist. While still in medical school, von Helmholtz investigated heat production in animals and found the law of conservation of energy. After working for six years as an army surgeon, he became an instructor of physiology in Königsberg, professor of anatomy and physiology in Bonn, then professor of physics in Berlin. Von Helmholtz wrote two books, covering similar ground as Smith before him, *Handbuch der physiologischen Optik* (Leipzig: L. Voss, 1867) and *Die Lehre von den Tonempfindungen als physiologische Grundlage für die Theorie der Musik* (Braunschweig: F. Vieweg & Sohn, 1863). He invented the ophthalmoscope and the keratometer; extended Young's theory of color vision; contributed to diffraction theory, thermodynamics, fluid dynamics, and meteorology; and showed that the character of a musical tone depends on harmonics present in addition to the fundamental.

Joseph Louis Lagrange (1736–1813), French mathematician. Lagrange is known for his work on the calculus of variations, which he summarized in his book *Analytical Mechanics*, Paris, 1788. He worked on perturbation theory (investigating three celestial bodies, the sun, Earth, and the moon) and introduced to optics the concept of conjugate elements. In 1793, after the French Revolution, Lagrange headed a commission to draw up a new system of weights and measures and, together with Pierre Simon Laplace and Adrien Marie Legendre, devised the metric system, now in use in most countries of the world.

Axial Magnification

So far we have always assumed that object and image are both flat, that they are two-dimensional, and that they lie in planes normal to the optical axis. But many objects also have *depth* (they are three-dimensional), and thus we should also consider the magnification along the axis, the *axial magnification*. Axial magnification, M_X, is defined as the ratio of a short length in the image, measured along the axis, to the conjugate length in the object:

$$M_X = \frac{\Delta s'}{\Delta s} \qquad [2\text{-}18]$$

To see how axial magnification relates to transverse magnification, we take the lens equation, solve it for *f*, and using the notation in Figure 2-17, apply it first to the *head* of the arrow and then to its *foot*:

$$f = \frac{s_1 s_1'}{s_1 - s_1'} = \frac{s_2 s_2'}{s_2 - s_2'}$$

$$s_1 s_1'(s_2 - s_2') = s_2 s_2'(s_1 - s_1')$$

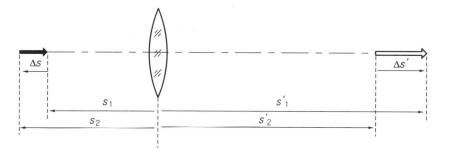

Figure 2-17   Axial magnification.

Multiplying and rearranging gives

$$s_1 s_2 s_2' - s_1 s_1' s_2 = s_1' s_2 s_2' - s_1 s_1' s_2'$$

$$(s_2' - s_1')(s_1 s_2) = (s_2 - s_1)(s_1' s_2')$$

and, since $s_2 - s_1 = \Delta s$ and $s_1' - s_2' = \Delta s'$,

$$\frac{s_2' - s_1'}{s_2 - s_1} = M_X = \frac{s_1' s_2'}{s_1 s_2} = M_{T1} M_{T2}$$

$$M_X = M_T^2 \qquad [2\text{-}19]$$

which shows that axial magnification is the *square* of transverse magnification.

The quadratic relationship between axial and transverse magnification accounts for the contraction in depth, the *foreshortening*, that we see on photographs taken with a telephoto lens. Consider the picture of a city street. Buildings at various distances appear in different sizes. Nearby buildings look larger and far-away buildings look smaller, in sizes predictable and familiar. But with a telephoto lens the far-away buildings are disproportionately larger and, since axial magnification is the square of the transverse, the street now gives the impression of an extraordinarily busy scene, with billboards and neon signs appearing close to each other.

Angular Magnification

Assume that initially the object is far away. The image formed on the retina is small. Then, as the object is brought closer, the image will grow larger. But there is a limit to how close the object can be brought, and still be seen in focus. This limit is a matter of age, but on the average the *distance of most distinct vision* is 25 cm [Figure 2-18 (top)].

But there is a way to bring the object closer (and hence see it larger) and *still keep it in focus*. This is possible by placing a converging lens in front of the eye [Figure 2-18 (center)]. A lens used that way is called a *magnifier*.

To find out how much the lens magnifies, again we make a *comparison*. Without the lens, the tangent of the angle subtended at the eye by the object is

$$\tan \theta = \frac{y}{-25} \qquad\qquad [2\text{-}20]$$

With the lens, the tangent is

$$\tan \theta' = \frac{y}{-f} \qquad\qquad [2\text{-}21]$$

The *angular magnification* (sometimes called "magnifying power"), therefore, is the ratio of these two tangents:

$$M_\theta = \frac{\tan \theta'}{\tan \theta} \qquad\qquad [2\text{-}22]$$

We insert Equations [2-21] and [2-20] in [2-22]; that gives

$$M_\theta = \frac{y/-f}{y/-25} = \frac{25}{f} \qquad\qquad [2\text{-}23]$$

with f given in centimeters. If instead of the focal length of the lens, we use its *power*, we have

$$\boxed{M_\theta = \tfrac{1}{4}P} \qquad\qquad [2\text{-}24]$$

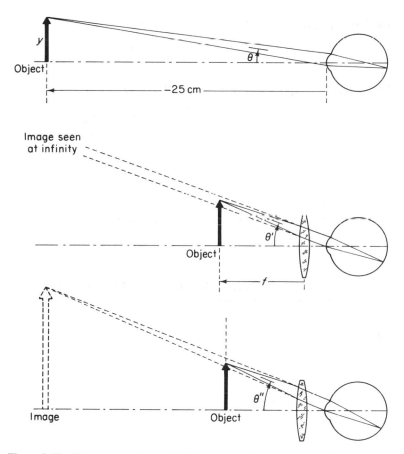

Figure 2-18 (*From top to bottom*) Object at the distance of most distinct vision, as viewed through magnifier and seen at infinity, and as viewed through magnifier and seen at −25 cm.

This is known as the "quarter-power equation." It applies when the magnifying lens is held close to the eye and the image is seen at infinity, *without accommodation*.

Next, let the observer *accommodate*. Accommodation means that the image is seen at the distance of most distinct vision,

$$s' = -25 \text{ cm}$$

As shown in Figure 2-18, bottom, the light entering the eye is now divergent, the same as when reading, and the object can be brought even closer, to within the focal length of the lens. This makes the object distance

$$s = \frac{s'f}{f - s'} = \frac{(-25)(f)}{f - (-25)}$$

so that, instead of $\tan \theta' = y/-f$, we have

$$\tan \theta'' = \frac{(y)(f + 25)}{-25f}$$

Inserting this expression in Equation [2-22] gives

$$M_\theta = \frac{y(f + 25)/-25f}{y/-25} = \frac{25}{f} + 1 \qquad [2\text{-}25]$$

or

$$\boxed{M_\theta = \tfrac{1}{4}P + 1} \qquad [2\text{-}26]$$

This means that, as the image is seen *with accommodation*, it is slightly more magnified than if seen without accommodation.

SUMMARY OF EQUATIONS

Single surface:

$$\frac{n}{s} + \frac{n' - n}{R} = \frac{n'}{s'} \qquad [2\text{-}3]$$

Surface power equation:

$$P = \frac{n' - n}{R} \qquad [2\text{-}5]$$

Lens-makers formula:

$$P_{\text{lens}} = (n_{\text{lens}} - n_0)\left(\frac{1}{R_1} - \frac{1}{R_2}\right) \qquad [2\text{-}8]$$

Thin-lens equation:

$$\frac{1}{s} + \frac{1}{f_2} = \frac{1}{s'} \qquad [2\text{-}10]$$

$$s = \frac{s'f}{f - s'} \qquad [2\text{-}11a]$$

$$s' = \frac{sf}{s + f} \qquad [2\text{-}11b]$$

$$f = \frac{ss'}{s - s'} \qquad [2\text{-}11c]$$

Transverse magnification:

$$M_T = \frac{y'}{y}$$

[2-12]

Angular magnification,
    without accommodation:

$$M_\theta = \tfrac{1}{4}P$$

[2-24]

    with accommodation:

$$M_\theta = \tfrac{1}{4}P + 1$$

[2-26]

PROBLEMS

2-1. Light coming out of a box has a vergence of -10 diopters. What is the vergence 90 cm farther away?

2-2. What is the vergence of light at a distance of 32 cm in front of an image point (that is, before it reaches the image)?

2-3. A point object is placed 10 cm in front of a convex surface of 2 cm radius of curvature, ground on a glass rod of $n' = 1.8$. How far from the surface is the image?

2-4. A glass sphere is used to form an image of a distant object. If the image is found to be located on the (back) surface of the sphere, what is the refractive index of the glass?

2-5. What radii of curvature must be ground onto a piece of glass of refractive index 1.64 to produce an equiconvex lens of 100 mm focal length?

2-6. A biconcave lens has a focal length of -25 cm. If the two surfaces have radii of curvature of 40 cm each, what is the index of refraction of the glass?

2-7. A thin lens of refractive index 1.5 has a focal length of 20 cm. If the radius of curvature of the back surface is -8 cm, what is the radius of the front surface?

2-8. A lens made of glass of $n = 1.5$ has, when in air, $+10.00$ diopters power. What is its power when immersed in a liquid of $n = 1.58$?

2-9. When an object is placed 75 mm in front of a converging lens, its image is three times as far away from the lens as when the object is at infinity. What is the focal length of the lens?

2-10. When an object is placed 15 cm in front of a thin lens, a virtual image is formed 5 cm away from the lens. What is the focal length of the lens?

2-11. An object is placed first at infinity and then 20 cm from the object-side focal plane of a converging lens. The two images thus formed are 8 mm apart from each other. What is the focal length of the lens?

2-12. An object is placed 30 cm from a lens having a focal length of +10 cm. Determine the image distance:

   (a) By the conventional form of the thin-lens equation.

   (b) By Newton's form of the lens equation.

2-13. Light is made to converge toward a certain point. If a −6.00-diopter lens is placed 25 cm ahead of this point, where will the light converge, or appear to come from?

2-14. An object is placed at a distance of one-half of the focal length in front of a converging lens. Where is the image located?

2-15. An object is located 1.25 m in front of a screen. Determine the focal length of a lens that forms on the screen a real, inverted image magnified four times.

2-16. If an object is placed 50 cm in front of a −3.00-diopter lens, what is the transverse magnification of the image?

2-17. If the real image formed by a lens is twice as large as the object, and if the distance between object and image is 90 cm, what is the power of the lens?

2-18. If a lens placed 4 cm to the right of an object produces a virtual image twice as large as the object, what is the power of the lens?

2-19. The object is a transparent cube, each edge 4 mm in length, placed 60 cm in front of a lens of +20 cm focal length. What is the transverse magnification of the rear surface of the cube (the face closer to the lens)?

2-20. Continue with the preceding problem and determine the image distance for the front surface of the cube. What is the *axial* magnification and how does it relate to the transverse magnification?

2-21. When a lens, held close to the eye and used without accommodation, magnifies three times, what is its power?

2-22. When an 8.5-diopter magnifying glass is held close to the eye and the image is seen at the distance of most distinct vision, what is its magnification?

SUGGESTIONS FOR FURTHER READING

FINCHAM, W. H. A., and M. H. FREEMAN. *Optics*, 10th ed. Boston: Butterworth, Inc., 1990.

HECHT, E. *Optics*, 2nd ed. Reading, Mass.: Addison-Wesley Publishing Company, Inc., 1987.

JENKINS, F. A., and H. E. WHITE. *Fundamentals of Optics*, 4th ed. New York: McGraw-Hill Book Company, 1976.

3

Thick Lenses and Combinations of Lenses

Customarily, lenses are divided into "thin lenses" and "thick lenses." But at what thickness does a thin lens end and a thick lens begin? That is entirely a matter of precision required for solving a given problem. The same lens may be considered "thin" for a preliminary, and "thick" for a rigorous solution. In addition, the term *lens*, especially when applied to a photographic camera, often means a *combination of lenses*, not a single lens. An example is the system shown in Figure 3-1.

THICK LENSES

As we saw in the preceding chapter, solving a thin-lens problem requires no more than simple thin-lens equations. Now let the light pass through a thick lens, or

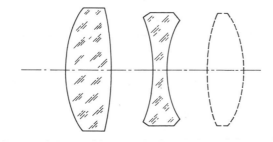

Figure 3-1 Introducing a combination of lenses.

49

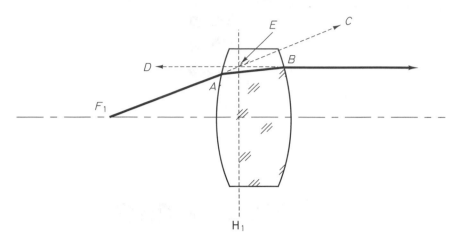

Figure 3-2   Using graphical construction to find the first principal plane, $\mathbf{H}_1$, of a thick lens, convex on both sides.

through a combination of lenses. Then the question arises from where to measure the focal length, the object distance, and image distance. The answer is: not from the center of the lens and not from either of its vertices but from certain hypothetical planes called *principal planes*. If we do this, we may *continue using convenient thin-lens equations*.

Assume, as shown in Figure 3-2, that a ray originates at the first focal point F_1 of a thick lens and that it is incident on the lens at a point A. At that point the ray is refracted toward a point B. There the ray is refracted again and, since the ray came from F_1, it emerges from the lens parallel to the optical axis.

Now we draw two extensions, a forward extension, AC, of the incident ray and a backward extension, BD, of the emerging ray. The two extensions intersect at point E. If we repeat the procedure for other rays, we find that points E all lie on the same plane, the *first principal plane*, $\mathbf{H}_1$.

Next consider the lens shown in Figure 3-3. Three rays come from an off-axis point on the left. Ray 1, the *parallel ray*, is first parallel (to the optical axis) and then passes through the second focal point, F_2. This ray locates the second principal plane, $\mathbf{H}_2$. Ray 2, the *focal ray*, goes through the first focal point, F_1, and then emerges parallel; it locates the first principal plane, $\mathbf{H}_1$. Ray 3, the *chief ray*, goes to the first principal point (the point where the axis intersects $\mathbf{H}_1$), is translated (shifted) along the axis to the second principal point, $\mathbf{H}_2$, and continues parallel to its initial direction.

We conclude that *the left-hand focal length, f_1, and the object distance, s, are measured from $\mathbf{H}_1$*, and that *the right-hand focal length, f_2, and the image distance, s', are measured from $\mathbf{H}_2$*.

Principal planes do not materially exist within a lens; they are merely where refraction is *assumed* to occur. They are conceptual planes. In a symmetrical lens,

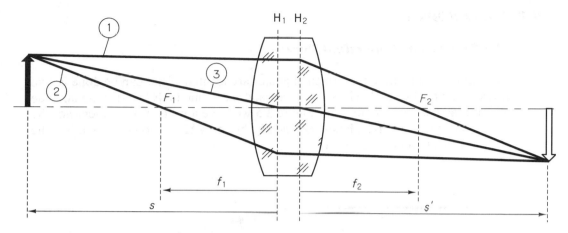

Figure 3-3 Principal planes, focal lengths, and object and image distances of a thick lens.

the principal planes, H_1 and H_2, lie at equal distances to both sides of the lens [Figure 3-4 (top left)]. In a plano-convex lens, and even more so in a meniscus, the two **H**-planes move to the side, and may even move outside the lens, toward the side of the steeper curvature. Their positions are invariant. They do not change with the object and image distances that occur in a particular case.

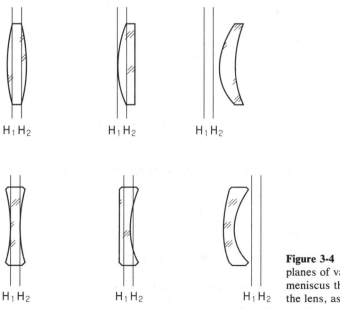

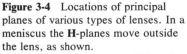

Figure 3-4 Locations of principal planes of various types of lenses. In a meniscus the **H**-planes move outside the lens, as shown.

EQUIVALENT POWER

The Concept of Equivalent Power

How can we find where the principal planes are located? That can be done in two ways, analytically (as I will show next) and experimentally, by the nodal slide method (as I will show later). In Figure 3-5 we have a system of two lenses. The system has two principal planes, H_1 and H_2, but only H_2 is shown.* Point F is the (second) focal point of the system.

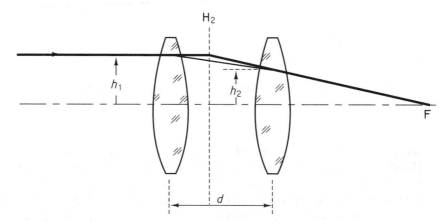

Figure 3-5   Deriving the equivalent-power equation.

Assume that a ray parallel to the optical axis is incident on lens 1 at a height h_1. At lens 2 the ray enters the lens at height h_2. Then consider the *slope angles* (not shown in the illustration) subtended by the ray and the axis. Behind (to the right of) the first lens the slope angle, u_1, is

$$u_1 = \frac{h_1}{f_1}$$

Since the reciprocal of the focal length of a lens is its *power*,

$$u_1 = P_1 h_1 \qquad \text{[3-1]}$$

In a similar way we find that behind (to the right of) the second lens the slope angle is

$$u_2 - u_1 = P_2 h_2$$

Adding these two equations gives

$$u_2 = P_1 h_1 + P_2 h_2 \qquad \text{[3-2]}$$

---

* With a single thick lens, H_1 lies to the left and H_2 to the right. But oddly enough, with a combination of two lenses, H_2 is to the left and H_1 to the right.

Next consider the two lenses together. Since the focal length of the combination, **f**, is measured from $\mathbf{H}_2$, slope angle u_2 is also

$$u_2 = \mathbf{P}h_1 \qquad\qquad [3\text{-}3]$$

where P is the power of the system. Substituting Equation [3-3] in [3-2] gives

$$\mathbf{P}h_1 = P_1h_1 + P_2h_2 \qquad\qquad [3\text{-}4]$$

From Figure 3-5 we also see that

$$u_1 = \frac{h_1 - h_2}{d}$$

or

$$h_2 - h_1 = -u_1 d$$

Substituting for u_1 Equation [3-1] gives

$$h_2 = h_1 - P_1 h_1 d$$

and substituting in Equation [3-4]

$$\mathbf{P}h_1 = P_1h_1 + P_2h_1 - P_1P_2h_1d$$

Canceling h_1 then gives the *equivalent power of the system,**

$$\boxed{\mathbf{P}_{\text{equivalent}} = P_1 + P_2 - P_1P_2d} \qquad\qquad [3\text{-}5]$$

Converting the powers again back into focal lengths, $P \to 1/f$, and multiplying the converted second and third terms by f_1f_2/f_1f_2 leads to

$$\mathbf{f}_{\text{equivalent}} = \frac{f_1f_2}{f_1 + f_2 - d} \qquad\qquad [3\text{-}6]$$

which is the *equivalent focal length of the system.*
If the two lenses are in contact, $d = 0$ and

$$\mathbf{P} = P_1 + P_2 \qquad\qquad [3\text{-}7]$$

---

* This equation was found by Alvar Gullstrand (1862–1930), Swedish ophthalmologist and professor of physiological and physical optics at the University of Uppsala. Gullstrand is best known for the *schematic eye* and for his development of the slit lamp. He also contributed to the tracing of skew rays, caustic surfaces (in aberrations), accommodation, and aspheric lenses for aphakics and to the propagation of light through gradient-index media (such as the crystalline lens of the eye). In 1911 Gullstrand received the Nobel Prize in physiology. A few years later he published a detailed exposition of geometrical-optical image formation, replete with numerous equations, "Das allgemeine optische Abbildungssystem" (General optical image formation), *K. Sven. vetenskapsakad. handl.* (4) **55** (1915–1916), 1–139.

which is the *approximate power* of the combination, and

$$\mathbf{f} = \frac{f_1 f_2}{f_1 + f_2} \qquad\qquad [3\text{-}8]$$

which is the *approximate focal length.*

   If, instead of two lenses, we have two *surfaces* separated by d, that is, if we have a rather *thick lens* (of thickness d and index n), then its equivalent power is

$$\mathbf{P} = P_1 + P_2 - P_1 P_2 \frac{d}{n} \qquad\qquad [3\text{-}9]$$

and its equivalent focal length

$$\mathbf{f} = \frac{f_1 f_2}{f_1 + f_2 - d/n} \qquad\qquad [3\text{-}10]$$

Example

   What happens if two positive lenses of equal power are first in contact and then are gradually pulled apart from each other?

   Solution: With the lenses in contact, the power of the combination is highest, the focal length least, and the two **H** planes are close together inside the system [Figure 3-6(a)]. Then, as we separate the lenses, the **H** planes separate too. For instance, when the distance between the lenses is equal to their focal length, the **H** planes coincide with the lenses (b). When the distance is larger than the focal length, the **H** planes have moved outside the lenses (c); and when the distance is equal to $2f$ and the system becomes *afocal* (zero power), $\mathbf{H}_1$ and $\mathbf{H}_2$ have moved out to infinity (d). Predictably, as we separate the lenses even farther, the power of the system becomes less than zero; that is, it becomes *negative*. Note that the ray that emerges on the right-hand side has continued to turn counterclockwise and now, in (e), is pointing *upward*, apparently refracted at $\mathbf{H}_2$, as it should.

   The question of whether a system has positive or negative power is easy to answer: If the emergent ray, after refraction at $\mathbf{H}_2$, points toward the axis and crosses it at $\mathbf{F}_2$ [(a), (b), and (c)], the system is positive. If the ray turns away from the axis, without crossing it (e), the system is negative.

   Assume, for example, that the two lenses have $+8.00$ diopters power each and that they are separated by a distance d. At which distance does the power of the combination turn from positive to negative?

   Numerical solution: We take the equivalent-power equation and set it equal to zero:

$$(+8.00) + (+8.00) - (+8.00)(+8.00)(d) = 0$$

Then

$$16 = 64d$$

and

$$d = \frac{16}{64} = 0.25m = \boxed{25 \text{ cm}}$$

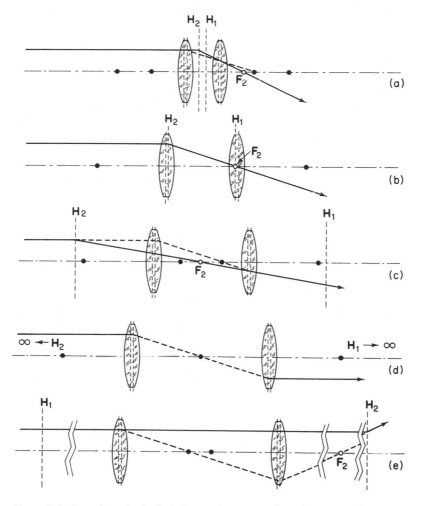

Figure 3-6 Location of principal planes of a system of two lenses as a function of distance between the lenses. Focal points of individual lenses are identified by dots, second focal points of system, **F**$_2$ by hollow circles.

Since $1/8.00 = 12.5$ cm, this is twice the focal length of either lens; at that distance the system is *afocal*.

Nodal Points

Consider a bundle of light that comes from a point somewhere off-axis. Within this bundle there is only one ray that has the *same direction* both *before* reaching the lens and *after* leaving it. That is the chief ray, shown in Figure 3-7. It intersects the optical axis at the two *nodal points*, N_1 and N_2.

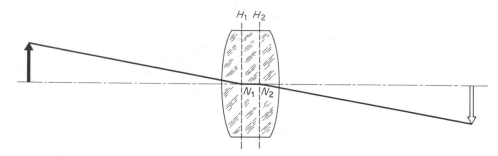

Figure 3-7   Nodal points. Note displacement, but no deviation, of the chief ray.

These two nodal points, together with the two focal points and the two principal points, H_1 and H_2, are known as the *cardinal points* of a lens or lens system. Whenever the refractive index in front of the lens is the same as the index behind it, the nodal points coincide with the principal points. If the refractive indices are different, the N points move away from the H planes, toward the side of the higher index. Thus, from knowing the locations of the N points we find the H planes (and these we like to know because from there the focal lengths and the object and image distances are measured). We find the N points using the *nodal slide method*.

In order to find the nodal points, we use a *nodal slide*. The nodal slide is a mechanical support for a lens or lens system that can be turned about a vertical axis. At the same time, the axis of rotation can be moved back and forth along the optical axis. First move the lens holders *on top of* the nodal slide along the optical axis in one direction and then move the slide *as a whole* an equal distance in the opposite direction. *Keep the image in focus all the time!* At the position where, on turning the slide, there is *no sideways displacement of the image*, the axis of rotation, as shown in Figure 3-8, goes through *the second nodal point, N_2*.

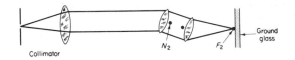

Figure 3-8   Schematic diagram of nodal slide (seen from the top). Rotation of lens system about N_2 will cause no image motion at $\mathbf{F}_2$.

VERTEX POWERS

Front Vertex Power and Back Vertex Power

Equivalent power is measured *at* the respective principal plane. This statement makes more sense when we realize that the reciprocal of equivalent power is the equivalent focal length and that this focal length is measured *from* the respective principal plane. Unfortunately, we often do not know where the principal planes are, and when we do know, they often are inaccessible within the lens.

What we need, therefore, are clearly defined and easily accessible reference points from which to measure. These reference points are the *vertices* of the two surfaces of the lens. For example, the *front vertex focal length* of a lens, f_{FV}, is measured from the front surface, and the *back vertex focal length*, f_{BV}, is measured from the back surface (Figure 3-9). The reciprocals of these focal lengths are the *front vertex power*, P_{FV}, and the *back vertex power*, P_{BV} (not shown).*

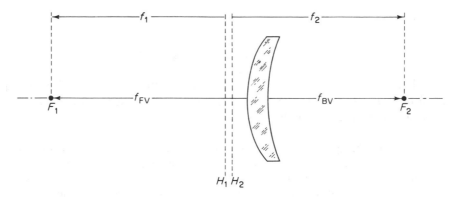

Figure 3-9 Conventional focal lengths, f_1 and f_2, and front vertex focal length, f_{FV}, and back vertex focal length, f_{BV}, of a positive meniscus.

With a lens of appreciable thickness, or with two lenses separated by an appreciable distance, we can no longer use the approximate-power equation, $\mathbf{P} = P_1 + P_2$. At the front vertex, P_1 would be measured correctly but, because of the thickness or separation, P_2 would not. Therefore, P_2 has to change.

To find out how much P_2 has to change, consider the vergence in two planes, 1 and 2 in Figure 3-10. In plane 1 the vergence is

$$\mathbf{V}_1 = \frac{n}{L}$$

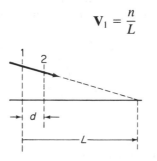

Figure 3-10 Change of vergence.

---

* Be sure to distinguish between back *surface* power and back *vertex* power. Back surface power is simply the power of the back surface. But back vertex power refers to the power of the whole lens (front surface plus back surface), *as measured at the back vertex.*

In plane 2 it is

$$\mathbf{V}_2 = \frac{n}{L - d}$$

Solving for L in the first equation, substituting in the second equation, and multiplying both numerator and denominator by $\mathbf{V}_1/n$ gives

$$\mathbf{V}_2 = \frac{n}{(n/\mathbf{V}_1) - d} = \frac{n(\mathbf{V}_1/n)}{(n/\mathbf{V}_1 - d)(\mathbf{V}_1/n)}$$

from where

$$\boxed{\mathbf{V}_2 = \frac{\mathbf{V}_1}{1 - \mathbf{V}_1(d/n)}} \qquad [3\text{-}11]$$

which is the *change-of-vergence equation*. Note that here, and in the equations to follow, n is the refractive index of the medium that surrounds the lens; it is *not* the refractive index of the lens as such.

The change-of-vergence equation, in essence, is also a *change-of-power equation*. As an example, assume that a three-diopter lens is placed in front of the eye but that it slides down the nose by 20 mm. Now what is its *effective power? Answer.* Moving the lens to the left, toward the object, makes d negative, $d = -0.02$, and therefore

$$P_2 = \frac{P_1}{1 - P_1 d} = \frac{3}{1 - (3)(-0.02)} = 2.83 \text{ diopters}$$

Now substitute P_2 in the approximate-power equation. This gives for the *front vertex power*

$$\boxed{\mathbf{P}_{FV} = P_1 + \frac{P_2}{1 - P_2(d/n)}} \qquad [3\text{-}12]$$

Back vertex power is the most important of all. Virtually all ophthalmic procedures, from determining visual acuity to writing lens prescriptions to "verifying" a lens using a lensometer, refer to the back vertex power. Similar to the derivation of Equation [3-12] we now replace the P_1 term; this gives

$$\boxed{\mathbf{P}_{BV} = \frac{P_1}{1 - P_1(d/n)} + P_2} \qquad [3\text{-}13]$$

Example

Two plano-convex lenses of $+5.00$ diopters power each, their plane sides facing each other, are separated by a distance of 8 cm.

(a) Determine the positions of the principal planes.

(b) Let the space between the lenses be filled with glass ($n = 1.523$), that is, consider a rather thick lens. Again find the principal planes.

Solution: (a) First, using Equation [3-13], determine the *back vertex power*:

$$\mathbf{P}_{BV} = \frac{P_1}{1 - P_1(d)} + P_2 = \frac{+5}{1 - (+5)(0.08)} + (+5) = +13.33 \text{ diopters}$$

The reciprocal is the *back vertex focal length*:

$$\mathbf{f}_{BV} = \frac{1}{+13.33}(100) = +7.5 \text{ cm}$$

Then, using Equation [3-6], determine the *equivalent focal length*:

$$\mathbf{f} = \frac{f_1 f_2}{f_1 + f_2 - d} = \frac{(20)(20)}{20 + 20 - 8} = +12.5 \text{ cm}$$

Since $\mathbf{f}_{BV}$ is measured from the last surface, and $\mathbf{f}$ from $\mathbf{H}_2$, the difference between the two,

$$12.5 - 7.5 = \boxed{5 \text{ cm}}$$

means that $\mathbf{H}_2$ is located 5 cm to the left of the last lens, or *1 cm to the left* of the midpoint between the two lenses.

(b) With the space between the lenses filled with glass, the back vertex power is

$$\mathbf{P}_{BV} = \frac{+5}{1 - (+5)(0.08/1.523)} + (+5) = +11.78 \text{ diopters}$$

the back vertex focal length,

$$\mathbf{f}_{BV} = +8.5 \text{ cm}$$

and the equivalent focal length,

$$\mathbf{f} = \frac{(20)(20)}{20 + 20 - (8/1.523)} = +11.5 \text{ cm}$$

This time, $\mathbf{H}_2$ is located

$$11.5 - 8.5 = \boxed{3 \text{ cm}}$$

to the left of the last lens, or 1 cm *to the right* of the midpoint, as it should be with a single lens.

MATRIX ALGEBRA

Describing a Lens Using Matrix Algebra

A matrix is an array of numbers, called *elements*, that are arranged in rows and columns. Matrix algebra as such has been known for years but, at least in optics, at first had not found much attention. The computer has changed all that.

Consider again the graphical construction that we had used to determine the principal planes of a thick lens (Figure 3-11). At both the front surface and the back surface of the lens, that is, at points *A* and *B*, the ray changes direction but it does not change in height; that, as we know, is called *refraction*. Between surfaces, the ray changes in height but it does not change direction; that is called *translation*.

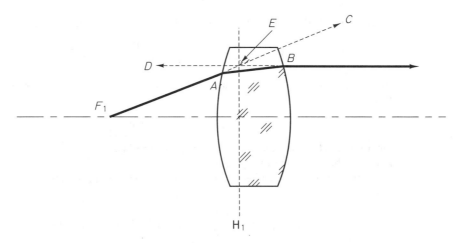

Figure 3-11   A ray passing through a thick lens undergoes both *refraction* and *translation*.

Consequently, as we follow the ray as it passes through the lens, we need two operators, one for the refraction process, the other for the translation. These two processes are linear transformations and as such can be handled very well by matrix algebra. The two operators are the *refraction matrix* and the *translation matrix*.

Use of such operators has many advantages. Most important perhaps is that, when we have a system of many lenses, calculating the various segments of a ray in the more traditional way would be tedious and time consuming. But if the radii and the separations of the surfaces are given in matrix form, a computer could do it quickly because all that is involved are multiplications and additions. A great many rays can then be traced from object to image much faster than what could be done otherwise. Matrices also connect readily to the concept of vergence. Finally, matrices also play an important role in physical optics such as in the design of multilayer filters and in polarization (Chapter 17).

Let us begin with the *refraction matrix*. We take Gauss' formula for refraction at a single surface, Equation [2-3],

$$\frac{n}{s} + \frac{n' - n}{R} = \frac{n'}{s'}$$

where n and n' are the refractive indices to the left and right of the surface, respectively, s is the object distance, s' the image distance, and R the radius of curvature of the surface. Replacing the second term by the power of the surface and rearranging gives

$$P + \frac{n}{s} = \frac{n'}{s'} \qquad \text{[3-14]}$$

The slope angles subtended by the ray and the optical axis, as we also have seen earlier, are

$$u = \frac{h}{s} \quad \text{and} \quad u' = \frac{h}{s'}$$

where h is the height of the ray as it enters the surface. Solving for s and s' and substituting in Equation [3-14] yields

$$\underset{\text{refraction}}{P} + \underset{\text{input}}{\frac{nu}{h}} = \underset{\text{output}}{\frac{n'u'}{h}} \qquad \text{[3-15]}$$

Note the sequence in which these terms are written, refraction *to the left* of input.

Now we convert Equation [3-15] to matrix form:

$$\underset{\text{refraction}}{\begin{bmatrix} 1 & P \\ 0 & 1 \end{bmatrix}} \; \underset{\text{input}}{\begin{bmatrix} nu \\ h \end{bmatrix}} = \underset{\text{output}}{\begin{bmatrix} n'u' \\ h \end{bmatrix}} \qquad \text{[3-16]}$$

with the angles, u and u', in radians. The input and output terms are 2×1 matrices (2 rows, 1 column). The left-hand term is the *refraction matrix*, R,

$$\mathsf{R} = \begin{bmatrix} 1 & P \\ 0 & 1 \end{bmatrix} \qquad \text{[3-17]}$$

If the surface is plane, the upper right-hand element becomes zero, $P = 0$, and we have a *unit matrix*.

In order to use such matrices, and to transform an initial set of coordinates into a new set of coordinates, the matrices are multiplied. Consider, for example, the two matrices

$$\begin{bmatrix} 4 & 3 \\ 6 & 5 \end{bmatrix}\begin{bmatrix} 1 \\ 2 \end{bmatrix}$$

Take the top row in the left-hand matrix, the shaded area containing 4   3, and multiply it by the right-hand matrix. Add the two products. This gives the *upper* element in the resultant matrix:

$$\begin{bmatrix} 4 & 3 \\ \cdot & \cdot \end{bmatrix}\begin{bmatrix} 1 \\ 2 \end{bmatrix} \rightarrow \begin{matrix} 4 \times 1 \\ 3 \times 2 \end{matrix} \rightarrow 4 + 6 \rightarrow \begin{bmatrix} \boxed{10} \\ \end{bmatrix}$$

Then take the bottom row. Multiply and add. This gives the *lower* element:

$$\begin{bmatrix} 4 & 3 \\ 6 & 5 \end{bmatrix}\begin{bmatrix} 1 \\ 2 \end{bmatrix} \rightarrow \begin{matrix} 6 \times 1 \\ 5 \times 2 \end{matrix} \rightarrow 6 + 10 \rightarrow \begin{bmatrix} 10 \\ \boxed{16} \end{bmatrix}$$

Example

A ray of light, emerging from an axial point object, subtends an angle of −6° with the optical axis. At a height $h = 2$ cm it intersects *a convex spherical surface, ground with a radius of +5 cm on glass of index n = 1.62. Find the angle the ray subtends with the axis after refraction.*

Solution: First we determine the power of the surface:

$$P = \frac{n' - n}{R} = \frac{1.62 - 1.00}{0.05} = 12.4 \text{ diopters}$$

Next we convert the angle into radians, $-6° = -0.1047$ rad. Then from Equation [3-16],

$$\begin{bmatrix} 1 & 12.4 \\ 0 & 1 \end{bmatrix}\begin{bmatrix} -0.1047 \\ 0.02 \end{bmatrix} = \begin{bmatrix} -0.1047 + 0.248 \\ 0.02 \end{bmatrix} = \begin{bmatrix} 0.1433 \\ 0.02 \end{bmatrix} = \begin{bmatrix} n'u' \\ h \end{bmatrix}$$

The angle subtended by the refracted ray is

$$u' = \frac{0.1433}{n'} = \frac{0.1433}{1.62} = 0.0885 \text{ rad} = \boxed{+5°}$$

Translation Matrix

Next we consider the change in height, Δh, of the ray as it proceeds from one surface to the next. If d is the distance (measured along the axis) between the two surfaces, then

$$\Delta h = h_1 - h_2 = d \tan u'$$

where u' is the angle (in radians) which the ray subtends with the axis after refraction. For paraxial rays,

$$h_1 - h_2 = du'$$

Dividing both sides by $n'u'$ and rearranging yields

$$-\frac{d}{n'} + \frac{h_1}{n'u'} = \frac{h_2}{n'u'} \qquad\qquad [3\text{-}18]$$
$$\text{translation} \quad \text{input} \quad \text{output}$$

which can be inverted and again be written in matrix form:

$$\begin{bmatrix} 1 & 0 \\ -d/n' & 1 \end{bmatrix} \begin{bmatrix} n'u' \\ h_1 \end{bmatrix} = \begin{bmatrix} n'u' \\ h_2 \end{bmatrix} \qquad [3\text{-}19]$$

$$\underbrace{}_{\text{translation}} \quad \underbrace{}_{\text{input}} \quad \underbrace{}_{\text{output}}$$

The left-hand term is the translation matrix, T:

$$\mathsf{T} = \begin{bmatrix} 1 & 0 \\ -d/n & 1 \end{bmatrix} \qquad [3\text{-}20]$$

System Matrix

Obviously, when a ray passes through a lens, especially a thick lens, (rather than through a single surface), refraction occurs twice, *at* each surface, and translation occurs once, *between* surfaces. Therefore, in order to fully describe the lens, we need two refraction matrices, R_1 and R_2, and one translation matrix, T. These matrices together, written again from right to left, form the *system matrix*, S,

$$\mathsf{R_2} \ \ \mathsf{T} \ \ \mathsf{R_1} = \mathsf{S}$$

and, substituting the complete refraction and translation matrices,

$$\begin{bmatrix} 1 & P_2 \\ 0 & 1 \end{bmatrix} \begin{bmatrix} 1 & 0 \\ -d/n & 1 \end{bmatrix} \begin{bmatrix} 1 & P_1 \\ 0 & 1 \end{bmatrix} = \mathsf{S} \qquad [3\text{-}21]$$

Multiplication of two 2×2 matrices is carried out as follows:

$$\begin{bmatrix} 6 & 5 \\ 8 & 7 \end{bmatrix} \begin{bmatrix} 2 & 1 \\ 4 & 3 \end{bmatrix}$$

Take the top row in the left-hand matrix, the shaded area containing 6 5, and multiply it with the right column, the shaded area containing 1 3, in the right-hand matrix. Add the products. This gives the upper right-hand element in the resultant matrix:

$$\begin{bmatrix} 6 & 5 \\ \cdot & \cdot \end{bmatrix} \begin{bmatrix} \cdot & 1 \\ \cdot & 3 \end{bmatrix} \rightarrow \begin{matrix} 6 \times 1 \\ 5 \times 3 \end{matrix} \rightarrow 6 + 15 \rightarrow \begin{bmatrix} & \boxed{21} \\ & \end{bmatrix}$$

Continue with the top row and multiply it with the left-hand column, 2 4. This gives the upper left-hand element:

$$\begin{bmatrix} 6 & 5 \\ \cdot & \cdot \end{bmatrix} \begin{bmatrix} 2 & 1 \\ 4 & 3 \end{bmatrix} \rightarrow \begin{matrix} 6 \times 2 \\ 5 \times 4 \end{matrix} \rightarrow 12 + 20 \rightarrow \begin{bmatrix} \boxed{32} & 21 \\ & \end{bmatrix}$$

Do the same with the bottom row of the left-hand matrix and the right-hand column of the right-hand matrix:

$$\begin{bmatrix} 6 & 5 \\ 8 & 7 \end{bmatrix}\begin{bmatrix} 2 & 1 \\ 4 & 3 \end{bmatrix} \quad \begin{matrix} 8 \times 1 \\ 7 \times 3 \end{matrix} \rightarrow 8 + 21 \rightarrow \begin{bmatrix} 32 & 21 \\ & \boxed{29} \end{bmatrix}$$

and again with the left-hand column:

$$\begin{bmatrix} 6 & 5 \\ 8 & 7 \end{bmatrix}\begin{bmatrix} 2 & 1 \\ 4 & 3 \end{bmatrix} \quad \begin{matrix} 8 \times 2 \\ 7 \times 4 \end{matrix} \rightarrow 16 + 28 \rightarrow \begin{bmatrix} 32 & 21 \\ \boxed{44} & 29 \end{bmatrix}$$

If we have two lenses, separated by a certain distance from each other, the system matrix is

$$\mathbf{R_4 \ T_3 \ R_3 \ T_2 \ R_2 \ T_1 \ R_1 = S} \tag{3-22}$$

where R_1, T_1, and R_2 refer to the first lens, T_2 refers to the translation between the lenses, and R_3, T_3, and R_4 refer to the second lens. The point is that any optical system, no matter how complex, can be fully described by one system matrix, just as such a system can be represented by one set of principal planes.

A system matrix contains four elements, a, b, c, d, known as *Gaussian constants*:

$$\mathbf{S} = \begin{bmatrix} b & a \\ d & c \end{bmatrix} \tag{3-23}$$

These constants represent various parameters of the system: Constant a is the *equivalent power*. Its reciprocal is the *equivalent focal length*,

$$\mathbf{f_1} = -\frac{n_1}{a} \quad \text{and} \quad \mathbf{f_2} = \frac{n_3}{a} \tag{3-24}$$

Constant b is the *angular magnification*. Constants a, b, and c together define the *front vertex focal length*, v_1,

$$v_1 = -\frac{bn_1}{a} \tag{3-25}$$

and the *back vertex focal length*, v_2,

$$v_2 = \frac{cn_3}{a} \tag{3-26}$$

Constant d is the (negative, reduced) *thickness* of the lens, $-d/n_2$.

Example

A lens has radii of curvature $R_1 = +8$ cm, $R_2 = -5$ cm, a center thickness $d = 1$ cm, and a refractive index $n = 1.6$. What is: (a) The system matrix? (b) The focal length?

Solution: (a) The surface powers are

$$P_1 = \frac{n' - n}{R} = \frac{1.6 - 1.0}{+0.08 \; m} = +7.5 \text{ diopters}$$

and

$$P_2 = \frac{1.0 - 1.6}{-0.05 \; m} = +12 \text{ diopters}$$

Thus the system matrix is

$$S = \begin{bmatrix} 1 & 12 \\ 0 & 1 \end{bmatrix} \begin{bmatrix} 1 & 0 \\ -0.01/1.6 & 1 \end{bmatrix} \begin{bmatrix} 1 & 7.5 \\ 0 & 1 \end{bmatrix} = \begin{bmatrix} 1 & 12 \\ 0 & 1 \end{bmatrix} \begin{bmatrix} 1 & 7.5 \\ -0.00625 & 0.953 \end{bmatrix}$$

$$= \begin{bmatrix} 0.925 & 18.936 \\ -0.00625 & 0.953 \end{bmatrix}$$

To check the result, calculate the *determinant*, multiplying the elements on the rightward arrow and subtracting the product of the elements on the leftward arrow:

$$\begin{bmatrix} 0.925 & 18.936 \\ -0.00625 & 0.953 \end{bmatrix}$$

$$(0.925)(0.953) - (18.936)(-0.00625) = 1$$

The result should be *unity*, as indeed it is.

(b) The focal length or, more precisely, the second, right-hand, equivalent focal length is simply the reciprocal of the equivalent power, a:

$$f_2 = \frac{1}{18.936} = 0.0528 = \boxed{+5.28 \text{ cm}}$$

The Matrix Description of Image Formation

The system matrix describes the properties of a lens or a combination of lenses, in a way similar to the lens-makers formula. Now we turn again to a lens-users equation. Consider the "typical" case of an axial point object, a converging lens, and a real image. Clearly, there is first a translation, from object to lens, then the system matrix, and then another translation, from the lens to the image.

The distances in these two translations are the object distance, s, and the image distance, s'. The sequence from object to image, therefore, can be written in the form of an *object-image matrix*,

$$\underbrace{\begin{bmatrix} 1 & 0 \\ -s' & 1 \end{bmatrix}}_{\substack{\text{image} \\ \text{distance}}} \underbrace{\begin{bmatrix} b & a \\ d & c \end{bmatrix}}_{\text{system}} \underbrace{\begin{bmatrix} 1 & 0 \\ -s & 1 \end{bmatrix}}_{\substack{\text{object} \\ \text{distance}}} \underbrace{\begin{bmatrix} u \\ h \end{bmatrix}}_{\text{input}} = \underbrace{\begin{bmatrix} u' \\ h \end{bmatrix}}_{\text{output}} \qquad \text{[3-27]}$$

Since much of image formation can be presented in terms of vergence, we also may use a *vergence matrix*, **V**. That is simply a 2 × 1 matrix which can be derived from the Smith-Helmholtz relationship. By definition, nu/h is the entrance vergence, **V**, and $n'u'/h$ is the exit vergence, **V'**. Multiplying by $1/h$ and substituting gives

$$\begin{bmatrix} 1 & P \\ 0 & 1 \end{bmatrix} \begin{bmatrix} V \\ 1 \end{bmatrix} = \begin{bmatrix} V' \\ 1 \end{bmatrix} \qquad [3\text{-}28]$$

The use of the vergence matrix, the same as vergences in general, is limited to points located on the optical axis. The vergence matrix is no substitute for exact ray tracing. Nevertheless, it is a useful approach because vergences, in the paraxial approximation, are independent of the height above the axis.

Determining Focal Length. There are many ways of determining the focal length of a lens or combination of lenses. We distinguish the following:

1. Measure the object distance and the image distance, and calculate the (equivalent) focal length using the *thin-lens equation*,

$$f = \frac{ss'}{s - s'}$$

2. *Autocollimation*. (Collimation means placing a source or target in the left-hand focal plane of a converging lens so as to obtain parallel light. Autocollimation means that the light, in addition, is reflected back along the same path and comes to a focus in the plane of the source. In a way, collimation is one-half of autocollimation.)

Place an illuminated slit or other target cut into a screen near the left-hand end of an optical bench and the lens to be measured to the right of the slit. Hold a plane mirror next to (the right of) the lens. (It is usually better to simply hold the mirror instead of mounting it in a lens holder, because then you can tell from the wiggly motion that the light is actually reflected by the mirror, rather than by the rear surface of the lens.) Slide the lens, together with the mirror, back and forth along the axis until the image of the slit is in focus on the screen (Figure 3-12). The distance from the screen to the lens is the focal length of the lens.

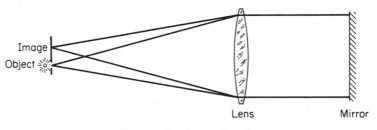

Figure 3-12   Autocollimation.

3. *Abbe's method*. Use an aperture as the object and form an image of it on a screen. Record the position of the object. Measure object size and image size. Determine the magnification M_1. Then, leaving the lens in place, move the object a short distance A to the right (which makes A positive) and focus again, moving the screen only. Determine the

new magnification M_2. Take the thin-lens equation, $1/s + 1/f = 1/s'$, multiply both sides by s, and replace s/s' by the inverse of the magnification. That gives

$$1 + \frac{s}{f} = \frac{1}{M}$$

Multiply by f and solve for s and do this twice, for s_1 and s_2:

$$s_1 = f\left(\frac{1}{M_1} - 1\right)$$

$$s_2 = f\left(\frac{1}{M_2} - 1\right)$$

Subtract one from the other, replace $s_2 - s_1$ by A, and solve for f; that gives *Abbe's equation*,

$$f = \frac{AM_1M_2}{M_1 - M_2} \qquad [3\text{-}29]$$

This method, preferred in some industrial laboratories, is convenient for thick lenses because no measurements need be made from the lens itself.

4. The focal length of a *minus lens* cannot be determined directly as it can with a plus lens. Therefore another approach is needed. For example, combine the minus lens with a sufficiently strong plus lens, but the two lenses need not be in contact. First form an image using the plus lens only. Note where the image is in focus; this is screen position 1 (Figure 3-13).

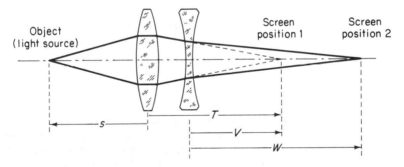

Figure 3-13 Determining the focal length of a minus lens.

Then insert the minus lens. Move the screen to the right to bring the image once again into focus (screen position 2). The earlier image (position 1) now acts as a virtual object for the minus lens. The former image distance T, less the separation of the two lenses, becomes the object distance V, and the distance from the minus lens to screen position 2 becomes the image distance W. Since all three distances, T, V, and W, are measured to the right, they are all positive. The focal length of the minus lens then follows from

$$f = \frac{VW}{V - W} \qquad [3\text{-}30]$$

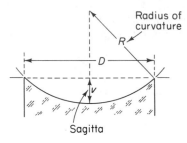

Figure 3-14 Deriving the vertex depth formula.

5. A *lens clock* actually measures the curvature of a surface, but for a given refractive index, it can be calibrated to show the surface *power*. The clock has two outer points a distance D apart, and a movable inner point. When the three points touch a plane surface, they form a straight line; on a curved surface, by contrast, the center point either protrudes, or is pushed in, by a certain distance v, called the *vertex depth or sagitta* (Figure 3-14). From Pythagoras' theorem

$$(R - v)^2 + \left(\frac{D}{2}\right)^2 = R^2$$

$$R^2 - 2Rv + v^2 + \left(\frac{D}{2}\right)^2 = R^2$$

$$2Rv = v^2 + \left(\frac{D}{2}\right)^2$$

Since $v \ll R$, v^2 may be neglected and

$$2Rv \approx \left(\frac{D}{2}\right)^2$$

Solving for R and substituting in the surface power equation then leads to

$$P \approx \frac{2(n' - n)v}{(D/2)^2} \qquad [3\text{-}31]$$

which shows that, D being constant, P is directly proportional to v.

6. A *lensometer* is used to measure the power, and other characteristics, of a lens. A lensometer consists of a light source, a target that can be moved back and forth along the axis, a standard lens, and a telescope focused for both object and image at infinity. If there is no other lens present in the path, the target is seen in focus [Figure 3-15 (top)].

Now assume the lens to be tested is placed between a standard lens and telescope (usually at a location 50 mm from the standard lens, at its right-hand focus). If this lens is a *plus* lens, the light entering the telescope is convergent, and the target is not seen in focus any longer (center). To make the light parallel again, the light to the left of the test lens must be divergent, and the light to the left of the standard lens must be even more divergent; hence, the target must be moved to the *right* (bottom).

The opposite holds true for a *minus* lens. This time the light entering the telescope is divergent, and to make it parallel again, the target must be moved to the *left*.

7. In recent years, fully automated *electronic lensometers* have come into use. The lens is simply inserted into the system and within seconds the built-in optics and subse-

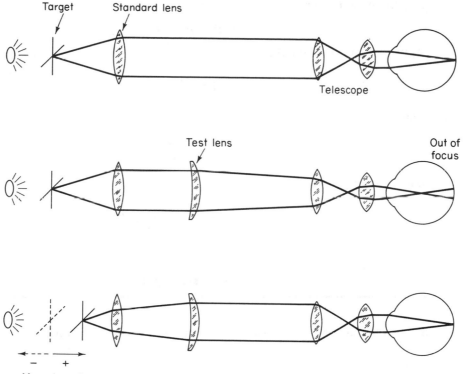

Figure 3-15   Ray diagram of a lensometer. (*Top*) Standard lens, with target at F_1, produces parallel light. Telescope is focused for infinity. Target is seen in focus. (*Center*) Adding unknown lens (plus lens in the example shown) will defocus image. (*Bottom*) Moving target brings image back into focus.

quent microprocessor produce the result, often in the form of both a numerical display and a printout.

SUMMARY OF EQUATIONS

Equivalent power:

$$\mathbf{P}_{\text{equivalent}} = P_1 + P_2 - P_1 P_2 d \qquad \text{[3-5]}$$

Equivalent focal length:

$$\mathbf{f}_{\text{equivalent}} = \frac{f_1 f_2}{f_1 + f_2 - d} \qquad \text{[3-6]}$$

Approximate power:

$$\mathbf{P} = P_1 + P_2 \tag{3-7}$$

Approximate focal length:

$$\mathbf{f} = \frac{f_1 f_2}{f_1 + f_2} \tag{3-8}$$

Change of vergence:

$$\mathbf{V}_2 = \frac{\mathbf{V}_1}{1 - \mathbf{V}_1(d/n)} \tag{3-11}$$

Back vertex power:

$$\mathbf{P}_{BV} = \frac{P_1}{1 - P_1(d/n)} + P_2 \tag{3-13}$$

PROBLEMS

3-1. In the lens shown in Figure 3-16, find the object point conjugate to image point S'.

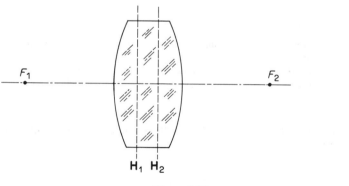

Figure 3-16

3-2. Draw the principal planes inside a thick equiconcave lens and show how a parallel ray, a focal ray, and a chief ray, all coming from the tip of an extended object, are (assumed to be) refracted at these planes.

3-3. Two thin lenses, one of 10 diopters power and the other of 5 diopters, are separated by a distance of 10 cm. What is the equivalent power of the combination?

3-4. Two lenses of -10.00 diopters power each are separated by a distance of 5 cm. What is the equivalent focal length of the system?

3-5. Two thin lenses of focal length +20 cm and +5 cm, respectively, are separated such that the system becomes *afocal*. What is the separation of the lenses?

3-6. A lens of +25 cm focal length is placed 15 cm in front of another lens. To make the system afocal, what must the power of the second lens be?

3-7. Two lenses, one with five times the power of the other, are separated by 60 cm. At this distance the combination is afocal. What are the powers of the two lenses?

3-8. Two thin lenses have a combined (approximate) power of +10.00 diopters. If they become separated by 20 cm, their equivalent power decreases to +6.25 diopters. What are the powers of the two lenses?

3-9. Nodal points are where rays, *without being refracted*, intersect the optical axis. Using the same definition, where is the nodal point of a *single refracting surface*?

3-10. Two +6.00-diopter lenses are placed a short distance apart from each other. The nodal points of the system, therefore, are located symmetrically between the lenses. If then one of the +6.00 lenses is replaced by a +10.00 lens, how do the nodal points change?

3-11. If a biconvex lens has surfaces of +4.00 diopters power each, and if the lens is 1 cm thick and made of glass of $n = 1.6$, what is its back vertex power?

3-12. A 10.6-mm-thick lens has +16.00 diopters front surface power and −6.00 diopters back surface power. If the lens is made of spectacle crown ($n = 1.523$), what is its back *vertex* power?

3-13. When two lenses of +4.00 diopters power each are separated by 5 cm, by how much does their back vertex power differ from their equivalent power?

3-14. Two thin lenses of +5.00 diopters each are separated by a distance of $6\frac{2}{3}$ cm. If the distance from the second lens to the focus is 10 cm, where is the second principal plane of the system?

3-15. Parallel light is incident on a surface of +6 diopters power. Using matrices, determine the (second) focal length of the surface.

3-16. A meniscus lens has surface powers $P_1 = +8$ 1/m and $P_2 = -5$ 1/m, glass of $n = 1.5$, and an axial thickness of $d = 20$ mm. Using matrices, determine first how the light is refracted on passing through the first surface.

3-17. Continue with Problem 3-16 and determine the *transfer* of the light to the second surface.

3-18. Continuing with the previous problem, determine the refraction of the light at the second surface. Therefore, what is the exit vergence of the light as it leaves the lens and what is its *back vertex focal length*?

3-19. Using Abbe's method, find the focal length of a lens from the following data: initial position of the target on the bench 12.0 cm; subsequent position 17.0 cm; target size 12 mm; image size first 6 mm, then 8 mm.

3-20. The outer points of a lens clock are 20 mm apart. If the center point is pushed in by 0.5 mm and the clock then reads +5.25 diopters, for what refractive index is it calibrated?

SUGGESTIONS FOR FURTHER READING

FANTONE, S. D., ed. *Optics Cooke Book*, 2nd ed. Washington, D.C.: Opt. Soc. Am. 1991.

HEAVENS, O. S., and R. W. DITCHBURN. *Insight into Optics*. New York: John Wiley and Sons, 1991.

HORNE, D. F. *Optical Production Technology*, 2nd ed. Bristol, England: Adam Hilger Ltd., 1983.

LONGHURST, R. S. *Geometrical and Physical Optics*, 3rd ed. New York: Longman, Inc., 1974.

MALACARA, D., ed. *Optical Shop Testing*, 2nd ed. New York: John Wiley and Sons, 1992.

PEDROTTI, F. L., and L. S. PEDROTTI. *Introduction to Optics*, 2nd ed. Englewood Cliffs, N.J.: Prentice-Hall, Inc., 1992.

4

Mirrors

In many ways, mirrors act like lenses. They focus light and form images, both real and virtual. They can be substituted for one another: a mirror can be used instead of a lens, and a lens instead of a mirror. The difference between them is that a mirror *returns* the light while a lens allows it to go on.

Because of the return of the light, perhaps we foresee some difficulties with the Cartesian sign convention, simply because a real object and a real image both lie on the same side, in front of the mirror. Note, however, that axial distances are measured *in the direction of propagation* of the light. Hence, as before, the distance of a real object is negative. But a real image, even though it lies to the left of the mirror, is formed along the return path and therefore its distance is *positive* (Figure 4-1).

PLANE MIRRORS

A plane mirror reflects the light, without focusing it. If you stand in front of a plane mirror, you see your image, of the same size as you are, behind the mirror. Indeed, the virtual image formed by a plane mirror is located at the same distance behind the mirror as the object is in front of it.

That is easy to explain. Assume that some light, as shown in Figure 4-2, emerges from a given point in the object. The light is divergent and, since the

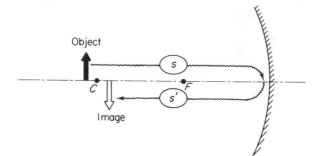

Figure 4-1   Mirror forming real image.

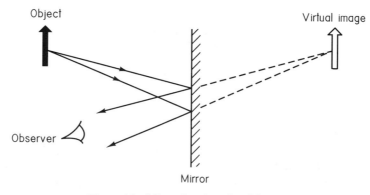

Figure 4-2   Mirror forming virtual image.

Figure 4-3   Looking at an object through a tunnel lined with four plane mirrors.

mirror is plane and has no power, after reflection it remains divergent: the rays never come together and form a real image. But if we extend the *reflected* rays to the other side of the mirror, these extensions *will* come together; they locate the virtual image. The same as with a lens, *no light actually reaches a virtual image.*

Consider now in more detail the *orientation* of such an image. The object may be the letter R. If the mirror is standing upright, like a dresser mirror, with its surface parallel to the straight line in the R, the images are upright but right and left are reversed (Figure 4-3, left and right). Such images, Я, are "upright backward"; they are *wrong-reading.*

If the mirror surface is horizontal, like a reflecting pool, the images, Я, are "inverted forward," and also wrong-reading. An image formed by a single plane mirror is always wrong-reading.

If a mirror is tilted (*rotated*) through a given angle, a beam reflected by the mirror will be rotated through twice that angle. This fact plays a role in optical levers, galvanometers, sextants, and similar instruments.

Multiple Plane Mirrors

Consider first *two* mirrors and assume that the rays and the normals to the two mirrors all lie in the same plane (Figure 4-4). The light is incident on the first mirror at an angle *A* and on the second mirror at an angle *B*. The two mirrors together subtend angle *C*, and the light is deviated through angle *D*.

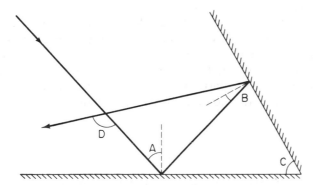

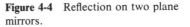

Figure 4-4 Reflection on two plane mirrors.

To find the angle of deviation, we remember that the exterior angle of a triangle is equal to the sum of the two opposite interior angles,

$$D = 2A + 2B \qquad\qquad [4\text{-}1]$$

Furthermore, since

$$(90° - A) + (90° - B) + C = 180°$$

$$A + B = C$$

We multiply both sides by 2 and substitute the result in Equation [4-1]. That gives

$$D = 2C \qquad\qquad [4\text{-}2]$$

which means that the angle of deflection is twice the angle subtended by the two mirrors. If the angle subtended is $C = 90°$, then no matter what the angle of incidence, the two mirrors together return the light by $D = 180°$, but only if the rays and the normals to the mirrors lie in the same plane.

Three plane mirrors, mounted at right angles to each other, form a *cube-corner retroreflector*. Such a reflector, shown in Figure 4-5, returns the light parallel to the direction from which it came.

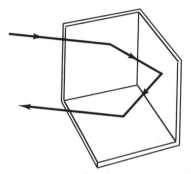

Figure 4-5   Three plane mirrors forming retroreflector.

These reflectors are often made by cutting a corner off a cube; consequently each has a plane front surface, the base, and a three-sided pyramid to the back of it, each face at right angles to the other two. Light entering the reflector passes first through the base. It then proceeds inside the block, is reflected on all three back surfaces, and passes again through the base. Cube-corner reflectors are widely used in surveying and in some interferometers; they were placed on the moon to determine with high precision the moon's distance from Earth.

SPHERICAL MIRRORS

According to our sign convention, the radius of curvature of a surface concave to the left is *negative*. That holds for any surface, refracting or reflecting. A concave mirror, which has a negative radius of curvature such as that shown in Figure 4-1, therefore acts like a converging lens. A convex mirror, whose radius of curvature extends to the right and is positive, acts like a diverging lens.

Assume now that a real object is placed some distance to the left of a concave mirror. If the object distance chosen is longer than the focal length, then the mirror, the same as a converging lens, forms a real image. As the light travels from the object to the mirror, it advances in the $+x$ direction and the velocity of the light, x/t, is positive:

$$\frac{+x}{+t} = +v$$

If the light travels in the $-x$ direction, then, because time can only go forward ($+t$, rather than $-t$), the velocity of the light is negative:

$$\frac{-x'}{+t} = -v'$$

We saw earlier that refractive index is defined as the ratio of the velocity of light in free space, c, to the light's velocity in a given medium, v,

$$n = \frac{c}{v}$$

Comparing two media of indices $n = c/v$ and $n' = c/v'$ shows that

$$\frac{n}{n'} = \frac{c/v}{c/v'} = \frac{v'}{v}$$

If after reflection the velocity of the light is negative, $-v'$, the refractive index is negative also,

$$n' = -n$$

All that is needed, then, is to substitute $-n$ for n' in the surface power equation:

$$P = \frac{n' - n}{R} = \frac{(-n) - n}{R} = \frac{-2n}{R} \qquad [4\text{-}3]$$

where P is now the *reflective power* of the mirror, R is its radius of curvature, and n is the refractive index of the medium, if any, in front of the mirror.

Next we substitute $P = n/f$, cancel n, and combine the result with the thin-lens equation solved for $1/f$. That gives

$$\boxed{\frac{1}{f} = \frac{1}{s'} - \frac{1}{s} = \frac{-2}{R}} \qquad [4\text{-}4]$$

which is the *spherical-mirror equation*. It holds for any mirror and for any object and image distance.

Taking just the first and last member of this equation shows that

$$f = -\frac{R}{2} \qquad [4\text{-}5]$$

which means that the focal length of a mirror is one-half the length of its radius of curvature.

Many characteristics of a mirror are equivalent to those of a lens. The *transverse magnification* produced by a mirror, for instance, is the same as that produced by a lens:

$$M = \frac{y'}{y} = \frac{s'}{s} \qquad\qquad [4\text{-}6]$$

In contrast to a lens, however, the refractive index in front of a mirror is (almost) irrelevant. True, a mirror in air has the same focal length as the same mirror under water. Only the *power* of the mirror changes, as we see from Equation [4-3].

Reflection Matrix

Since a mirror is so much like a lens, a reflection matrix is very similar to a refraction matrix. We only need to substitute

$$P = \frac{-2n}{R}$$

which is part of Equation [4-3], in the refraction matrix, Equation [3-17]. That leads to the *reflection matrix*, **M**,

$$\mathbf{M} = \begin{bmatrix} 1 & P \\ 0 & 1 \end{bmatrix} = \begin{bmatrix} 1 & -2n/R \\ 0 & 1 \end{bmatrix} \qquad [4\text{-}7]$$

As before, n is the refractive index of the medium in front of the mirror and R is the radius of curvature. If the mirror were turned around to face the other way, both the index and the radius of curvature would change signs, but the matrix itself would remain the same.

Example 1

An object is located 25 cm to the left of a concave mirror of 10 cm radius of curvature. Using vergence terminology determine the image distance.

Solution: First we determine the entrance vergence,

$$\mathbf{V} = \frac{1}{s} = \frac{1}{-0.25} = -4 \text{ diopters}$$

and the power of the mirror,

$$P = \frac{-2}{R} = \frac{-2}{-0.1} = +20 \text{ diopters}$$

Adding both gives the exit vergence,

$$\mathbf{V'} = (-4) + (+20) = +16 \text{ diopters}$$

The image, as we see from the plus sign, is real and located

$$s' = \frac{1}{\mathbf{V'}} = \frac{1}{16} = +0.0625 = \boxed{6.25 \text{ cm}}$$

to the *left* of the mirror.

Example 2

With an object 2 m away, a concave mirror forms a real image 50 cm away. Using matrices, find the radius of curvature of the mirror.

Solution: Convert the known object and image distances into vergences and these into vergence matrices:

$$V = \frac{n}{s} = \frac{1}{-2} = -0.5 \text{ diopters}, \quad V = \begin{bmatrix} V \\ 1 \end{bmatrix} = \begin{bmatrix} -0.5 \\ 1 \end{bmatrix}$$

$$V' = \frac{1}{0.5} = 2 \text{ diopters}, \quad V' = \begin{bmatrix} 2 \\ 1 \end{bmatrix}$$

Then, writing the sequence from object to mirror to image,

$$\begin{matrix} \text{M} & \text{V} & = \text{V}' \end{matrix}$$

$$\begin{bmatrix} 1 & -2/R \\ 0 & 1 \end{bmatrix}\begin{bmatrix} -0.5 \\ 1 \end{bmatrix} = \begin{bmatrix} 2 \\ 1 \end{bmatrix}$$

$$\begin{bmatrix} (-0.5) + (-2/R) \\ 1 \end{bmatrix} = \begin{bmatrix} 2 \\ 1 \end{bmatrix}$$

$$-0.5 - \frac{2}{R} = 2$$

$$\frac{-2}{R} = 2.5$$

$$R = \frac{-2}{2.5} = -0.8 = \boxed{-80 \text{ cm}}$$

Graphical Construction

In principle very similar to refraction, the rays are reflected as follows:

The *parallel ray* (1), after reflection, goes through the focal point.
The *focal ray* (2) is reflected back parallel to the optical axis.
The *chief ray* (3) passes through the center of curvature of the mirror; it is returned in its own initial path.

The object in Figure 4-6, for example, lies *outside the center of curvature*, farther away from the mirror than twice its focal length. The image is real, inverted, and reduced in size.

In Figure 4-7, the object is *inside the focal length*. That makes the image virtual, upright, and magnified. *Convex* mirrors produce only virtual, upright, and reduced images. The principal points of a mirror coincide at the vertex, and the nodal points coincide at the center of curvature.

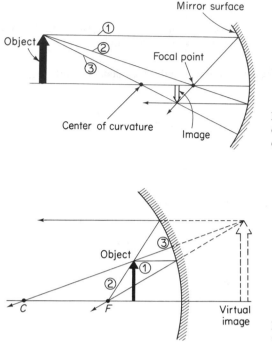

Figure 4-6 Image formed by a spherical mirror, with the object *outside* the center of curvature.

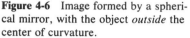

Figure 4-7 Image formed by a mirror, with the object *inside* the focal length.

Applications

Reflection of light on a convex mirror occurs in the *keratometer*. That is an instrument used to determine the radius of curvature, and therefore the power, of the cornea of the eye.

A keratometer consists of a back-lit object, usually two "mires" separated by a distance *a* (Figure 4-8). The mires produce two virtual images, seen on the cornea, which are separated by a distance *b*.

To locate the images, draw two rays (3) from the mires toward the center of curvature, *C*. These rays hit the surface at normal incidence, and hence are reflected back in their initial paths. Then draw the two focal rays (2) from the mires toward the focal point, *F*. By definition, these rays are reflected parallel to the optical axis, as shown. Extend the rays to the right; their intersections define the two images.

Now consider the triangle formed by the two sources and by *F*, and the triangle formed by the two images as seen *on the cornea* and also by *F*. The triangles are similar and thus

$$\frac{a}{s+f} = \frac{b}{f}$$

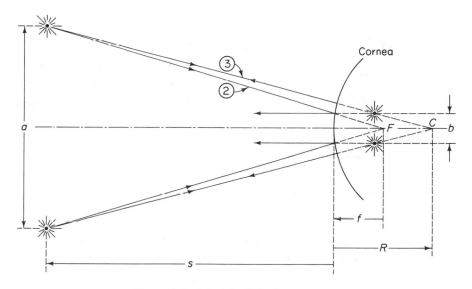

Figure 4-8 Principle of the keratometer.

Cross-multiplication gives

$$af = bs + bf$$

$$f(a - b) = bs$$

The focal length of a mirror, neglecting the sign, is one-half its radius of curvature and therefore

$$R = 2f = 2\frac{bs}{a - b} \qquad [4\text{-}8]$$

In a given keratometer, the distances a and s are usually fixed and cannot be changed; thus, we only need to measure the separation, b, of the two reflections to find the radius of curvature, R, of the cornea.

To convert the radius into power, we use the surface power equation, $P = (n' - n)/R$, and arbitrarily set $n' = 1.3375$. Consequently, a radius of 7.5 mm, for example, means a power of 45 diopters.

ASPHERICAL MIRRORS

A *spherical* mirror will reflect a point object into a perfect point image only when the object and the image both lie at the center of curvature of the mirror. Needless to say, that rarely happens. With other distances, a perfect image will result only if we have an *aspherical mirror* (Figure 4-9).

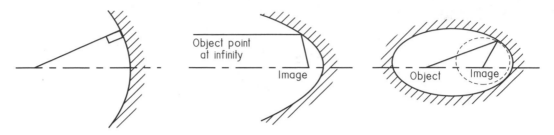

Figure 4-9 (*From left to right*) Spherical mirror, paraboloidal mirror, ellipsoidal mirror.

A *paraboloidal* mirror, for example, is used when either the object is at infinity (as with an astronomical telescope) or the image is at infinity (as with a searchlight).

The equation of a parabola is particularly simple if the vertex is at the origin and the focus lies on one of the coordinate axes. The function

$$y = x^2$$

describes a parabola open at the top and symmetric about the *y*-axis. To turn the parabola so that its axis is horizontal, interchange *x* and *y* and, with the vertex placed at the origin, include the focal length, *f*,

$$y^2 = 4fx \qquad\qquad [4\text{-}9]$$

A paraboloid is a three-dimensional figure of revolution of a parabola about its axis of symmetry.

When both the object and the image are at a finite distance, the mirror should be *ellipsoidal* [Figure 4-9 (right)]. Such mirrors have two conjugate foci: light originating at one focus is projected into the other focus.

Determining the Focal Length of a Mirror

1. The power of a mirror can be determined using a *lens clock*. Assume that the clock, when held against the mirror, gives a reading of *x* diopters. Then, using the surface power equation, $P = (n' - n)/R$, setting $n = 1$ and $P = x$, and solving for *R* gives $R = (n' - 1)/x$. Substituting this in the mirror equation yields $P = -2/R = -2x/(n' - 1)$ and, if the lens clock is calibrated for $n' = 1.5$,

$$P = -4x \qquad\qquad [4\text{-}10]$$

which means that the mirror has four times the power, and the opposite sign, of that shown by the clock.

2. With a *concave* mirror, place an object-image screen (a transilluminated target) in front of the mirror at a distance such that the target is focused back into the plane of the target. The distance from the target to the mirror is then equal to the radius of curvature. One-half that distance is the focal length.

3. Focus a telescope for infinity and then aim at the (concave) mirror. A small object such as a pin is mounted, slightly off-axis, between mirror and telescope. Moving the

object back and forth along the axis, find the position where the image is seen in focus. Light coming from the object and reflected by the mirror must now be parallel and the object be located in the mirror's focus.

4. A *convex* mirror, the same as a diverging lens, does not form a real image. Proceed as with a minus lens, first forming an image with the plus lens only. Focus on the mirror surface (dashed curve on the right in Figure 4-10). Then move the mirror to the left (solid curve) until an image is formed in the plane of the target. The distance between the two positions of the mirror is equal to the radius of curvature. The same procedure can be used with a concave mirror, moving the mirror *away* from the lens.

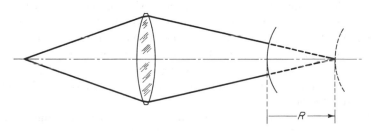

Figure 4-10   Determining the radius of curvature of a mirror.

SUMMARY OF EQUATIONS

Mirror equation:

$$\frac{1}{f} = \frac{1}{s'} - \frac{1}{s} = \frac{-2}{R}$$

[4-4]

PROBLEMS

4-1.   How long must a (vertical) mirror be for a man who stands 1.82 m tall to see his full height? How far must the man be from the mirror?

4-2.   Two plane mirrors subtend an angle of 35°. At what angle must light be incident on one mirror so that, after reflection at the other mirror, the light exactly retraces its path?

4-3.   An object is placed 15 mm in front of a concave mirror of 4 cm radius of curvature. What is the image distance?

4-4.   With an object 2 m away, a concave mirror forms a real image 50 cm away. What is the radius of curvature of the mirror?

4-5.   An object 2 cm high is placed 6 cm in front of a concave mirror. If the mirror has a radius of curvature of 16 cm, how high is the image?

4-6. How far from a concave mirror of 10 cm radius of curvature must a real object be placed so that its image is real and four times the size of the object?

4-7. If a concave mirror of 60 cm radius of curvature forms a real image twice as far away as the object, what is the object distance?

4-8. If an object is placed halfway between the focal point and the vertex of a concave mirror, how much will the image be magnified?

4-9. A small object is placed between the center of curvature and the focus of a concave mirror. What is the type (real or virtual), orientation (inverted or upright), and magnification of the image?

4-10. When an object is placed at a distance of $1\frac{1}{2}$ times the focal length in front of a *convex* mirror, what is the type, orientation, and magnification of the image?

4-11. A −2.00-diopter lens is held in contact with a mirror of +80 cm focal length and of the same diameter as the lens. What is the power of the combination?

4-12. A symmetrical (equiconvex) lens has +10 diopters power and is made of glass of $n = 1.5$. If the back surface of the lens has a reflective coating, what is the (total) power of the combination?

4-13. If you want to look at your own eyes, without accommodation, using a concave mirror of 60 cm radius of curvature, how far in front of your eyes must the mirror be?

4-14. When a polished steel ball 15 mm in diameter is observed in a keratometer, it gives a reading of +48.00 diopters. For what index is the keratometer calibrated?

4-15. A +6.00-diopter lens is placed 25 cm from a transilluminated target. If a convex mirror, as in Figure 4-10, is set 40 cm from the lens, it returns the light in its original path. What is the power of the mirror?

4-16. A lens clock, held against a mirror, gives a reading of −2.50 diopters. If this mirror forms a real image at a distance one-half the object distance, what is the image distance?

SUGGESTIONS FOR FURTHER READING

ELMER, W. B. *Optical Design of Reflectors*, 2nd ed. New York: John Wiley and Sons, Inc., 1980.

GREENSLADE, T. B., JR. "Multiple Images in Plane Mirrors," *Phys. Teacher* **20** (1982), 29–32.

HOPKINS, R. E. "Mirror and Prism Systems," in R. Kingslake, ed., *Applied Optics and Optical Engineering*, vol. III, pp. 269–308. New York: Academic Press, Inc., 1965.

Aberrations

When we trace a series of rays through a lens with conventional spherical surfaces, we find that the rays do not converge to a point as we would like. Instead, the peripheral rays, as shown in Figure 5-1, come to a focus closer to the lens, and the central rays come to a focus farther away. That is an example of an *aberration*.

Aberrations are due to inherent shortcomings of a lens, even a lens made of the best glass and free from manufacturing and other defects. Some aberrations occur already with monochromatic light; they are *monochromatic aberrations* (1 through 5 in the list below). Other aberrations occur only with light that contains at least two wavelengths; these are *chromatic aberrations* (6). We distinguish the following:

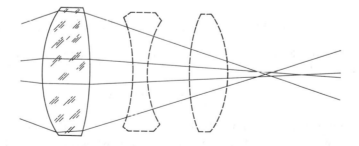

Figure 5-1 Spherical aberration.

1. Spherical aberration
2. Coma
3. Oblique astigmatism
4. Curvature of field
5. Distortion
6. Chromatic aberration

In addition, spherical aberration and chromatic aberration both have a longitudinal (axial) and a transverse (lateral) variety, each with its own causes and corrections.

Remember how we derived the refraction of light passing through a spherical surface. Instead of using Snell's law,

$$n \sin I = n' \sin I'$$

which is rigorously correct, we assumed that the angles were small, no more than a few degrees, and that their sines were equal to the angles themselves, in radians, and therefore we thought that

$$nI = n'I'$$

That simplification, however, holds true only for rays that are close to the optical axis, that is, for *paraxial rays*.

But now assume that the angles are large. Look at Figure 5-2 where we see, side by side, the sine of an angle and the angle as such. As the angle is small, $\sin I$ is nearly equal to I (in radians) but, as the angle grows larger, it is not.

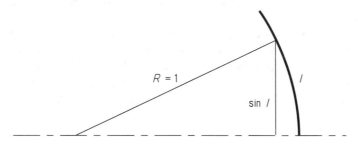

Figure 5-2 Comparing the sine of an angle with the angle as such.

When the rays subtend larger angles with the optical axis, the sine must be represented by a series,

$$\sin x = x - \frac{x^3}{3!} + \frac{x^5}{5!} - \cdots \qquad [5\text{-}1]$$

The factorials, such as 3!, stand for $3! = 1 \cdot 2 \cdot 3$, and so on.

If we use only the first term (that is, if we assume that $\sin I = I$), we have *first-order* or *Gaussian optics*. If we include the next term and assume that $\sin x =$

$x - I^3/3!$, we deal with *third-order optics*. Aberrations of that kind are often called *von Seidel aberrations*.* Sometimes, even higher orders, based on sin $x = x - x^3/3! + x^5/5!$, are considered also.

SPHERICAL ABERRATION

Spherical aberration is the deficiency that rays passing through a lens farther away from the optical axis (as shown earlier in Figure 5-1) are refracted more strongly and come to a focus closer to the lens than paraxial rays. That is an example of *longitudinal spherical aberration* and a lens acting that way is said to be *undercorrected* (Figure 5-3). If the curve were to lean over to the right, the lens would be *overcorrected*. With a large-diameter lens, therefore, image details that are formed by paraxial rays are surrounded by a diffuse patch formed by peripheral rays, and vice versa, which accounts for the blur of the image.

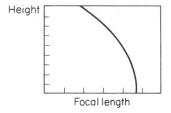

Figure 5-3 Plot of focal length as a function of height in spherical aberration. Rays farther away from the optical axis (as on *top* of the plot) come to a focus *closer* to the lens than rays at the *bottom*.

Besides longitudinal spherical aberration, LSA, we also have a *transverse* spherical aberration, TSA. The transverse variety is defined as the distance, above or below, from the optical axis at which a peripheral ray intersects the plane of the paraxial focus (as shown in Figure 5-4). Since the transverse aberration is measured *in the plane* of the image, perhaps it is a more meaningful measure of image blur than the longitudinal type. Calling U' the slope angle that a peripheral ray after leaving the lens subtends with the optical axis, both the transverse and the longitudinal type are connected as

$$\text{TSA} = \text{LSA} \tan U' \qquad\qquad [5\text{-}2]$$

---

* Named after Ludwig Philipp von Seidel (1821–1896), German mathematician and astronomer, professor at the University of Munich. After working on divergent and convergent series, von Seidel became interested in stellar photometry, probability theory, and the method of least squares, studied the trigonometry of skew rays, and became the first to establish a rigorous theory of monochromatic third-order aberrations, which led to the construction of much improved astronomical telescopes. His principal work is L. Seidel, "Ueber die Entwicklung der Glieder 3ter Ordnung, welche den Weg eines ausserhalb der Ebene der Axe gelegenen Lichtstrahles durch ein System brechender Medien bestimmen," *Astron. Nachr.* **43** (1856), 289–304, 305–20, 321–32.

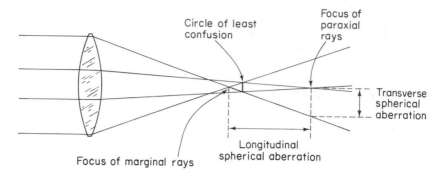

Figure 5-4   Longitudinal and transverse spherical aberration.

Correction for Spherical Aberration

With a single spherical lens, spherical aberration cannot be eliminated. But at least it can be minimized. That is possible by making the two surfaces of the lens contribute equally, similar to setting a prism to the angle of minimum deviation (Figure 5-5). Giving the lens the proper shape is called *bending* the lens.

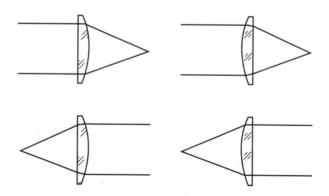

Figure 5-5   If, with light coming from infinity, one surface contributes nothing to the refraction (*top left*), spherical aberration is worse than if both surfaces contribute about equally (*top right*). With parallel light emerging from the lens, it is still the same: If both surfaces contribute equally (*bottom left*), that is better than one surface contributing nothing (*bottom right*).

To further reduce spherical aberration, the radii should be as long as practical, which means using *high-index* glass. Complete elimination of spherical aberration is possible only by making either or both surfaces *aspherical*, or by using a *gradient-index lens* whose index of refraction is higher in the center than in the periphery (see Chapter 7). Finally, in a lens *system*, the undercorrection of one lens can be compensated for by the overcorrection of another lens.

The shape of a lens at which it has the least amount of spherical aberration is called its *best form*. To find the best form, we need to consider two factors. One is the *Coddington shape factor*,* S. It describes the degree of bending as a function of the two radii of curvature, R_1 and R_2:

$$S = \frac{R_2 + R_1}{R_2 - R_1} \qquad \text{[5-3]}$$

For example, when $S = 0$, the lens is symmetric, either equiconvex or equiconcave. When $S = -1$, the first surface is plane and the second surface is either convex or concave. When $S = +1$, the second surface is plane. When S is less than -1, or greater than $+1$, the lens is of the meniscus type.

Whereas S is determined by the physical shape of the lens, the *Coddington position factor*, P, depends on the object and image distances actually used:

$$P = \frac{s' + s}{s' - s} \qquad \text{[5-4]}$$

When $P = -1$, the incident light is parallel. When $P = 0$, object distance and image distance are equal, except for the sign; and when $P = +1$, the emergent light is parallel.

To find the best form, we first determine the position factor. Substituting in Equation [5-4] the thin-lens equation, solved for s', gives

$$P = \frac{sf + s(s + f)}{sf - s(s + f)} = -\left(\frac{2f}{s} + 1\right)$$

and solved for s,

$$P = \frac{s'(f - s') + s'f}{s'(f - s') - s'f} = 1 - \frac{2f}{s'}$$

Then from the lens-makers formula,

$$\frac{1}{f} = (n - 1)\left(\frac{1}{R_1} - \frac{1}{R_2}\right) = \frac{n - 1}{R_1} - \frac{n - 1}{R_2}$$

$$R_1 = f(n - 1) - \frac{fR_1(n - 1)}{R_2} \qquad \text{[5-5]}$$

Next we take the shape factor, Equation [5-3], and solve for R_2:

$$R_2 = R_1 \frac{S + 1}{S - 1} \qquad \text{[5-6]}$$

---

* Henry Coddington (born around 1800, died 1845), English mathematician and cleric. Coddington had many interests, spoke several languages, and was a good musician, draftsman, and botanist. He wrote two books on optics, the more important one the two-volume *A System of Optics*. Part I, *A Treatise on the Reflexion and Refraction of Light* (Cambridge: Cambridge University Press, 1829), contains a thorough investigation of reflection and refraction. In volume 2, Coddington discusses the eye and the theory and construction of various types of eyepieces, telescopes, and microscopes.

Substituting Equation [5-6] in [5-5] gives

$$R_1 = f(n - 1) - \frac{f(n - 1)(S - 1)}{S + 1}$$

Multiplying the $f(n - 1)$ term by $(S + 1)/(S + 1)$ and simplifying leads to

$$R_1 = \frac{2f(n - 1)}{S + 1} \qquad [5\text{-}7]$$

and similarly,

$$R_2 = \frac{2f(n - 1)}{S - 1} \qquad [5\text{-}8]$$

Differentiation with respect to the shape factor and setting equal to zero show that *minimum spherical aberration* occurs when

$$\boxed{S = - \frac{2(n^2 - 1)}{n + 2} P} \qquad [5\text{-}9]$$

Example

Determine the radii of curvature of a lens of $f = +10$ cm, $n = 1.5$, which for parallel incident light has minimum spherical aberration.

Solution: First, determine the position factor:

$$P = \frac{s' + s}{s' - s} = \frac{(+10) + (-\infty)}{(+10) - (-\infty)} = -1$$

Substitute $n = 1.5$ and $P = -1$ in Equation [5-9]:

$$S = - \frac{(2)(1.5^2 - 1)}{1.5 + 2} (-1) = +0.714$$

Then from Equations [5-7] and [5-8],

$$R_1 = \frac{2f(n - 1)}{S + 1} = \frac{(2)(10)(1.5 - 1)}{(+0.714) + 1} = \boxed{+5.83 \text{ cm}}$$

$$R_2 = \frac{2f(n - 1)}{S - 1} = \frac{(2)(10)(1.5 - 1)}{(+0.714) - 1} = \boxed{-35 \text{ cm}}$$

Dividing R_1 by R_2 shows that to attain minimum spherical aberration for parallel incident light, the radii of curvature of a lens should be related as $+1 : -6$. Such a lens is nearly plano-convex, the convex side facing the light. If the lens were placed halfway between object and image, the best form would be symmetric.

COMA

Coma is similar to spherical aberration but in addition the rays come from *off-axis* points. As shown in Figure 5-6 (top), the object may be a circular opening of diameter D. The lens forms an image of diameter D'. Actually, it is only rays 2,

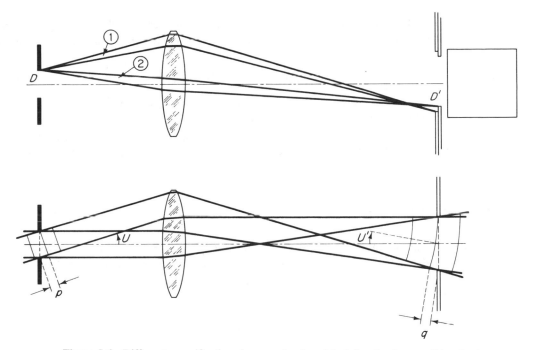

Figure 5-6 Different magnifications in coma (*top*) and deriving the sine condition for its elimination (*bottom*).

because they are paraxial rays, that form such an image; rays 1, which are peripheral rays, form an image slightly *larger* than D'. That results in the raindrop-like figure of coma.* The diffuse tail of the comatic flare may lie in a direction away from the axis ("outward" coma), as shown, but, depending on the type of lens used, it may also lie toward the axis ("inward" coma).

To correct for coma, we need to make the different images coincide, as shown in Figure 5-6 (bottom). This time let the light come from an extended source, placed far to the left of the opening. Again some of the light proceeds along the axis. Other light is more oblique, subtending an angle U with the axis. This angle relates to the path difference p as

$$\sin U = \frac{p}{D}$$

At the image, the conjugate angle is U' and

$$\sin U' = \frac{q}{D'}$$

---

* The term *coma* comes from the Greek κόμη, long hair, referring to the long tail of a comet. Coma is easy to see. Hold a magnifying glass in the path of sunlight and tilt it: The image of the sun will elongate into the cometlike shape characteristic of coma.

For the different images to coincide, elemental lengths of p must equal conjugate lengths of q. Taking into account the refractive indices on both sides, then, if

$$nD \sin U = n'D' \sin U' \qquad\qquad [5\text{-}10]$$

is satisfied for all zones of the system, there is no coma, a relationship known as *Abbe's sine condition*.

Correction for coma is possible by bending, or by using a combination of lenses symmetric about a central stop. A lens or system free of both spherical aberration and coma is said to be *aplanatic*.

OBLIQUE ASTIGMATISM

Oblique astigmatism is another off-axis aberration.* It is characterized as an object *point* with oblique rays that is drawn out into two image *lines*. These lines lie on (hypothetical) surfaces, the *tangential* surface and the *radial* surface (Figure 5-7).

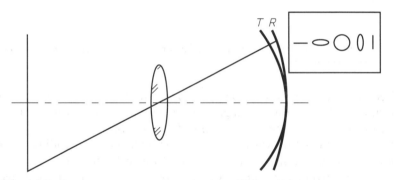

Figure 5-7   Tangential surface (T) and radial surface (R) as they occur in positive oblique astigmatism. Insert shows cross sections through conoid.

The tangential surface is called tangential because the lines seen on it are *tangent* to a circle drawn around the optical axis. In positive oblique astigmatism the tangential surface is closer to the lens (Figure 5-8). The radial surface is called radial because lines seen on it are oriented radially, like *radii*, with the axis in the center. The radial surface is farther away from the lens. Because of astigmatism, an off-axis point object will, in short, be imaged as a tangential line at T and as a radial line at R [Figure 5-8(a)].

---

* The term astigmatism comes from the Greek, 'α = alpha privative, meaning "not," and στίγμα = point, meaning that a point object is no longer imaged as a point. A system free of astigmatism is called an *anastigmat*.

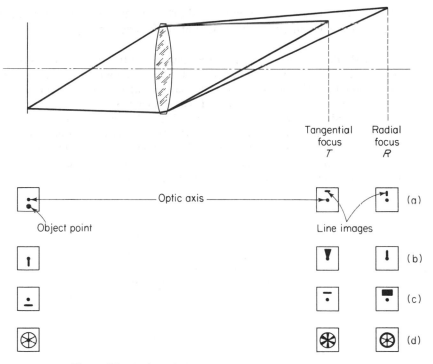

Figure 5-8 Astigmatic images of a sequence of objects.

To understand why an object *point* forms an image *line*, oriented either tangentially or radially, consider the following. A cone of light may originate at an off-axis point source, some distance away from the optical axis. When this cone is obliquely incident on a surface, plane or curved, it forms an ellipse. The major axis of the ellipse is always radial, the minor axis is always tangential. But rays in the major axis encounter a lens whose diameter seems to be *less* than the diameter really is, and consequently its power is *higher*. This light, therefore, comes to a focus *closer* to the lens, forming the tangential focus, *T*. Rays that lie in the minor axis form the radial focus, *R*.

Between the two surfaces the tangential line widens into an ellipse and then into a circle. Subsequently the circle contracts, first into an ellipse and again into a line, but this line is now radial. The three-dimensional figure between the two lines is called a *conoid*. At the optical axis the two surfaces touch: for axial rays there is no astigmatism.

If the object were a radial line, its image would be blurred at the tangential focus, *T*, and sharp at the radial focus, *R* [Figure 5-8(b)]. If the object were a tangential line, its image would be sharp at *T*, and blurred at *R* (c). If the object were a combination of radial and tangential lines such as a spoked wheel, the rim of the wheel (which is the sum of short tangential lines) will be in focus at *T* and the spokes (which are radial lines) at *R* [Figure 5-8(d)].

Cylinder Lenses. Cylinder lenses have properties that remind us of astigmatism. But *oblique* astigmatism is an off-axis aberration. Cylinder lenses have *axial astigmatism*. While oblique astigmatism is a nuisance, axial astigmatism is often induced on purpose; it is used in the correction of certain visual deficiencies.

Consider the cylinder lens shown in Figure 5-9. Assume that two pieces of cardboard are held to either side of the lens, with the light grazing along their surfaces: on the cardboard we see a *band of light*. In the orientation shown, the light continues straight through the lens with *no refraction* because the thickness of the lens, in this *axis meridian*, is constant; the lens merely acts as a plane-parallel slab of glass.

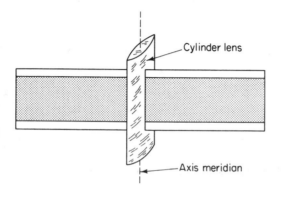

Figure 5-9 Biconvex cylinder. Band of light in the axis meridian.

Imagine, however, that the cardboard is turned through 90°, as in Figure 5-10. The band of light then lies in the *power meridian*, 90° away from the axis meridian. In that orientation the light *is* refracted, and comes to a focus the same as with a conventional plus lens. A *negative cylinder*, not shown, is comparable to a minus lens. Both types of cylinder lenses are *simple cylinders*; they have power in their power meridians but no power in their axis meridians.

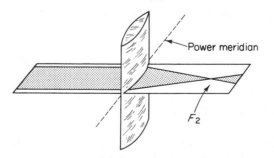

Figure 5-10 Cylinder with band of light in the power meridian.

A *spherocylinder* is a combination of a conventional lens and a simple cylinder. Consider the two lenses illustrated in Figure 5-11. The spherical lens alone, in the meridian

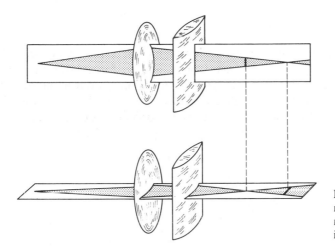

Figure 5-11 Spherocylinder (schematic), with bands of light in the axis meridian (*top*) and in the power meridian (*bottom*).

shown in the upper diagram, projects a point object into a point image; the cylinder, in this meridian, *does not contribute.*

In the power meridian, however, *both lenses contribute*, causing the light to come to a focus *closer* to the lens (bottom). Obviously, the light does not stop at the foci; at the first focus to the right of the lenses (lower diagram) it continues, and by the time it has reached the second focus, it has spread out into a *line*, oriented in the direction of the power meridian. On the other hand (upper diagram), when the light is at the first focus, it has *not yet* converged into a point; it still is a line. (Note that the first line focus is always parallel to the meridian of lesser power.) In reality, the two separate lenses shown in Figure 5-11 are fused into one, a *spherocylinder*. A spherocylinder, in other words, has one surface that is spherical and another that is cylindrical.

The two lines on the right are a certain distance apart called the *interval of Sturm.** Within this interval there is one cross-section where the bundle of light is circular; this is the *circle of least confusion*. It is located at the *dioptric midpoint*, a little closer to the lens than the "linear midpoint."

A *toric surface* has two radii of curvature (different in the two principal meridians), but in contrast to a simple cylinder, even the axis meridian has some power other than zero. Toric surfaces are used in ophthalmic lenses (which are menisci) whenever an additional cylinder component is needed. Usually, only one surface is toric, but *bitoric* lenses can be made also.

---

* Charles-François Sturm (1803–1855). Born in Switzerland, Sturm became a French citizen, mathematician and physicist, and professor of analysis and mechanics at the École Polytechnique in Paris. He is known for the theorem that bears his name (referring to the number of roots of an algebraic equation) and for his contributions to projective geometry, propagation of sound in water, and the optics of cylinder lenses. Much of Sturm's scientific work is contained in lecture notes, published posthumously as two books, *Cours d'analyse de l'École Polytechnique* (Paris, 1857–1859) and *Cours de mécanique de l'École Polytechnique* (Paris, 1861).

CURVATURE OF FIELD

To light that comes from off-axis, a positive lens appears to be thicker than it really is and its diameter appears to be less. Consequently, for oblique rays the power of the lens is higher and its focal length shorter. Points on a plane object then form a curved image, a deficiency called *curvature of field*. Shown schematically in Figure 5-12, curvature of field is especially objectionable in cameras, enlargers, and slide projectors, where the image is expected to be flat.

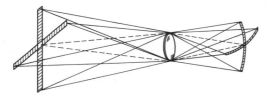

Figure 5-12   Curvature of field.

Correction for curvature of field, called "field flattening," is possible by using a combination of at least two lenses that meet the *Petzval condition*:*

$$n_1 f_1 + n_2 f_2 = 0 \qquad\qquad\qquad [5\text{-}11]$$

Example

The Petzval condition can be met even if the two lenses have the same index of refraction. Consider a combination of two lenses that have the same power but opposite signs, and set them some distance d apart. Assume that $P_1 = +10 \text{ m}^{-1}$, $P_2 = -10 \text{ m}^{-1}$, $n = 1.55$, and $d = 25$ mm. This satisfies the Petzval condition,

$$\frac{1.55}{+10} + \frac{1.55}{-10} = 0$$

but is there any power left? (Work Problem 5-11 to find out.)

Obviously, it would be better to make the plus lens of glass of higher index than the minus lens. But this is just the opposite of what is needed for correction of chromatic aberration. It is only by the use of newer types of high-index low-

---

* Josef Max Petzval (1807–1891), Hungarian mathematician and professor of mathematics at the University of Vienna. Petzval is known for his eloquent, colorful lectures; he wrote two volumes on the integration of linear differential equations and extended Gaussian optics beyond the paraxial approximation by including higher powers. For diversion he liked fencing and other forms of physical activity. Soon after Louis Jacques Maudé Daguerre (1787–1851) had announced, in 1839, his process of photography, Petzval set out to calculate, with the help of eight artillery men familiar with arithmetic and loaned to him by Archduke Ludwig of Austria, a portrait camera lens, of much larger aperture (to allow shorter exposures) and much better correction than any other lens known at that time (1840). Unfortunately, when in 1859 burglars broke into his country home on the Kahlenberg near Vienna looking for valuables, they destroyed part of the manuscript and only a short version remains: J. Petzval, *Bericht über die Ergebnisse einiger dioptrischer Untersuchungen* (Pest, 1843).

dispersion glass that the conditions for achromatism and flatness of field can be met at the same time.

DISTORTION

Distortion is the transverse counterpart of curvature of field. Like curvature of field, distortion also is limited to oblique rays; these rays become displaced, either toward or away from the optical axis, causing a change of magnification.

We distinguish two types of distortion, *pincushion distortion* and *barrel distortion*. In pincushion distortion, the transverse magnification increases with increasing obliquity of the rays; thus, the corners of a square, for example, are drawn out like a pillow. In barrel distortion the magnification decreases and the corners retract (Figure 5-13).

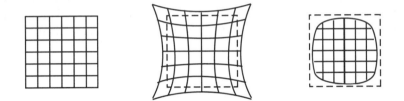

Figure 5-13 Undistorted image (*left*), pincushion distortion (*center*), and barrel distortion (*right*).

Look at a sheet of millimeter paper from a distance of about 50 cm. Look through a +20.00 diopter lens, holding the lens first fairly close to the paper and then farther away from it. You will have no difficulty distinguishing the two types of distortion.

Distortion is readily explained as follows. The lens in Figure 5-14 forms an image of object 1–2 on the screen on the right. While point 1 is imaged (correctly) into 1′, point 2, because of astigmatism and curvature of field, is imaged into 2′, which lies to the left of the

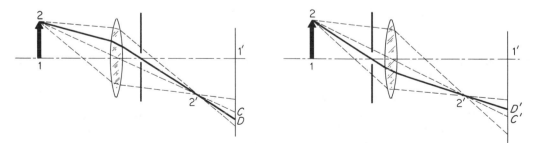

Figure 5-14 Rays causing pincushion distortion (*left*) and barrel distortion (*right*).

screen. But when the rays forming 2' have reached the screen, they have expanded into a blur circle with the chief ray at its center, *C*. If a circular aperture is placed to the right of the lens, the blur circle contracts and its center moves to *D* (solid line), which is farther away from the axis. This results in pincushion distortion.

On the other hand, if the stop is placed to the left of the lens, *C* moves to *D'*, which is closer to the axis; this results in barrel distortion. To eliminate distortion, the stop should be placed *between* two lenses. Systems free of both curvature of field and distortion are called *orthoscopic*.

CHROMATIC ABERRATION

The index of refraction of glass, and of matter in general, varies as a function of wavelength; it is always higher for blue than for red. A single lens, therefore, has different focal lengths for different colors: blue light comes to a focus closer to the lens than red light (Figure 5-15). The horizontal distance between the two images is called *longitudinal chromatic aberration*, or "longitudinal color" for short.

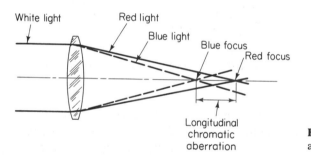

Figure 5-15   Longitudinal chromatic aberration.

The images produced by different colors are also of different sizes; they have different transverse (lateral) magnifications. That is called *lateral chromatic aberration*, or "lateral color." An image formed in the blue focus is closer to the lens and smaller; at the red focus it is farther away and larger.

Correction for Chromatic Aberration

There are several ways of correcting for chromatic aberration. The best known is to use two lenses in contact, one made of crown, the other of flint (Figure 5-16). The crown lens is given more plus power than necessary. Its dispersion is moderate. Flint, on the other hand, has high dispersion. The two dispersions are then made equal but of opposite sign so that they cancel. Therefore, because of the higher dispersion of flint, the flint component can have less minus power than the crown has plus power, the combination has excess plus power, and the result is a

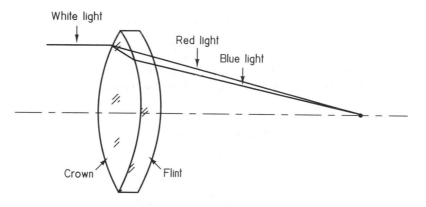

Figure 5-16 Dichromatic doublet. White light is entering the lens from the left. The crown as such refracts blue more than red, but in the end, the two rays come together at a common focus.

positive dichromat (the former *achromat*).* A system corrected for three colors is called a *trichromat*.

Consider a combination of two lenses in contact to be corrected for blue and red. We know that the approximate power of a combination is the sum of the powers of the elements:

$$\mathbf{P} = P_1 + P_2 \qquad\qquad [5\text{-}12]$$

We substitute for the two *P* terms (twice) the lens-makers formula, once for the crown element and once for the flint,

$$\mathbf{P} = (n - 1) \underbrace{\left(\frac{1}{R_1} - \frac{1}{R_2}\right)}_{\textit{crown}} + (n - 1) \underbrace{\left(\frac{1}{R_1} - \frac{1}{R_2}\right)}_{\textit{flint}}$$

---

* This method of correcting for chromatic aberration goes back to 1729, when Chester Moor-Hall (1704–1771), British justice of the peace and amateur astronomer, designed the first achromatic contact doublet. Apparently, Moor-Hall had discovered that glass containing lead oxide, of the type used to make fine table glassware, had higher dispersion than window glass. To keep his discovery secret, he ordered one element for his doublet from a certain lens-maker in London, the other from another. As it happened, both men subcontracted the work to a third optician, who, on finding that both lenses were for the same customer and had one radius in common, placed them in contact and saw that the image was free of color. News of this discovery slowly spread to other opticians, among them John Dollond (1706–1761), whose son Peter (1739–1820) urged him to apply for a patent so that he could collect royalties. Naturally, the other London opticians objected and took the case to court, producing Moor-Hall as a witness. The court agreed that indeed Moor-Hall was the inventor, but in a much-quoted decision, the judge, Lord Camden, ruled in favor of Dollond, saying: "It is not the person who locked up his invention in his scritoire that ought to profit by a patent for such invention, but he who brought it forth for the benefit of the public."

For convenience we replace the terms containing the R's by K's, writing

$$\mathbf{P} = (n_1 - 1)K_1 + (n_2 - 1)K_2 \qquad [5\text{-}13]$$

To correct for chromatic aberration, the power of the combination at the two colors must be the same. We denote blue by the subscript F' and red by C'; thus

$$(n_{1F'} - 1)K_1 + (n_{2F'} - 1)K_2 = (n_{1C'} - 1)K_1 + (n_{2C'} - 1)K_2$$

Multiplying out and canceling gives

$$\frac{K_1}{K_2} = -\frac{n_{2F'} - n_{2C'}}{n_{1F'} - n_{1C'}} \qquad [5\text{-}14]$$

Note the minus sign in front of the right-hand term.

We repeat the process for green (subscript e),

$$P_{1e} = (n_{1e} - 1)K_1 \quad \text{and} \quad P_{2e} = (n_{2e} - 1)K_2$$

Dividing one by the other gives

$$\frac{K_1}{K_2} = \frac{P_{1e}(n_{2e} - 1)}{P_{2e}(n_{1e} - 1)} \qquad [5\text{-}15]$$

Then setting Equation [5-15] equal to [5-14] and solving for P_{1e}/P_{2e} yields

$$\frac{P_{1e}}{P_{2e}} = -\frac{(n_{2F'} - n_{2C'})/(n_{2e} - 1)}{(n_{1F'} - n_{1C'})/(n_{1e} - 1)}$$

Both the numerator and the denominator in the right-hand term are the inverse of Abbe's number,

$$V = \frac{n_e - 1}{n_{F'} - n_{C'}}$$

(which describes the dispersion) and therefore

$$\frac{P_1}{P_2} = -\frac{V_1}{V_2} \qquad [5\text{-}16]$$

Since Abbe's number can only be positive, the minus sign means that if one of the lenses is positive, the other must be negative.

Finally, we take Equations [5-12] and [5-16], solve them for P_2, and set them equal to each other by eliminating P_2:

$$P_2 = \mathbf{P} - P_1 = -\frac{P_1 V_2}{V_1}$$

We solve the two right-hand terms for P_1:

$$\mathbf{P}V_1 - P_1 V_1 = -P_1 V_2$$
$$\mathbf{P}V_1 = P_1 V_1 - P_1 V_2$$

$$\mathbf{P}V_1 = P_1(V_1 - V_2)$$

$$P_1 = \mathbf{P}\left(\frac{V_1}{V_1 - V_2}\right) \qquad [5\text{-}17a]$$

Similarly,

$$P_2 = -\mathbf{P}\left(\frac{V_2}{V_1 - V_2}\right) \qquad [5\text{-}17b]$$

Example

Design a crown-flint doublet of +10 cm focal length corrected for blue and red, using the refractive indices listed in Chapter 1.

Solution: First, we determine Abbe's numbers. We find for crown

$$\nu_1 = \frac{n_d - 1}{n_F - n_C} = \frac{1.5230 - 1.0}{1.5293 - 1.5204} = 58.7640$$

and for flint

$$\nu_2 = \frac{1.7200 - 1.0}{1.7378 - 1.7130} = 29.0323$$

Inserting these figures in Equations [5-17] gives

$$P_1 = (+10)\left(\frac{58.7640}{58.7640 - 29.0323}\right) = +19.7648 \text{ diopters}$$

and

$$P_2 = -(+10)\left(\frac{29.0323}{58.7640 - 29.0323}\right) = -9.7648 \text{ diopters}$$

The combined power of the two lenses is + 10 diopters, which serves as a check on our calculations so far.

Knowing the focal lengths required of the two lenses, we are ready to choose their radii. For reasons of economy, the converging lens is made symmetric, equiconvex. Also, the two lenses are to be in contact. Thus $R_1 = -R_2 = -R_3$. For the first lens, from the lens-makers formula,

$$P = (n_{\text{lens}} - 1)\left(\frac{1}{R_1} - \frac{1}{R_2}\right)$$

$$+19.7648 = (1.523 - 1.000)\left(\frac{2}{R_1}\right) = \frac{1.046}{R_1}$$

and thus

$$R_1 = 0.0529 = \boxed{+5.29 \text{ cm}}$$

The next two surfaces have radii of

$$R_2 = R_3 = \boxed{-5.29 \text{ cm}}$$

and for the last surface,

$$-9.7648 = (0.7200) \left(\frac{1}{-0.0529} - \frac{1}{R_4} \right)$$

$$R_4 = -\frac{0.7200}{3.8458} = -0.1872 = \boxed{-18.72 \text{ cm}}$$

Another method of making a system achromatic is to use two positive lenses made of the same type of glass and separated by a distance equal to one-half the sum of their focal lengths. To see why this approach works, we start out from the equivalent power equation,

$$\mathbf{P} = P_1 + P_2 - P_1 P_2 d$$

Then, following Equation [5-13],

$$\mathbf{P} = (n - 1)K_1 + (n - 1)K_2 - (n - 1)K_1(n - 1)K_2 d$$
$$= (n - 1)(K_1 + K_2) - (n - 1)^2 K_1 K_2 d$$

For the combination to be achromatic, $\mathbf{P}$ must remain constant even if the wavelength changes; thus by differentiation

$$\frac{d\mathbf{P}}{dn} = K_1 + K_2 - 2(n - 1)K_1 K_2 d = 0$$

We multiply the two K's by $(n - 1)$ and substitute the corresponding P's for $(n - 1)K$:

$$P_1 + P_2 = 2 P_1 P_2 d$$

$$d = \frac{P_1 + P_2}{2 P_1 P_2}$$

which is equal to

$$\boxed{d = \frac{1}{2} (f_1 + f_2)} \qquad\qquad [5\text{-}18]$$

Spaced doublets of this type are used in eyepieces (the Huygens and Ramsden eyepiece); these we will discuss in Chapter 9.

Concluding Remarks

Are some aberrations more important than others? That depends on the application. A high-speed camera lens must be well corrected for spherical aberration; distortion is not as critical. For a lens used for reproducing topographic maps it is just the opposite. In any case, there is no way to eliminate all aberrations at the same time; in fact, eliminating some of them will usually make others worse.

Table 5-1   Summary of Aberrations

Aberration	Character	Correction
1. Spherical aberration	Monochromatic, on- and off-axis, image blur	Bending, high index, aspherics, gradient index, doublet
2. Coma	Monochromatic, off-axis only, blur	Bending, spaced doublet with central stop
3. Oblique astigmatism	Monochromatic, off-axis, blur	Spaced doublet with stop
4. Curvature of field	Monochromatic, off-axis	Spaced doublet
5. Distortion	Monochromatic, off-axis	Spaced doublet with stop
6. Chromatic aberration	Heterochromatic, on- and off-axis, blur	Contact doublet, spaced doublet

All that we can do is try to *balance* them. A summary of the various aberrations and what to do about them is found in Table 5-1.

SUMMARY OF EQUATIONS

Coddington shape factor:

$$S = \frac{R_2 + R_1}{R_2 - R_1} \qquad\qquad \text{[5-3]}$$

Coddington position factor:

$$P = \frac{s' + s}{s' - s} \qquad\qquad \text{[5-4]}$$

Abbe's sine condition:

$$nD \sin U = n'D' \sin U' \qquad\qquad \text{[5-10]}$$

Petzval condition:

$$n_1 f_1 + n_2 f_2 = 0 \qquad\qquad \text{[5-11]}$$

Correction for chromatic aberration:

$$P_1 = \mathbf{P}\left(\frac{V_1}{V_1 - V_2}\right) \quad \text{and} \quad P_2 = -\mathbf{P}\left(\frac{V_2}{V_1 - V_2}\right) \qquad \text{[5-17a and b]}$$

$$d = \frac{1}{2}(f_1 + f_2) \qquad\qquad \text{[5-18]}$$

PROBLEMS

5-1. If a lens of 50 mm diameter and 25.4 cm focal length has a longitudinal spherical aberration of 4 mm, what is its transverse spherical aberration?

5-2. A thin meniscus of index 1.60 has radii $R_1 = +15$ cm and $R_2 = +30$ cm. Determine:
(a) The Coddington shape factor.
(b) The position factor for an object 1.0 m away.

5-3. A lens of index 1.6 forms of an object 40 cm away an image 8 cm away. The front surface of the lens has a radius of curvature of $+120$ mm. Find the shape and position factors.

5-4. A lens made of flint of index 1.72 has a focal length of $+5$ cm. For parallel incident light and for the lens to have minimum spherical aberration, determine the position and shape factors and the two radii of curvature necessary.

5-5. If the object consists of a series of dots arranged in a circle concentric with the optical axis, what does the image look like:
(a) In the tangential focus?
(b) In the radial focus?

5-6. A target in the form of a large + sign serves as the object for a positive spherocylinder, axis vertical. What does the image look like in the two principal image planes?

5-7. When a conventional (spherical) lens is turned 22.5° about the vertical axis, a point object 60 cm in front of the lens gives two line images, one 25 cm and the other 40 cm behind the lens. How much cylinder (power) has been induced?

5-8. A large-diameter spherocylinder forms one line focus at a distance of 21.7 cm and another line focus at a distance of 23.8 cm. To what diameter must an iris diaphragm close to the lens be stopped down in order to produce a circle of least confusion 0.8 mm in diameter?

5-9. A spherocylinder 70 mm in diameter produces two line foci whose distances from the lens are 12.5 cm and 20 cm, respectively. How large is the circle of least confusion?

5-10. A conventional thin lens forms of a pinhole 40 cm to the left of the lens a real image 20 cm to the right of the lens. On adding a cylinder to the lens, a horizontal line image is formed 10 cm to the right of the two lenses. What is the power and the orientation of the cylinder?

5-11. To meet the Petzval condition for the elimination of curvature of field, two lenses, of $+10.00$ and -10.00 diopters power and both of $n = 1.55$, are placed 25 mm apart. What is the power of the combination?

5-12. If a lens, made of glass of index 1.5, has $+9.00$ diopters power, what is the power of a lens of the same shape but made of glass of index 1.6?

5-13. A certain lens has, for yellow light, an index of refraction of 1.6500 and a focal length of 62 cm. For red light, the focal length becomes 62.5 cm. What is the refractive index for that light?

5-14. A $+6.1$-diopter crown lens is to be combined with a flint lens to make a contact doublet achromatic for blue and red. Using the V-numbers from Table 1-2, determine the power of the flint lens.

5-15. What are the powers of two thin lenses, one made of crown, the other of dense flint, that must be placed in contact to obtain a +6.00-diopter doublet achromatic for blue and red? Use the refractive indices from Table 1-2.

5-16. An achromatic doublet consists of two positive lenses separated by 8 cm. The first lens has a focal length of 12 cm. What is the focal length:
(a) Of the second lens?
(b) Of the whole system?

SUGGESTIONS FOR FURTHER READING

KINGSLAKE, R. *Lens Design Fundamentals.* New York: Academic Press, Inc., 1978.

PRICE, W. H. "The Photographic Lens," *Sci. Am.* **235** (Aug. 1976), 72–83.

WELFORD, W. T. *Aberrations of Optical Systems.* New York: American Institute of Physics, 1986.

6

Stops and Pupils

In any optical system the light passing through is limited in cross section, either by the finite diameter of the lenses or by additional diaphragms called *stops* (Figure 6-1). Stops, as well as *pupils*, are as important as lenses or mirrors. The exit pupil of a system, for instance, should be placed so as to coincide with the pupil of the observer's eye. Only then will the observer be comfortable looking through the system, an aspect just as necessary as seeing the image in focus.

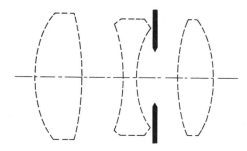

Figure 6-1 Aperture stop placed in a three-element lens system.

The Slide Projector

A good example that shows why it is so important to use lenses of the right size is the *slide projector*. When we set out to build a slide projector, perhaps we feel that all we have to do is take a light source, place the object (the *slide*) in front of it, and

106

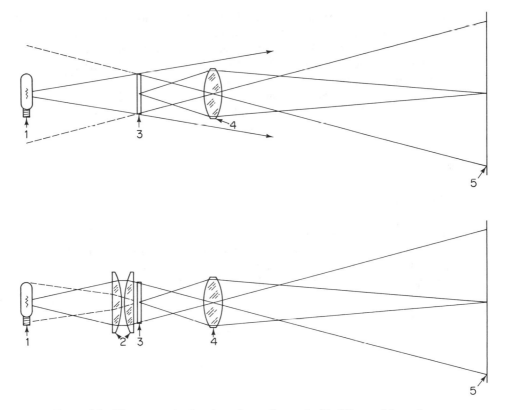

Figure 6-2 Wrong way (*top*) and good way (*bottom*) of building a slide projector: 1, light source; 2, condenser; 3, transparency; 4, projection lens; 5, screen.

project a magnified image onto a distant screen [Figure 6-2 (top)]. Most likely, the image will be disappointing. Even while it may be "in focus," probably only the center of the image will be bright; the corners will be dim or even be cut off entirely, an effect called *vignetting* (vig net' ting). Placing a sheet of ground glass between light source and transparency would help but that is hardly an elegant solution.

Obviously, if all of the transparency is to receive light that then goes on to the projection lens, a source of unreasonably large size would be required. Instead of trying to find such a source, we use a *condensing lens*. That lens is placed just ahead of the transparency; it should be slightly larger than the diagonal across the transparency, but since the condenser does not contribute directly to image formation, it need not be of high quality.

The purpose of the condenser is to project an image of the light source into, or close to, the projection lens (bottom). In other words, its purpose is to *fill the aperture of the projection lens*. The rays then continue and fill the screen. The projection lens, in turn, forms an image (of the slide). But with the light source relatively small, *its* image will be small too and hence the projection lens can be small also—which makes it less expensive to design and produce (but certainly it should be of high quality). The image, as it appears on the screen, is now fully and evenly illuminated, with no vignetting.

Clearly, the proper design of an optical system includes the correct choice of diameters of the various lenses, and of additional diaphragms. That is the subject of our discussion of *stops and pupils*.

STOPS

Aperture Stop

The first type of a stop that we discuss is an *aperture stop*. Often the aperture stop is simply the rim of a lens, or its mount. If it is a separate (mechanical) stop, it is placed *close to a lens*. In a camera the aperture stop is usually an iris diaphragm (which can be varied in size). Making the aperture stop smaller, as in Figure 6-3, limits the amount of light that reaches the image and therefore makes the image dimmer (but it does not change the focal length of the lens, or the location or magnification of the image, and it does not restrict the field of view).

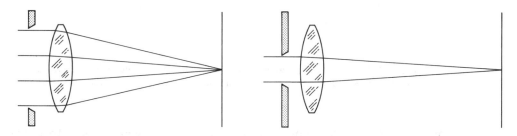

Figure 6-3   Aperture stop. Making the aperture stop smaller (*right*) limits the amount of light that reaches the image.

The ratio of the focal length of the lens to the aperture stop's diameter is called the *f-stop number*,

$$f\text{-stop number} = \frac{\text{focal length}}{\text{diameter of aperture}} \qquad [6\text{-}1]$$

For objects nearby and image distances, s', appreciably longer than the focal length, f, the *effective f*-stop number is s'/D, rather than f/D.

As a practical example, consider a camera lens of 50 mm focal length with an aperture 12.5 mm in diameter. At this setting, the *f*-stop number is 50 mm/12.5 mm = 4, which customarily is written $f/4$. Opening the aperture by one f-stop *doubles* the amount of light that reaches the film.

While an aperture stop has no effect on the focal length of a lens or the location of the image, it does affect the *depth of focus*. A smaller aperture *increases* the depth of focus. The concept of depth of focus is closely tied to the structure of the detector. The retina of the eye, for example, has a particular discrete structure. Photographic film, likewise, has a certain graininess. Two image points that are *closer* to one another than two adjacent elements of the retina, or two silver grains of photographic film, cannot be resolved. Resolution, therefore, even if the image were infinitely sharp, is limited.

In addition, the finite *thickness* of the retina or film limits the resolution. Ideally, the detector, retina or film, should be infinitesimally thin. But in reality, it is not, with light focusing on the rear surface causing a blur of diameter b on the front surface, the distance between the two planes representing the depth of focus (Figure 6-4). These two factors, inhomogeneity and depth of focus, account for a certain unsharpness of the image that in practice we can, or must, tolerate.

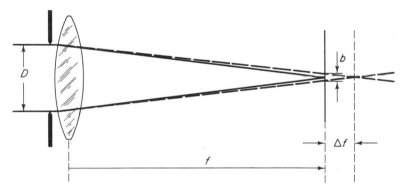

Figure 6-4   Defining depth of focus.

Assume, for example, that the light comes from infinity. If D is the diameter of the aperture, b the blur circle, and Δf the depth of focus, then from similar triangles and assuming that $f \gg \Delta f$,

$$\frac{D}{f} \approx \frac{b}{\Delta f} \qquad [6\text{-}2]$$

A blur circle 30 μm in diameter or 1/1000 the focal length of the lens, whichever is larger, is considered acceptable.

Depending on the depth of focus, Δf or $\Delta s'$, there is a conjugate distance, Δs, in the object space (not shown). That distance is called the *depth of field*. The same as with the depth of focus, a smaller aperture will increase the depth of field.

But to reach the largest possible depth of field, it would not be wise to focus at infinity because some depth of field, "beyond infinity," would be wasted. Instead, we focus at the *hyperfocal distance*; the depth of field then extends from one-half that distance out to infinity. Use the focal length of the lens, in mm, and divide it by the *f-stop*: that gives the hyperfocal distance, in meters:

$$\text{hyperfocal distance} = \frac{\text{focal length}}{f\text{-stop}} \qquad [6\text{-}3]$$

For example, with a 50-mm lens set at *f*/8, focus at 6.25 m: the depth of field then extends from 3.125 m to infinity.

Field Stop

The second type of a stop is the *field stop*. It limits the (angular) *field of view*. The field stop may either be a (mechanical) stop as such, or it may be the aperture of another lens placed in the focal plane of the first lens. In a camera, the iris diaphragm is the aperture stop, and the film size is the field stop. As the field stop is made smaller, the field of view becomes correspondingly smaller (as in Figure 6-5). A field stop is often placed in the image plane (as shown) but it could also be located in the object plane (as in a slide projector where the opaque mount of the slide limits the field).

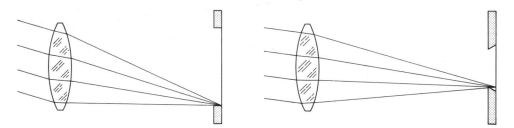

Figure 6-5 Field stop. Making the field stop smaller (*right*) limits the field of view.

PUPILS

The *pupil* of an optical system is simply the image of a stop. Assume, for example, that a stop is placed between an object and a lens, as in Figure 6-6. Its diameter is less than the diameter of the lens. Rays from all points in the object pass through the stop *completely mixed*, with no preferential spatial separation. If we partially occlude the stop with the point of a pencil and at the same time look at the image, the point can hardly be seen (except that the image becomes still dimmer).

It is this mixing of the rays, without spatial separation, that is characteristic of a pupil. The lens in this case does *not* act as a pupil: rays from the tip of the

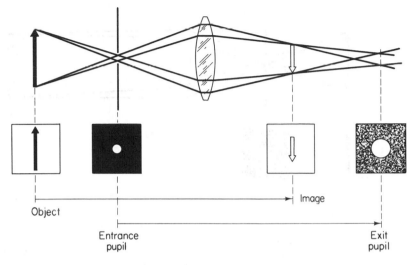

Figure 6-6 Entrance pupil and exit pupil. Horizontal arrows indicate two independent processes of image formation.

object pass through the lower part of the lens, and rays from the foot pass through the upper part. If we hold the pencil close to the lens, the point *could be seen*, forming a shadow superimposed on the image.

We also see from Figure 6-6 that actually there is an *entrance pupil* and an *exit pupil*. The entrance pupil is the stop as such. Conjugate to it is the exit pupil. This conjugate relationship shows us that the lens handles two processes of image formation that go on side by side, independent of each other: (1) the lens projects the object into its image, and (2) the lens also projects the entrance pupil into the exit pupil.

If we wish, we can *move* the entrance pupil, to a position either farther away from, or closer to, the lens. If we move the stop closer to the lens, its image, the exit pupil, moves away. If the entrance pupil is moved into the left-hand focal plane, the exit pupil moves out to infinity on the right. The system is then called *telecentric on the image side* (Figure 6-7). Conversely, if the stop is moved into the right-hand focal plane, the system is *telecentric on the object side*. Such systems are useful in precision measurements, whenever the (transverse) size of an object is to be compared with a scale that cannot be brought into contact with the object.

For comfortable viewing, as I said earlier, the exit pupil should coincide with the pupil of the eye. But that is not easy to accomplish: the size of the pupil depends on the amount of light. A telescope designed for looking at the stars, even if all other characteristics including magnification were the same, is not necessarily the best also for viewing a landscape during the day. At night, the pupil of the eye may be 7 mm in diameter while during the day it may be 3 mm. Using a

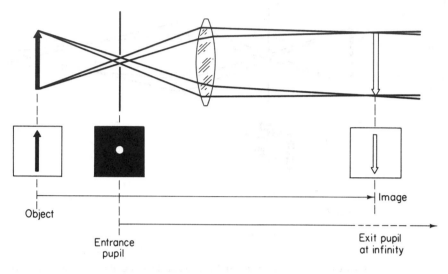

Figure 6-7 System telecentric on the image side.

daylight telescope at night means that the eye could accept a wider bundle of light, and therefore the telescope's front lens could be larger, more suitable for low light levels. On the other hand, using such a telescope during the day does not utilize all the light that passes through the front lens.

Lab Experiment. Take a filter one-half of which is green, the other half red. Hold the filter in front of a diffuse light source and, using a converging lens, form an image of the filter on a screen. The image will show both colors clearly separate. Then take a white card, hold the card first next to the screen, and then move it slowly toward the lens: the colors will fuse into white. This shows that the *lens* is the pupil (*where all rays mix*).

Remove the filter, without changing anything else. Place a stop halfway between the light source and the lens. Hold a pencil point close to the stop: the point's image can hardly be seen on the screen, which means that now the *stop* has become the pupil.

Move the screen farther away from the lens, to a position where you see an image of the stop. Again hold the pencil point close to the stop. Now the image of the stop with the point inside can be seen easily on the screen: While the stop is the *entrance pupil*, the stop's image is the *exit pupil*.

Numerical Example

A stop 8 mm in diameter is placed halfway between an extended object and a large-diameter lens of 9 cm focal length. The lens projects an image of the object onto a screen 14 cm away. What is the diameter of the exit pupil?

Solution: Consider again Figure 6-6. From the known focal length and the image distance, determine the object distance.

$$s = \frac{s'f}{f - s'} = \frac{(14)(9)}{9 - 14} = -25.2 \text{ cm}$$

One-half that distance is the distance of the stop (from the lens).

$$\frac{-25.2 \text{ cm}}{2} = -12.6 \text{ cm}$$

With the diameters given, we have little doubt that it is the stop, rather than the lens, that is the entrance pupil. The image of the stop, hence, is the exit pupil. It is formed

$$s'' = \frac{(-12.6)(9)}{(-12.6) + 9} = 31.5 \text{ cm}$$

to the right of the lens. The diameters of the two pupils are proportional to their distances.

$$\frac{D_{\text{EP}}}{D_{\text{XP}}} = \frac{12.6}{31.5}$$

Therefore, the diameter of the exit pupil is

$$D_{\text{XP}} = \frac{(0.8)(31.5)}{12.6} = \boxed{2 \text{ cm}}$$

Pupils of Combinations of Lenses

When a stop is placed ahead of a lens, that stop becomes the entrance pupil. But if another lens is placed ahead of the stop, the light does not reach the stop directly. Instead, the light "sees" the stop the same way we see newsprint through a magnifying glass. In other words, the light comes to an *image* of the stop, rather than to the stop itself, and *this image*, by definition, *is the entrance pupil*. Therefore, the entrance pupil of a combination of lenses is the image of the stop formed by all lenses *preceding* the stop. Conversely, the exit pupil of a combination of lenses is the image of the stop formed by all lenses *following* the stop.

If we are not sure whether it is the stop or one of the lenses that limits the light, the rule is that the aperture that subtends the *smallest angle* at the center of the object is the entrance pupil. We should also note that the exit pupil may be located to the right of the entrance pupil, or to the left of it, or it may even coincide with it.

In Figure 6-8, for example, we have placed a stop halfway between two lenses. By definition, the entrance pupil is the image of the stop formed by the lens, or lenses, preceding the stop. In this case this is lens 1. With the distances shown, the image is virtual, and located to the right of the stop. Conversely, the exit pupil is the image of the stop formed by the lens, or lenses, following the stop; that is lens 2. This image is virtual also, but located to the left of the stop.

Example

Two lenses, a +8.00-diopter lens and a minus lens of unknown power, are mounted coaxially and 8 cm apart. The system is afocal, that is, light entering the system parallel at one side emerges parallel at the other. If a stop 15 mm in diameter is placed halfway between the lenses:

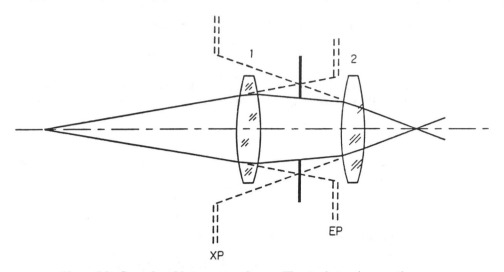

Figure 6-8   Stop placed between two lenses. The stop's two images, the entrance pupil, *EP*, and the exit pupil, *XP*, are both virtual.

(a) Where is the entrance pupil?
(b) Where is the exit pupil?
(c) What are their diameters?

Solution: (a) For the system to be afocal, the focal points of the two lenses must coincide (at *F* in Figure 6-9). Therefore, if the first lens has +8.00 diopters power, or (⅛)(100) = 12.5 cm focal length, and if the two lenses are 8 cm apart, the second lens must have −(12.5 − 8) = −4.5 cm focal length.

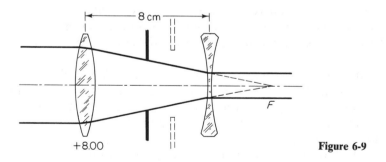

Figure 6-9

The *entrance pupil* is the image of the stop formed by the first lens. This time, contrary to our usual custom, the object (that is, the stop) lies to the right of the lens, the rays that form the image proceed from right to left, and the f_2 focal length is on the left. Hence, the image of the stop (that is, the entrance pupil) lies

$$s' = \frac{sf}{s+f} = \frac{(4)(-12.5)}{(4)+(-12.5)} = \boxed{+5.88 \text{ cm}}$$

to the right of the (first) lens.

(b) The *exit pupil* is the image of the stop formed by the second lens. It is

$$s' = \frac{(-4)(-4.5)}{(-4)+(-4.5)} = \boxed{-2.12 \text{ cm}}$$

to the left of the (second) lens. In other words, the two pupils coincide—which is typical of an afocal system.

(c) However, the pupil *sizes* are different. The diameter of the entrance pupil is

$$D_{\text{EP}} = \frac{D_{\text{STOP}} \ s'}{s} = \frac{(15)(58.8)}{40} = \boxed{22 \text{ mm}}$$

and the diameter of the exit pupil

$$D_{\text{XP}} = \frac{(15)(21.2)}{40} = \boxed{8 \text{ mm}}$$

Now consider a system of two lenses that are placed farther apart from each other, and a stop located close to the second lens (Figure 6-10). In that case, the stop acts more like a field stop. By definition, the image of a field stop, formed by the lenses preceding the stop, is the *entrance window*, EW. Conversely, the image of the stop, formed by the lenses following it, is the *exit window* (not shown). These are very appropriate terms: If we are inside a room and look out, it is the *window*, and its size and position relative to the observer, that limits the field of view.

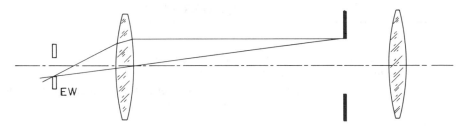

Figure 6-10   Field stop and its object-side image, the entrance window.

In conclusion, we note that

1. *Aperture stops relate to pupils.* The image of an aperture stop, formed by the lens or lenses preceding the stop, is the *entrance pupil*. The image formed by the lens or lenses following the stop is the *exit pupil*.
2. *Field stops relate to windows.* The image of a field stop, formed by the lenses preceding the stop, is the *entrance window*. The image formed by the lenses following the stop is the *exit window*.

3. At the exit pupil, the bundle has the *least diameter* because an aperture stop is by necessity farther away from the last lens than a field stop. That has important consequences: In many optical systems, such as in a telescope or microscope, the entrance pupil is the objective lens (no separate stop is needed). The exit pupil then is the image of this lens formed by the eyepiece. This image is where the density of the light is highest and where, as we have shown, the pupil of the observer's eye should be placed to assure comfortable viewing and avoid vignetting.

4. Finally, the light source, or in place of it a condenser lens, should be large enough to fill the entrance pupil. A source or lens that does not fill the entrance pupil cannot fill the aperture stop, and if it does not, the amount of light reaching the image will be less than it could be. This is why, in designing an optical system, we need to make sure that the lenses and other elements not only have the right focal lengths, but also the right *diameters*.

SUMMARY OF EQUATIONS

$$f\text{-stop number} = \frac{\text{focal length}}{\text{diameter of aperture}} \qquad [6\text{-}1]$$

PROBLEMS

6-1. Consider a lens of 100 mm focal length, together with an aperture stop 12.5 mm in diameter. What is the *f*-stop number?

6-2. If a camera lens of 50 mm focal length is set at *f*/8 and focused at an object 15 cm away, what is the effective *f*-stop number?

6-3. If a 50-mm lens is set at *f*/8, what is:
(a) The hyperfocal distance?
(b) The depth of field?

6-4. Using a camera with an 80-mm lens, you may want to have everything in focus from a distance of $2\frac{1}{2}$ meters out to infinity. What *f*-stop should you use and at which distance should you focus?

6-5. If a pinhole camera has the shape of a cube, how large an angular field of view, measured horizontally, will it cover?

6-6. A photographic negative taken on 35-mm film is 36 mm wide (with the camera held horizontally). Using a lens of 80 mm focal length, how wide is the angular field?

6-7. With the light coming from the left, a stop is placed two focal lengths to the right of a converging lens. Where is:
   (a) The entrance pupil of the system?
   (b) The exit pupil?

6-8. A stop is placed 10 cm to the right of an object, and a +10.00 diopter lens 20 cm to the right of the stop. How far from the lens is the exit pupil?

6-9. If a stop is placed 3 cm to the left of a lens of 50 mm focal length, where is the exit pupil located?

6-10. An object is placed 25 cm in front of a 60-mm-focal-length lens and a stop 2 cm behind the lens. If the lens is 50 mm in diameter, and the stop 20 mm, determine:
   (a) The position of the entrance pupil relative to the lens.
   (b) The diameter of the entrance pupil.

6-11. A lens of 50 mm focal length and 36 mm diameter is used as a magnifier. If a stop 12 mm in diameter is placed 10 mm to the right of the lens:
   (a) How far from the lens is the entrance pupil?
   (b) What is the diameter of the entrance pupil?

6-12. Two thin lenses, 5 cm in diameter each and of focal lengths +10 cm and +6 cm, respectively, are placed 4 cm apart. An aperture stop 2 cm in diameter is set halfway between the lenses. Find the diameters of the entrance pupil and the exit pupil.

6-13. A lens of +4 cm focal length and 2 cm diameter is combined with a second lens of −2 cm focal length and 1 cm diameter, and the two lenses are separated by 2 cm. By graphical ray tracing locate the:
   (a) Entrance pupil.
   (b) Exit pupil.
   (c) Field stop.
   (d) Entrance window.

6-14. A lens of +5 cm focal length is mounted 6 cm in front of another lens but of −7 cm focal length. When a stop 5 mm in diameter is placed halfway between the two lenses, what is:
   (a) The location and diameter of the entrance pupil?
   (b) The location and diameter of the exit pupil?

7

Gradient-Index, Fiber, and Integrated Optics

Gradient-index optics has become a prominent part of modern optics. The term *gradient index*, or *GRIN* for short, refers to local variations of refractive index. For example, plane-parallel circular plates can be made with a refractive index that is higher in the center than at the periphery. Such a plate acts like a converging lens. If the plate also has curved surfaces, like an ordinary lens, it is equivalent to a combination of two lenses. Other applications of GRIN optics extend to waveguides (*fiber optics*) and to miniaturized systems (*integrated optics*).

Atmospheric Refraction

Gradient-index phenomena are very common. They occur when hot air mixes with cold air, when warm water is poured into cold water, when liquids are separated as in electrophoresis or ultracentrifugation, and in many other processes that involve temperature, pressure, or density gradients.

The atmosphere, for example, has a refractive index that decreases with height; higher up its density is less. This causes light to proceed in a curved path, an effect known as *regular atmospheric refraction* [Figure 7-1 (top)]. A similar effect, called *looming*, is seen when looking across a body of cold water; it makes

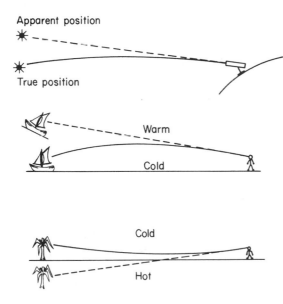

Apparent position

True position

Warm

Cold

Cold

Hot

Figure 7-1 Regular atmospheric refraction (*top*), looming (*center*), and mirage (*bottom*). Dashed lines indicate apparent positions, full lines true positions.

objects on the surface appear to be lifted up (center).* The opposite is the *mirage*; it occurs when looking across an expanse of hot desert. Here, the air directly above the ground is hotter than the air higher up, its refractive index is lower, and distant objects appear below the horizon as though reflected in a pool of water (bottom). *Random atmospheric refraction* is due to turbulence; it causes the twinkling, or scintillation, of the stars.

THEORY OF REFRACTIVE GRADIENTS

Imagine a medium whose index of refraction is lowest at the top, increasing from there down toward the bottom. Let the light, as shown in Figure 7-2, be incident from the left. Outside the gradient field the light has a wavelength λ_0. At the top, the index is n and the wavelength λ. At the bottom, the index is n' and the wavelength λ'. If there are N wavefronts within the volume shown, the upper arc has length

$$\Delta L = N\lambda = N\frac{\lambda_0}{n} \qquad [7\text{-}1]$$

and the lower arc

---

* Some time ago people believed that big four-masted brigs, after having been lost at sea, kept on sailing through the air. Richard Wagner wrote an opera ''The Flying Dutchman'' referring to such a ship. But perhaps this may not all be fantasy. As shown, vessels on the sea can indeed appear as if lifted up into the air.

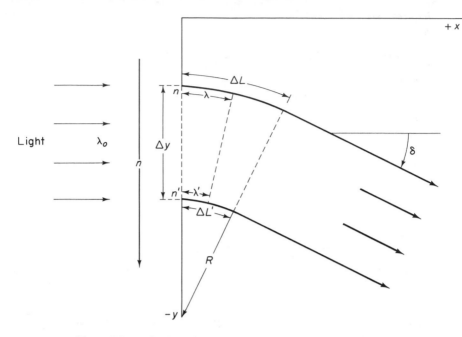

Figure 7-2 Deflection of light passing through a gradient-index field.

$$\Delta L' = N\lambda' = N\frac{\lambda_0}{n'}$$

The arc lengths depend on the radius of curvature, R, of the rays and on the angle of deviation, δ, measured in radians:

$$\Delta L = R\delta \qquad\qquad [7\text{-}2]$$

and

$$\Delta L' = (R - \Delta y)\delta \qquad\qquad [7\text{-}3]$$

Subtracting Equation [7-3] from [7-2] gives

$$\Delta L - \Delta L' = \Delta y\delta \qquad\qquad [7\text{-}4]$$

Since

$$\Delta L - \Delta L' = N\frac{\lambda_0}{n} - N\frac{\lambda_0}{n'} = N\lambda_0\left(\frac{n'-n}{nn'}\right)$$

we find that

$$\Delta L - \Delta L' = n\Delta L\left(\frac{n'-n}{nn'}\right)$$

and

$$\delta = \frac{1}{\Delta y} \left(1 - \frac{n}{n'}\right) \Delta L \qquad [7\text{-}5]$$

For light proceeding horizontally, in the $+x$ direction, it is sufficient to resolve δ into two components, δ_y and δ_z. The y component, in a first approximation, is given by

$$\delta_y = \int \frac{1}{n} \frac{\partial n}{\partial y} \, dL \qquad [7\text{-}6]$$

where the integral is taken over the length, L, traversed by the light. If this length is relatively short and the medium homogeneous in that direction,

$$\boxed{\delta_y = \frac{1}{n} \frac{\partial n}{\partial y} L} \qquad [7\text{-}7]$$

The angle of deflection, hence, is a function of the gradient $\partial n/\partial y$, of the index n, and of the length of path traversed, L. Inserting Equation [7-7] into [7-2] gives

$$\Delta L = R \frac{1}{n} \frac{\partial n}{\partial y} L$$

and setting $\Delta L = L$ and solving for R,

$$R = \frac{n}{\partial n/\partial y} \qquad [7\text{-}8]$$

Thus *the steeper the gradient, the shorter the radius of curvature through which the light is bent.*

Example

Assume that a cuvette is filled with a salt solution whose refractive index varies from $n_1 = 1.4$ at the top to $n_2 = 1.6$ at the bottom. If the distance between top and bottom is 5 cm and the path through the solution 1 cm:

(a) Determine the *angle of deflection* of the light.
(b) Determine the *radius of curvature* through which the light is bent.
(c) Determine the *apex angle of a prism* (made of glass of $n = 1.5$) which would give the same deflection.

Solution: (a) The angle of deflection is found from Equation [7-7], using for n the average refractive index, $(1.4 + 1.6)/2 = 1.5$:

$$\delta = \frac{1}{n} \frac{\partial n}{\partial y} L = \frac{1}{1.5} \frac{0.2}{0.05} \, (0.01) = 0.02667 \text{ rad} = \boxed{1.53°}$$

(b) The radius of curvature follows from Equation [7-8],

$$R = \frac{1.5}{0.2/5} = \boxed{37.5 \text{ cm}}$$

(c) Note that the 1.53° angle refers only to deflection within the solution. Outside, after passing through the rear surface of the cuvette, the angle, from Snell's law, becomes

$$I' = \arcsin(1.5)(\sin 1.53°) = 2.29°$$

The apex angle of a (thin) prism that gives the same deflection is found from Equation [1-8].

$$A = \frac{D}{n-1} = \frac{2.29°}{1.5-1} = \boxed{4.58°}$$

GRADIENT-INDEX LENSES

In a conventional lens, the refractive index is the same throughout. Refraction takes place only at the surfaces of the lens. In a *gradient-index lens*, refraction also takes place within the lens. That has major advantages. A single GRIN lens can be corrected for various aberrations using sometimes no more than a single element. Hard-to-grind aspherics can be replaced by (spherical) GRIN lenses. In complex optical systems the number of elements can be reduced, without sacrificing performance, by replacing a number of homogeneous lenses by a lesser number of GRIN lenses.

GRIN lenses come in three forms. One form has a *radial* gradient (of cylindrical symmetry), which means that the index of refraction varies as a function of distance *from* the optical axis.* The endfaces of such lenses can either be plane, or curved to give additional power.

Another type of GRIN lens contains an *axial* gradient. Here the refractive index varies as a function of distance *along* the axis: the surfaces of constant index are plane and normal to the axis. Axial refractive gradients are particularly useful for the correction of spherical aberration, replacing aspherical surfaces.

In Figure 7-3 (top), we have an example of spherical aberration, similar to the one we saw earlier in Chapter 5: the marginal rays come to a focus closer to

---

* GRIN lenses of radial symmetry were made as early as 1905 by Robert Williams Wood (1868–1955), professor of experimental physics at Johns Hopkins University. Wood used a mixture of glycerol and gelatin, and let it gel between two parallel plates, forming a cylinder a few centimeters in diameter. When the cylinder is soaked in water, the water slowly diffuses into the jelly, replacing (part of) the glycerol, and lowering the index. The cylinder is then cut into several slices, each slice representing a GRIN lens of radial symmetry—which forms an image just as a conventional (converging) lens does. Wood also made notable contributions also to UV and IR spectroscopy and was a superb lecturer with a flair for showmanship. For some of the spectroscopic experiments he did at his farm on Long Island, he used a grating at the end of a long section of sewer pipe; he cleaned the pipe of cobwebs by letting a cat run through it. Wood wrote two books, one of them a collection of nonsense poetry, *How to Tell the Birds from the Flowers* (New York: Dover Publications, Inc., 1959). When he sent a copy of it to President Theodore Roosevelt, the President asked for more of his writings, so Wood sent him a copy of his other book, *Physical Optics*. Even today, *Physical Optics* is worth reading for its wealth of experimental detail.

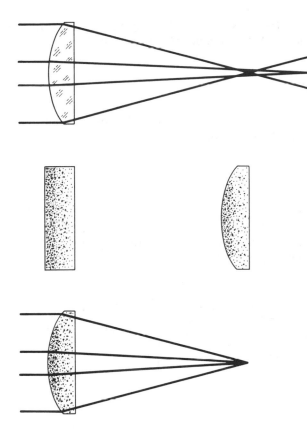

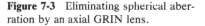

Figure 7-3 Eliminating spherical aberration by an axial GRIN lens.

the lens than the paraxial rays. Now consider a GRIN-lens blank containing an axial gradient (center left). The refractive index is highest near the front surface of the lens. If the front surface is ground convex (center right), some of the peripheral high-index material is removed and the marginal rays are bent less. With the gradient profile chosen correctly, all rays passing through come together at the same point, eliminating spherical aberration (bottom).

The third type is the *spherical* GRIN lens, where the index of refraction varies symmetrically about a point: The surfaces of constant index are *spheres*. An example of such a lens is the crystalline lens of the human eye.

A persistent problem in GRIN optics is how to produce the gradients. There are several methods available, neutron irradiation, chemical vapor deposition, polymerization, and ion stuffing, but the most promising, it seems, is *ion exchange*. Usually the process starts with an aluminosilicate glass. The sodium ions in such glasses are only loosely bound to the glass structure by weak interactions with the silicon and oxygen atoms. The Na^+ ions are then exchanged with other ions known to increase the refractive index, most notably silver, Ag^+. A lens blank is suspended in a bath of molten AgCl at a temperature of about 500°C for some 40 h. During that time, while the bath is continuously being stirred, the Ag^+

ions diffuse into the glass, replacing the Na^+ ions and raising the index. Theoretically, the index difference can be as high as 0.15, but in practice Δn is limited to about 0.05.

EXPERIMENTAL GRIN OPTICS

Under this heading we discuss various ways of testing lenses or otherwise showing the typical properties of GRIN elements.*

Toepler's Method

In Toepler's schlieren method, the light, as shown in Figure 7-4, comes from a source (1) and is focused by two lenses (2 and 4) onto a knife edge (5). In addition, the GRIN field (3) is imaged by another lens (6), often the objective lens of a camera, onto a screen or photographic film (7).

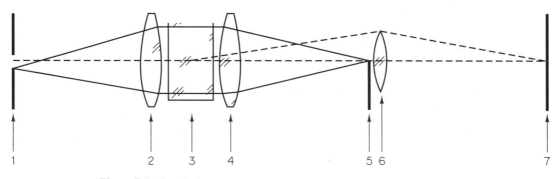

Figure 7-4 Toepler's method. 1, light source; 2 + 4, schlieren head; 3, gradient-index field; 5, knife edge; 6, camera lens; 7, image plane.

The two lenses 2 and 4 together are called the *schlieren head*. They must be large enough to cover the field. They also must be of high quality and free from

---

* It all began with Robert Hooke (1635–1703), English physicist, at that time curator of experiments of the Royal Society of London. In Observation LVII of his *Micrographia*, 1665, Hooke reports that he took a large convex lens and placed it at some distance from a candle. Another candle was set nearby. Looking through the lens at the first candle, he saw how the flame (of the second candle) was "encompassed with a stream of liquor, which seemed to issue out of the wick, and to ascend up in a continual current . . . unmixt with the antient air," an observation that even today is a most impressive sight.

Some 200 years later came Léon Foucault (1819–1868), French physicist, with his *knife edge test*, and August Toepler (1836–1912), German scientist. In his book, *Beobachtungen nach einer neuen optischen Methode* (Bonn: Maximilian Cohen und Sohn, 1864), Toepler tells how difficult it is to produce large lenses of good quality. Often the craftsman labors for weeks grinding the lens, only to find after polishing that it contains some inhomogeneities, since called *schlieren*.

polishing and other defects; otherwise all such defects would show. Sometimes, the two lenses are replaced by concave first-surface mirrors.

   As in all schlieren methods, there are two processes that take place side by side. If the sampling field is homogeneous and of uniform refractive index, the *light passing through* is blocked by the knife edge and no light reaches the screen. But if there are inhomogeneities present, these gradients will show as *images*, as bright streaks on a dark background.

FIBER OPTICS

Optical fibers come in two types: *step-index fibers* and *GRIN fibers*. Step-index fibers were developed first.* They consist of a transparent core of glass or plastic of a given refractive index surrounded by a cladding of lower index. Most of the light travels inside the core and is contained there by total internal reflection.

   Assume that n_0 in Figure 7-5 is the refractive index of the medium outside the fiber, n_1 the index of the core of the fiber, and n_2 the index of the cladding. From the construction we see that

$$\sin I'' = \frac{n_2}{n_1} = \frac{c}{a}$$

Since

$$a^2 = b^2 + c^2$$

$$1 = \frac{b^2}{a^2} + \frac{c^2}{a^2}$$

$$\left(\frac{b}{a}\right)^2 = 1 - \left(\frac{c}{a}\right)^2 = 1 - \left(\frac{n_2}{n_1}\right)^2$$

$$\frac{b}{a} = \sqrt{1 - \left(\frac{n_2}{n_1}\right)^2} = \sin I'$$

Then, from Snell's law,

$$n_0 \sin I = n_1 \sin I' = n_1 \sqrt{1 - \left(\frac{n_2}{n_1}\right)^2} = \sqrt{n_1^2 - n_2^2} \qquad [7\text{-}9]$$

where $n_0 \sin I$ is the *numerical aperture*, NA, of the fiber. It is a measure of how wide a cone of light the fiber can accept.

---

   * At night, when we look at a water fountain illuminated from below, the light seems to follow the curved streams of water. Actually, the light follows a zigzag path, because of total internal reflection. The first to show this effect, using a stream of water flowing from a tank, was John Tyndall, "On Some Phenomena Connected with the Motion of Liquids," *Proc. Roy. Inst.* **1** (1854), 446–48. J. L. Baird extended the principle to fibers to convey an image, *An Improved Method of and Means for Producing Optical Images*, Brit. Patent 285,738, Feb. 15, 1928.

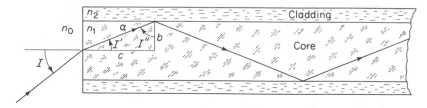

Figure 7-5   Propagation of light in a step-index fiber.

Gradient-index, or *GRIN*, *fibers* have a refractive index that is highest along the axis and lower farther away from it. Ordinarily, the light moves along near the center of the fiber. But if the ray is incident obliquely, making an angle with the fiber's axis as shown in Figure 7-6, the ray encounters the gradient as if it were a prism and through it is bent back toward the axis. The ray never even touches the surface of the fiber. Obviously, that is different from a step-index fiber which relies on total internal reflection. The advantage of a GRIN fiber is that the rays inside the fiber all have the same path length so that a pulse of light injected at one end retains its shape when it emerges at the other.

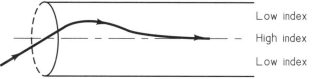

Low index

High index

Low index

Figure 7-6   Propagation of light within a GRIN fiber.

Applications

The most elementary application of fiber optics is the *conduction of light*, either to illuminate hard-to-reach places or to conduct light out of such places, perhaps to a photocell located at a more accessible site. In an instrument-laden cockpit, for instance, numerous individual light bulbs have given way to fiber bundles and a single light bulb that can easily be replaced when needed. In a similar way, it is easy to isolate a sensitive photoelectric detector from electronic noise by shielding the detector and routing the signal to it by way of a fiber.

Of great practical importance is the use of waveguides in the field of *communications*. Compared with electrical conductors, optical fibers are lighter in weight, less expensive, equally flexible, not subject to electrical interference, and more secure from interception. Their main advantage, though, is their enormous *bandwidth*. That makes it possible for one fiber to carry many more signals than any metal wire can.

A wave "carries" information by modulation. For example, a radio wave with a frequency of 30×10^6 Hz may have superimposed on it a variation of its amplitude with a frequency of 440 Hz (the sound of the note A in the middle of the keyboard of a piano). The wave then transmits the signal to a receiver and speaker which convert it back into sound. This type of transmission is called *analog*.

Today, though, more and more signals are processed by *digital* conversion: the signal is converted into numbers, such as a series of 1s and 0s. In either case, analog or digital, the time variation of the signal must be much slower than the frequency of the carrier. If the carrier has a higher frequency, more digits can be transmitted per second. For example, if we use *light* (which for reddish-orange has a frequency of 5×10^{14} Hz) it can carry information at a rate more than 10 million times greater than radio waves.

The first transatlantic telegraph cable was installed in 1858. The first analog transatlantic *telephone* cable, TAT-1, was built in 1955; it could handle 36 simultaneous calls. The transatlantic *fiber* cable TAT-9 was installed in 1991; it contains six glass fibers, each of them 1.3 μm thick, two pairs for messages and one pair for backup. It has a capacity of over 8000 digital channels and can handle 80,000 simultaneous calls, voice, electronic data, and video. Repeaters, powered from the shores, are placed every 70 km. At each repeater the signals are changed from optical (infrared) to electrical pulses, amplified, and changed back to optical pulses that pass on to the next repeater.

Finally, there is the *transmission of images*. A good example is the *flexible fiberscope*. As shown in Figure 7-7, some of the fibers conduct light into the cavity to be examined, while others, sometimes up to 140,000 of them, one for each pixel, carry the image back to the observer. Other types have only a few fibers (and thus are even more flexible) and at their front ends carry a tiny image tube with a scanning mechanism. Fiberscopes are used extensively in engineering as well as in medicine. They make it possible to inspect just about any cavity in the human body, from the respiratory to the digestive tract, and to look into the heart while it beats.

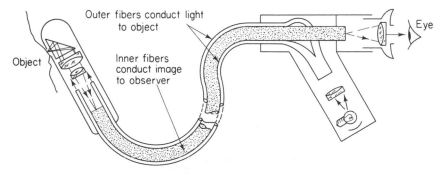

Figure 7-7 Flexible fiberscope.

INTEGRATED OPTICS

In recent years, fiber optics has developed further into complex systems of minia-ture dimensions. This is the field of *integrated optics*. Like fibers, the individual elements can be made in the form of step-index, or as gradient-index, elements.

Lenses used in integrated optics can be made in either of two forms. One of them resembles a conventional lens, with the light passing through approximately in direction of the axis of the lens. Figure 7-8 shows an example of two plano-convex lenses of this type, forming a collimator. These lenses can be made very small, with diameters no more than 0.5 mm in size and focal lengths less than 1 mm.

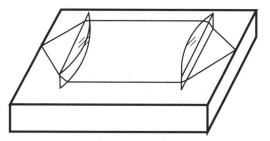

Figure 7-8   Two plano-convex lenses forming a collimator.

In other types of lenses the light passes through in the direction of the lenses' diameter. One type, called *mode-index lenses*, can perhaps be understood best by considering a prism made out of a thin film whose thickness varies. In a thinner film the effective velocity of the light is higher (because its zigzag path is stretched out longer), in a thicker film it is lower. If we add another layer of transparent material to the film, through a mask with a triangular opening, the light that passes through the base of the triangular-shaped, thicker film is delayed and, as in a conventional prism, deviates from its initial direction.

A third type of lenses used in integrated optics is the gradient-index lens, which is similar to a mode-index lens. The best known of these is the *Luneburg*

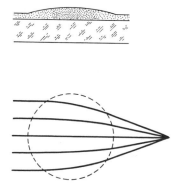

Figure 7-9   Cross-section of a thin-film Luneburg lens (*top*), with ray trace for paraliel light shown below.

lens. It looks like a flat, circular mound (Figure 7-9). Its index is highest in the center of the lens and decreases toward the periphery. Interestingly, light incident from any direction within the plane of the film comes to a focus without aberrations except curvature of field.

Manufacturing such elements is a rapidly evolving art. Systems containing only *passive* elements such as prisms and lenses are often made out of polymethyl methacrylate. If *active* components are also part of the system, often gallium arsenide, GaAs, is used. Gallium arsenide is well known from the construction of lasers. Such lasers serve as tiny, highly efficient light sources that, together with other passive and active elements, can be incorporated into small *monolithic* systems, often no larger than a few millimeters in size. Figure 7-10 shows an example.

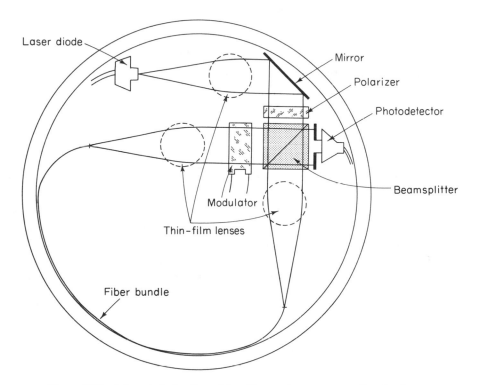

Figure 7-10 Integrated circuit used in a fiber-optics gyroscope. For details on gyros see Chapter 24.

Other uses of integrated optics include: spectrum analyzers (to tell the pilot of a military aircraft whether his plane is being tracked by enemy radar); multiplexed lasers operating at different wavelengths; Doppler velocimeters; readheads for computer, video, and audio disks; and, again, optical communication systems where we now have networks that are thousands of kilometers long and permit data rates up to gigabits per second and up to 80,000 telephone conversations at the same time.

SUMMARY OF EQUATIONS

Angle of deflection in GRIN medium:

$$\delta = \frac{1}{n} \frac{\partial n}{\partial y} L \qquad\qquad [7\text{-}7]$$

Radius of curvature in GRIN medium:

$$R = \frac{n}{\partial n / \partial y} \qquad\qquad [7\text{-}8]$$

Numerical aperture of step-index fiber:

$$n_0 \sin I = \sqrt{n_1^2 - n_2^2} \qquad\qquad [7\text{-}9]$$

PROBLEMS

7-1. A slab of glass, 14 mm thick and 60 mm long, has a refractive index of 1.51 at one end and 1.55 at the other. When placed in a bundle of collimated light, how much will the light be deflected on a screen 10 m away?

7-2. A palm tree is seen across 1 km of desert that has become so hot that directly above the ground the refractive index of the air is 1.00029, whereas 2 m higher up it is 1.0003. How much farther down does the tree seem to be compared with where it really is?

7-3. A round, plane-parallel plate of glass is 20 mm thick. If 20 mm from the center of the plate the refractive index is 1.5 and if it decreases, along the radius, by 0.003/mm, by how much does a ray of light, entering the plate 20 mm from the center, change in direction (before it again leaves the plate)?

7-4. A plane-parallel GRIN plate of radial symmetry is 40 mm in diameter and 10 mm thick. Its refractive index varies from 1.6 in the center to 1.5 in the periphery. What is its focal length?

7-5. Imagine a solid sphere of gradient-index material whose index is highest in the center and 1.4 at the periphery.
   (a) Assume that light is injected tangentially into the sphere such that it travels inside, next to the surface. What is the *lowest-possible index* in the center to accomplish that?
   (b) What must be the *diameter* of the sphere?

7-6. Some GRIN material is made in the form of a doughnut. The inside diameter (the diameter of the hole in the doughnut) is 34 cm and the outside diameter 46 cm; the cross section of the ring, therefore, is 6 cm in diameter.

   (a) If the average index is 1.6, what radial gradient is needed to keep a beam of light traveling along exactly in the center of the ring?

   (b) What should be the highest, and the lowest, refractive index of the material?

7-7. Assume that the front surface of an optical fiber is polished flat and its core has a refractive index of 1.7. How wide a cone of light can propagate inside the fiber?

7-8. With a larger difference between the refractive indices of core and cladding, a fiber will accept more light. For example, if the core has an index of 1.55 and the cladding 1.53, what is the largest (total) angle at which light can enter the fiber?

7-9. Parallel light enters a fiber of diameter $D = 0.1$ mm (Figure 7-11). Determine the least radius R through which the fiber may be bent, assuming that the core has an index of 1.54, the cladding 1.52.

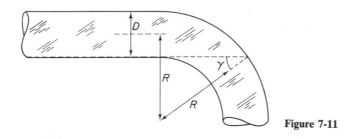

Figure 7-11

7-10. Assume that a series of object points lie on a circle concentric with a thin-film Luneburg lens. If the circle has a radius 1.5 times the focal length of the lens, where do we find the conjugate image points?

7-11. How could you possibly construct a *diverging* Luneburg lens?

7-12. Three individual fibers enter an integrated circuit, and three other fibers leave it. A switching element can be designed such that any incoming fiber is connected to any outgoing fiber, but no incoming fiber is connected to more than one fiber, nor is any fiber *not* connected to any other fiber. How many permutations are possible?

SUGGESTIONS FOR FURTHER READING

CHEO, P. K. *Fiber Optics and Optoelectronics*, 2nd ed. Englewood Cliffs, N.J.: Prentice-Hall, Inc., 1990.

HUNSPERGER, R. G. *Integrated Optics: Theory and Technology*, 3rd ed. New York: Springer-Verlag, 1991.

MEYER-ARENDT, J. R. *Schlieren Optics*. Bellingham, Wash.: The International Society for Optical Engineering, 1992.

YOUNG, M. *Optics and Lasers Including Fibers and Optical Waveguides*, 4th ed. New York: Springer-Verlag, 1992.

8

Lens Design

After we have become familiar with the individual lenses and stops, we follow a ray as it passes through the system. That is called *ray tracing*. Ray tracing can be practiced to different degrees of precision. *Paraxial ray tracing* is limited to rays that are close to the optical axis. *Trigonometric ray tracing* includes larger angles but still it applies only to rays that lie in a meridional plane (which also contains the optical axis). *Skew rays* proceed in three dimensions, outside of meridional planes. The modern way of ray tracing is by *computer*. An example is shown in Figure 8-1.

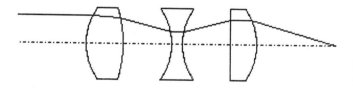

Figure 8-1 Triplet drawn by computer graphics.

TRIGONOMETRIC RAY TRACING

Assume that we have a combination of two lenses in contact, a *doublet* as shown in Figure 8-2. The first lens is equiconvex; its surfaces have radii of curvature of +50 mm and −50 mm, respectively. The second lens is a negative meniscus; its

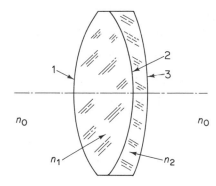

Figure 8-2 Example of a doublet. To scale.

surfaces have radii of -50 mm and -87 mm. Lens 1 has an axial thickness of 20 mm and is made of glass of index 1.50. Lens 2 has an axial thickness of 5 mm and an index of 1.80. A ray of light enters the system parallel to the axis, at a height of 22 mm. After passing through the two lenses, how far from the back vertex does the ray intersect the axis?

When entering the first surface, the ray proceeds as shown in Figure 8-3, at a height of

$$h = 22 \text{ mm}$$

Initially, the *slope angle*, subtended by the ray and the horizontal, is zero.

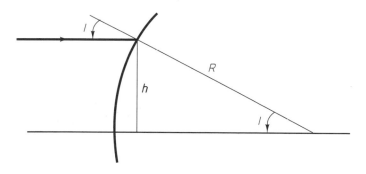

Figure 8-3 Ray incident on the first surface.

First we determine the angle of incidence. Since we know the height of the ray and the radius of curvature of the surface, we can find the sine of the angle of incidence,

$$\sin I = \frac{h}{R} = \frac{22}{50} = 0.44$$

But angles of incidence are measured *from* the surface normal and thus, since angle *I* turns counterclockwise, it is *negative*,

$$I = -26.1°$$

The angle of refraction is found from Snell's law,

$$n \sin I = n' \sin I'$$

and therefore

$$\sin I' = \left(\frac{1.0}{1.5}\right) (0.44)$$

$$I' = -17.1°$$

The difference between the two angles, as we see from Figure 8-4, is the slope angle after refraction,

$$U' = I - I'$$

$$= 26.1° - 17.1° = 9.0°$$

Next, instead of the height of the ray above the axis, we consider the perpendicular *vertex-ray distance*. Distance Q', as shown in Figure 8-4, consists of two parts, $a + b$. Since

$$\sin I' = \frac{a}{R}$$

and

$$\sin U' = \frac{b}{R}$$

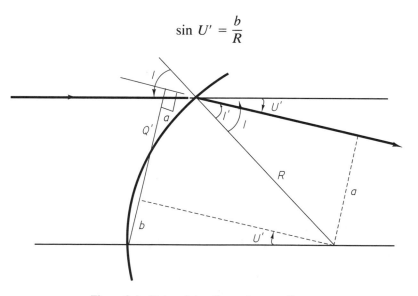

Figure 8-4   Determining the vertex-ray distance.

the total distance is

$$Q' = R \sin I' + R \sin U' \qquad [8\text{-}1]$$

$$= (50)(\sin 17.1°) + (50)(\sin 9.0°)$$

$$= 22.53 \text{ mm}$$

Now we are ready to consider the *transfer* of the (refracted) ray from the first surface to the second. If d is the distance between the two vertices, as illustrated in Figure 8-5,

$$\sin U' = \frac{Q_1 - Q_2}{d}$$

$$d \sin U' = Q_1 - Q_2$$

$$Q_2 = Q_1 - d \sin U' \qquad [8\text{-}2]$$

$$= 22.53 - (20)(\sin 9.0°)$$

$$= 19.38 \text{ mm}$$

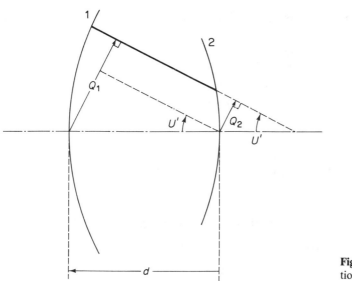

Figure 8-5 Deriving the transfer equation.

At the second surface, angle I_2 is turning clockwise and thus, compared with Equation [8-1], its sign, and that of the radius, changes,

$$\sin I_2 = \frac{Q_2}{R_2} + \sin U_2 \qquad [8\text{-}3]$$

$$= \frac{19.38}{50} + \sin 9.0° = 0.5449$$

$$I_2 = 33.0°$$

Again we use Snell's law to find the angle of refraction,

$$\sin I_2' = \left(\frac{1.5}{1.8}\right)(0.5449) = 0.4541$$

$$I_2' = 27.0°$$

The slope angle after refraction at the second surface is

$$U_2 = U_1 - I_2 + I_2' \qquad [8\text{-}4]$$

$$= 9.0° - 33.0° + 27.0°$$

$$= 3.0°$$

For the transfer from surface 2 to 3, again using Equation [8-2], we have

$$Q_3 = Q_2 - d_{23}\sin U_2$$

$$= 19.38 - (5)(\sin 3.0°)$$

$$= 19.12 \text{ mm}$$

and for the angles of incidence and refraction at the third surface, from Equation [8-3],

$$\sin I_3 = \frac{Q_3}{R_3} + \sin U_2$$

$$= \frac{19.12}{87} + \sin 3.0° = 0.2727$$

$$I_3 = 15.8°$$

$$\sin I_3' = \left(\frac{1.8}{1.0}\right)(0.2727) = 0.4908$$

$$I_3' = 29.4°$$

The slope to the right of the third surface, following Equation [8-4], is

$$U_3 = U_2 - I_3 + I_3'$$

$$= 3.0° - 15.8° + 29.4°$$

$$= 16.6°$$

Finally, we determine the *axial intercept*, the distance from the back vertex of the last lens to the point where the refracted ray crosses the optical axis. That distance is

$$L' = \frac{\text{final } Q}{\text{final } \sin U}$$

$$= \frac{19.12}{\sin 16.6°} = \boxed{66.9 \text{ mm}}$$

SKEW RAYS

As discussed earlier, meridional rays lie in planes that also contain the optical axis. Skew rays proceed in three-dimensional space, outside of meridional planes.

For the tracing of skew rays, and in lens design in general, it is best to use x and y as the coordinates normal to the optical axis and to use z for distances along the axis (Figure 8-6).

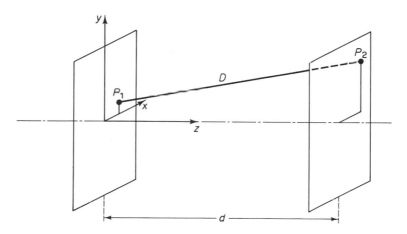

Figure 8-6 Tracing a skew ray from one surface to the next.

Assume that the ray comes from a point P_1, which has the coordinates (x_1, y_1, z_1), and proceeds to a point P_2, which has the coordinates (x_2, y_2, z_2). The distance between the two points is

$$D = \sqrt{(x_2 - x_1)^2 + (y_2 - y_1)^2 + (z_2 - z_1)^2} \qquad [8\text{-}5]$$

This ray is a *vector*; its three components are the projections of the ray on the three coordinate axes: $(x_2 - x_1)$, $(y_2 - y_1)$, and $(z_2 - z_1)$. The ray subtends with the axes a set of three angles whose cosines are the *direction cosines*, X, Y, and Z,

$$X = \frac{x_2 - x_1}{D}$$

$$Y = \frac{y_2 - y_1}{D} \qquad [8\text{-}6]$$

$$Z = \frac{z_2 - z_1}{D}$$

We note that an arbitrary ray in space is completely specified by its direction cosines and by the coordinates of the point at which it intersects a given surface. For example, if point $P_1(x_1, y_1, z_1)$ is given and we know the cosines X, Y, and Z, we can find the coordinates of point P_2 using Equations [8-6],

$$x_2 = x_1 + DX$$

$$y_2 = y_1 + DY$$

and if $z_1 = 0$,

$$z_2 = DZ$$

The sum of the squares of the direction cosines is unity,

$$X^2 + Y^2 + Z^2 = 1$$

which shows that two of the angles are independent variables and one is dependent. If both x and X are zero, the ray is a meridional ray and arccos Z is the slope of the ray, U.

Now assume that the second surface is part of a sphere of radius R, with its center, C, on the optical axis (Figure 8-7). To find the point where the ray intersects the sphere, we use the vertex depth formula,

$$(R - v)^2 + h^2 = R^2$$

where v is the z-component of the distance of point P_2 from the vertex. Solving for v gives

$$v = R - \sqrt{R^2 - h^2} \tag{8-7}$$

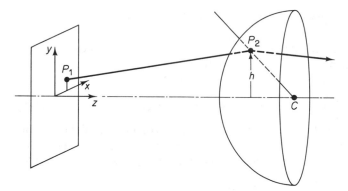

Figure 8-7   Tracing the ray through a spherical surface.

At P_2 we erect a *surface normal*, again a vector. Its direction cosines (which we identify by a degree sign, °), follow Equations [8-6], except that we change the subscripts from 1 to 2 (because we are at point P_2) and distance D to radius R,

$$X° = -\frac{x_2}{R}$$

$$Y° = -\frac{y_2}{R} \tag{8-8}$$

$$Z° = \frac{z_C - z_2}{R}$$

Next we determine the *angle of incidence I*, at P_2. We note that the cosine of an angle subtended by two lines in space is given by the direction cosines of the two lines,

$$\cos I = XX° + YY° + ZZ° \qquad [8\text{-}9]$$

Substituting the terms in Equations [8-8] for $X°$, $Y°$, and $Z°$ we obtain

$$\cos I = (X)\left(-\frac{x_2}{R}\right) + (Y)\left(-\frac{y_2}{R}\right) + (Z)\left(\frac{z_C - z_2}{R}\right)$$

$$= -\frac{1}{R}[x_2 X + y_2 Y - (z_C - z_2)Z] \qquad [8\text{-}10]$$

Then we use *Snell's law* to find the angle of refraction, I'. Ordinarily, Snell's law is written

$$n \sin I = n' \sin I'$$

but since Equation [8-10] refers to cos I, it is more practical to convert Snell's law to cosine terms. To do so, we square both sides,

$$n^2 \sin^2 I = n'^2 \sin^2 I'$$

use the identity

$$\sin^2 I + \cos^2 I = 1$$

solve for $\sin^2 I$, and substitute in the preceding equation:

$$n^2(1 - \cos^2 I) = n'^2(1 - \cos^2 I')$$

$$n^2 - n^2 \cos^2 I = n'^2 - n'^2 \cos^2 I'$$

$$n'^2 \cos^2 I' = n'^2 - n^2 + n^2 \cos^2 I$$

$$\cos^2 I' = 1 - \left(\frac{n}{n'}\right)^2 + \left(\frac{n}{n'}\cos I\right)^2$$

$$\cos I' = \sqrt{1 - \left(\frac{n}{n'}\right)^2 + \left(\frac{n}{n'}\cos I\right)^2} \qquad [8\text{-}11]$$

Now comes the essential point. Since both the incident ray and the surface normal are vectors, we anticipate that the refracted ray is a vector as well. To show that these three vectors all lie in the same plane, we use *vector cross products*. Generally, the cross product of two vectors is another vector, pointing in a direction *normal* to the plane defined by the two initial vectors. The *direction* of $\mathbf{A} \times \mathbf{B}$, for example, is found from the right-hand rule: Rotate $\mathbf{A}$ toward $\mathbf{B}$ with the fingers of the right hand; the thumb will point in the direction $\mathbf{A} \times \mathbf{B}$ [Figure 8-8 (left)].

If $\mathbf{A}$ and $\mathbf{B}$ are unit vectors, the resultant $\mathbf{A} \times \mathbf{B}$, likewise, is a unit vector. But if $\mathbf{A}$ and $\mathbf{B}$ subtend an angle less than 90°, the resultant, $\mathbf{C}$, becomes less than

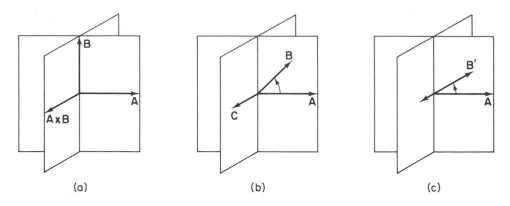

(a)   (b)   (c)

Figure 8-8   (*From left to right*) As the angle subtended by **A** and **B** decreases, the vector cross product **A** × **B** also decreases.

one [Figure 8-8 (center and right)]. In the example of light passing through a surface, **A** is the surface normal, **N**°; **B** is the incident ray, **R**; and **B′** is the refracted ray, **R′**. This makes the angle subtended by **N**° and **R** the angle of incidence, I, and the angle subtended by **N**° and **R′** the angle of refraction, I',

$$\mathbf{N}^\circ \times \mathbf{R} = \sin I$$

$$\mathbf{N}^\circ \times \mathbf{R'} = \sin I'$$

Substituting these terms in Snell's law,

$$n \sin I = n' \sin I'$$

gives

$$\boxed{n(\mathbf{N}^\circ \times \mathbf{R}) = n'(\mathbf{N}^\circ \times \mathbf{R'})}$$ [8-12]

which is the *vector form of Snell's law.*

   The cross products, though, do not mean that the light actually goes to where the resultant vectors point; the resultants merely connect the two sides of an equation. Their ratios, however, are significant; they are equal to n'/n, and also

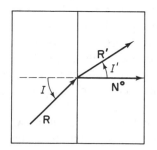

Figure 8-9   Eliminating the resultant leads again to the conventional form of Snell's law.

equal to sin I/sin I'. Consequently, we can redraw Figure 8-8, retaining **R**, **R′**, and **N°** and eliminating the cross-product resultants and the plane that contains them. The three remaining vectors, **R**, **R′**, and **N°**, in other words, lie in the same plane, the *plane of incidence*, as indeed they should (Figure 8-9).

From the vector form of Snell's law, from the definition of the angle of incidence, Equation [8-9], and from a similar definition of the angle of refraction it follows that

$$n(X - X° \cos I) = n'(X' - X° \cos I')$$

$$nX - nX° \cos I = n'X' - n'X° \cos I'$$

$$n'X' = nX - nX° \cos I + n'X° \cos I'$$

$$X' = \left(\frac{n}{n'}\right) X - \left(\frac{1}{n'}\right)(n \cos I - n' \cos I')(X°)$$

Substituting Equation [8-8a], $X° = -x_2/R$, and repeating the process for Y' and Z', we obtain the *direction cosines of the refracted ray*:

$$
\begin{array}{l}
X' = \dfrac{n}{n'} X - \dfrac{1}{n'R}(n' \cos I' - n \cos I)x_2 \\[2mm]
Y' = \dfrac{n}{n'} Y - \dfrac{1}{n'R}(n' \cos I' - n \cos I)y_2 \\[2mm]
Z' = \dfrac{n}{n'} Z - \dfrac{1}{n'R}(n' \cos I' - n \cos I)(z_2 - z_C)
\end{array}
\qquad [8\text{-}13]
$$

To sum up the pertinent steps: Presumably we know the coordinates of two points of intersection, P_1 and P_2. From the coordinates of these points we find the distance between them:

$$D = \sqrt{(x_2 - x_1)^2 + (y_2 - y_1)^2 + (z_2 - z_1)^2} \qquad [8\text{-}5]$$

Then we calculate the direction cosines:

$$X = \frac{x_2 - x_1}{D}$$

$$Y = \frac{y_2 - y_1}{D} \qquad [8\text{-}6]$$

$$Z = \frac{z_2 - z_1}{D}$$

If the second surface is part of a sphere, we need to know its radius of curvature, R, and the location of the center, C. This gives us the angle of incidence, I:

$$\cos I = -\frac{1}{R}[x_2X + y_2Y - (z_C - z_2)Z] \qquad [8\text{-}10]$$

Knowing $\cos I$ and the refractive indices on both sides of the surface, n and n', we find the angle of refraction, I':

$$\cos I' = \sqrt{1 - \left(\frac{n}{n'}\right)^2 + \left(\frac{n}{n'}\cos I\right)^2} \qquad [8\text{-}11]$$

With I' known, we determine the refraction cosines:

$$X' = \frac{n}{n'} X - \frac{1}{n'R}(n'\cos I' - n\cos I)x_2$$

$$Y' = \frac{n}{n'} Y - \frac{1}{n'R}(n'\cos I' - n\cos I)y_2 \qquad [8\text{-}13]$$

$$Z' = \frac{n}{n'} Z - \frac{1}{n'R}(n'\cos I' - n\cos I)(z_2 - z_C)$$

At the next surface, the refracted ray becomes the incident ray and the process is repeated until the ray reaches the image plane.

Example

Trace a ray through a spherical surface that has a radius of curvature of $+26$ mm and separates air ($n = 1.0$) on the left from glass ($n' = 1.8$) on the right. The ray originates at an object placed 100 mm in front of the surface, at a point P_1 that has the coordinates $(2, 2, 0)$; it enters the surface at a point P_2 that has the coordinates $x_2 = 6$, $y_2 = 8$. Determine the direction cosines before and after refraction and the angles of incidence and refraction.

Solution: First we determine the distance from P_1 to P_2. So far we know only the distance measured parallel to the axis (100 mm) to the *tangent plane*, the plane tangent to the surface at the vertex. The (axial) distance from there to P_2 is found from the vertex depth formula, but for using that formula we need to know h, the distance of P_2 from the axis:

$$h = \sqrt{x_2^2 + y_2^2} = \sqrt{6^2 + 8^2} = 10$$

The additional axial distance follows from Equation [8-7],

$$v = R - \sqrt{R^2 - h^2} = 26 - \sqrt{26^2 - 10^2} = 2$$

and the total distance, from P_1 to P_2, from Equation [8-5]:

$$D = \sqrt{(x_2 - x_1)^2 + (y_2 - y_1)^2 + (z_2 - z_1)^2}$$
$$= \sqrt{(6 - 2)^2 + (8 - 2)^2 + (102 - 0)^2} = 102.25$$

Now we determine the direction cosines of the incident ray, using Equations [8-6]:

$$X = \frac{x_2 - x_1}{D} = \frac{6 - 2}{102.25} = \boxed{0.039}$$

$$Y = \frac{8 - 2}{102.25} = \boxed{0.059}$$

$$Z = \frac{102 - 0}{102.25} = \boxed{0.998}$$

The cosine of the angle of incidence is found from Equation [8-10]:

$$\cos I = -\frac{1}{R} [x_2 X + y_2 Y - (z_C - z_2)Z]$$

Note that the center of curvature, z_C, is located at the z-coordinate of the vertex plus R, $z_C = 100 + 26 = 126$; thus

$$\cos I = -\frac{1}{26} [(6)(0.039) + (8)(0.059) - (126 - 102)(0.998)] = 0.894$$

Next we determine the cosine of the angle of refraction, using Equation [8-11]:

$$\cos I' = \sqrt{1 - \left(\frac{n}{n'}\right)^2 + \left(\frac{n}{n'} \cos I\right)^2}$$

$$= \sqrt{1 - \left(\frac{1}{1.8}\right)^2 + \left(\frac{1}{1.8} 0.894\right)^2} = 0.9685$$

and the direction cosines of the refracted ray, using Equations [8-13]:

$$X' = \frac{n}{n'} X - \frac{1}{n'R} (n' \cos I' - n \cos I)x_2$$

$$= \frac{n}{n'} X - (K)(x_2)$$

$$= \frac{1}{1.8} (0.039) - \frac{1}{(1.8)(26)} [(1.8)(0.9685) - 0.894](6)$$

$$= \boxed{-0.087}$$

$$Y' = \frac{n}{n'} Y - (K)(y_2)$$

$$= \frac{1}{1.8} (0.059) - (K)(8)$$

$$= \boxed{-0.112}$$

$$Z' = \frac{n}{n'} Z - (K)(z_2 - z_C)$$

$$= \frac{1}{1.8} (0.998) - (K)(102 - 126)$$

$$= \boxed{+0.990}$$

The angle of incidence is easily found from its cosine:

$$I = \text{arccos } 0.894 = \boxed{26.6°}$$

and the angle of refraction either from extending Equation [8-11]:

$$I' = \text{arccos } 0.9685 = \boxed{14.4°}$$

or from Snell's law:

$$\sin I' = \frac{n}{n'} \sin I = \frac{1}{1.8} \sin 26.6°$$

$$I' = \boxed{14.4°}$$

This completes our example.

COMPUTER-AIDED LENS DESIGN

Presumably, it takes a skilled person with a desk calculator several minutes to trace a ray through an optical system. A computer can do it in perhaps one thousandth of a second and, what is more, it will do it ray after ray, always with the same undiminished enthusiasm. Hence, the ascent of *computer-aided lens design*, CALD.*

The personal computer has revolutionized the way we work and live. Properly used, it can greatly increase our productivity and on occasion even our creativity. The nature of geometrical optics offers a good opportunity to use the potential of such computers. Computers also make it possible to learn interactively; this greatly increases the depth of our understanding of optics.

A good introduction to CALD is provided by the use of *spreadsheets*. Their name comes from the sheets of ledger paper used by accountants, with a set of categories across the top and another set down the left-hand margin. With a computer, they are easy to use. It is merely necessary to enter numbers or text directly into their "cells," the numbers entered either by the user or by the computer.

As we have already seen in detail, in geometrical optics we use *rays* to describe the path of light. These rays are straight lines until something, such as a refractive surface, changes their direction. Our goal, therefore, is to account for such rays as they propagate through the system. Mathematically, a line is completely specified if at any point both its height and slope are known. In ray tracing, the height is given by *y* and the slope by *u*, hence the expression *y-u ray tracing*.

Consider a ray that progresses from the base of an object to the base of its image. In general, the equation of a line is

---

* Much of the material that follows was written by Professor James K. Boger, Albuquerque, New Mexico.

$$y = ux + b \qquad [8\text{-}14]$$

where x is the distance along the (optical) axis and b is the height of an intercept, for example a lens.

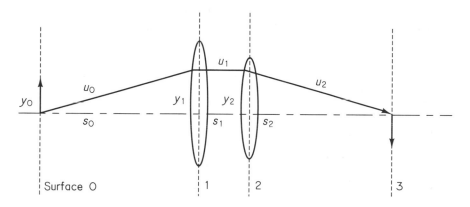

Figure 8-10 A sequence of surfaces as used in ray tracing.

As illustrated in Figure 8-10, surface 0 is the plane of the object. The other surfaces follow in sequential order. The various terms have subscripts depending on the surfaces from which they are measured, thus s_0 is measured from surface 0, s_1 from surface 1, and so on. In the plane of lens 1, therefore, we have

$$y_1 = u_0 s_0 + y_0 \qquad [8\text{-}15]$$

and at lens 2

$$y_2 = u_1 s_1 + y_1 \qquad [8\text{-}16]$$

We begin with Equation [8-15] where u_0 is chosen arbitrarily. But u_1 changes depending on the power of the lens. Similarly to what we have seen before (in Chapter 3), we find that

$$u_1 = u_0 - y_1 P_1 \qquad [8\text{-}17]$$

The two equations [8-15] and [8-17] are the basic y-u equations needed to trace a ray through a lens. In more general terms,

$$y_{i+1} = y_i + u_i s_i \qquad [8\text{-}18]$$

and

$$u_{i+1} = u_i - y_{i+1} P_{i+1} \qquad [8\text{-}19]$$

Next we turn to the *spreadsheet* (Table 8-1). First we enter the basic parameters of the two lenses, including their radii (one-half their diameters) to see which lenses may act as stops.

Table 8-1 Example of *y-u* ray-tracing spreadsheet

Ray Surface		Object Surface O	Lens 1 Surface 1 (F.S.)	Lens 2 Surface 2 (A.S.)	Image Surface 3
System	s	100	10	100	
	p		0.01	0.01	
	r		5	3	
Axial	y	0	1	1	0 Closing trace
	u	0.01	0	-0.01	
	r/y		5	3	
Chief	y	-10	-1	0	10
	u	0.09	0.1	0.1	0.1
	r/y		5		
Parallel	y	1	1	0.9	-1
	u	0	-0.01	-0.019	-0.019
Focal	y	0	0.9	1	1
	u	0.019	0.01	0	0

		Relative to surface			Relative to surface			Relative to surface
efl =	52.632	h1 h2	image =	100	2	ExW =	11.111	2
ffl =	47.368	1	fov =	96.911	degrees	EnPH =	3.3333	
bfl =	47.368	2	mag =	-1		ExPH =	3	
h1 =	5.2632	1	EnP =	11.111	1	EnWH =	5	
h2 =	-5.263	2	ExP =	0	2	ExWH =	-5.556	
f/# =	7.8947		EnW =	0	1			

The most basic operation, presumably, is to find the location of the image. This can be done by tracing just one ray. Generally two rays are needed, in graphical as well as in numerical ray tracing. Yet the ray shown in Figure [8-10] is sufficient, simply because the axis acts as the second ray.

We begin our ray trace by entering in the table at the object (surface 0) a height of zero and an arbitrary slope. We follow the ray to surface 1, using Equation [8-18] and setting the height = 1. The distance between surfaces 2 and 3 is the image distance we seek. At the image the height of the ray is again 0. Inserting this zero is sometimes called *closing the trace*.

The single ray also helps us find the *aperture stop*. We merely calculate the ratio r/y at each surface to find the least of these ratios. From the table, and from Figure 8-11, we see that it is lens 2 that acts as the aperture stop.

Knowing the aperture stop we trace the *chief ray*. It passes through the center of the aperture stop. We choose a height of 0 and an arbitrary slope at the aperture stop. Using Equations [8-18] and [8-19], the ray is traced forward to the image plane and backward to the object plane. Again look for the least r/y ratio; it will identify the *field stop*.

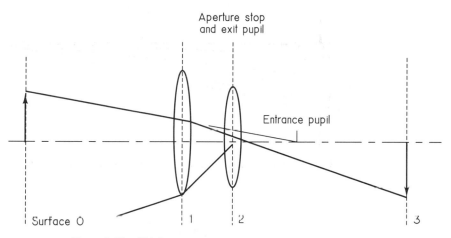

Figure 8-11   Chief ray, aperture stop, and exit pupil.

With both aperture stop and field stop located, we turn to the *pupils*. The exit pupil, for example, is the image of the aperture stop as seen from the side of the image. Since in our case there is no other element between the last lens and the eye of the observer, the exit pupil is right at the aperture stop.

Clearly, it is most important to find the *second principal plane*, H_2, and the *second focal point*, F_2. The axial distance between the two is the *equivalent*, or effective, *focal length* (efl). To find H_2, we let a parallel ray enter the system. A parallel ray has an arbitrary height and a 0.0 slope. As shown in Figure 8-12, the projections of the ray, both back and forth, show us where the H_2 plane is located.

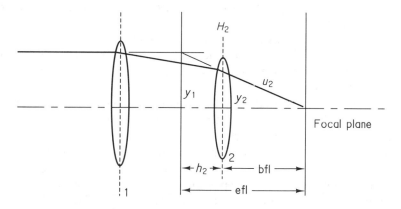

Figure 8-12   A parallel ray and the H_2 principal plane.

The last slope, u_2, may be defined in either of two ways. First, u_2 consists of y_2 as the rise and the back vertex focal length (bfl) as the run. This defines the back vertex focal length,

Figure 8-13   Computer graphics showing rays passing through a four-lens system.

$$\text{bfl} = -\frac{y_2}{u_2}$$

Instead, we could describe u_2 in terms of y_1 and the effective focal length as

$$\text{efl} = -\frac{y_1}{u_2}$$

The difference between both tells us where H_2 is located, measured from the last surface of the lens,

$$\text{bfl} - \text{efl} = h_2 \qquad\qquad [8\text{-}20]$$

The next step would be to realize that no lens is really negligibly thin. That means we need to include the various indices of refraction and the dispersions. We can then analyze thick elements as well as estimate the degree of *chromatic aberration.*

Most programs draw, whenever needed or desired, a *graphical* layout of the system. They also draw a series of rays including skew rays. If the results look promising, we choose some of the data entered in the spreadsheet and use them as variables.

Suppose we make a small change in the radius of curvature of the first surface. That must be accompanied by a change in some other curvature (frequently, but not always, the last) to hold the focal length constant. If this leads to an improvement, we continue in the same direction until the trend reverses, at each step tracing a number of rays to see how that will affect the image. Again we include the indices and dispersions. If we change these variables within preset limits, in effect, we make the computer respond to the results of its own computations; that is called *optimization.* Figure 8-13 shows an example, tracing five rays through a four-element camera lens.*

---

* For this demonstration I have used Stellar Software's *BEAM TWO* lens design program.

Conclusions

There is no question that computers greatly facilitate any lens design. Just imagine how Petzval or Conrady, in bewilderment perhaps but also with appreciation, would look at a computer churning out ray traces in rapid succession. Still, in spite of all the publicity, the role of the computer is probably overrated. After all, it is people who design lenses, come up with the concept, and tell the computer what to do.

Even optimization is not the answer. In fact, some systems have not one but several best solutions. There are also cases where by making some changes in the basic layout (even at the price of a temporary deterioration) we find new avenues that lead to a superior design. Perhaps we even find a completely different solution, far removed from the initial "optimized" design. The role of the computer is to help the designer make decisions and to take the repetitiveness out of the work. *Lens design as such is still an art.*

SUMMARY OF EQUATIONS

Direction cosine of a skew ray:

$$X = \frac{x_2 - x_1}{D} \qquad \text{[8-6a]}$$

Vector form of Snell's law:

$$n(\mathbf{N}° \times \mathbf{R}) = n'(\mathbf{N}° \times \mathbf{R}') \qquad \text{[8-12]}$$

Direction cosine of refracted ray:

$$X' = \frac{n}{n'} X - \frac{1}{n'R} (n' \cos I' - n \cos I)x_2 \qquad \text{[8-13a]}$$

PROBLEMS

8-1. A ray, parallel to the optical axis and at a height of 20 mm, passes through a single surface with a radius of curvature of +64 mm separating air on the left from crown ($n = 1.523$) on the right. What is the angle of incidence?

8-2. Continue with Problem 8-1 and determine the angle of refraction and the slope of the ray after refraction.

8-3. A parallel bundle of light proceeds inside a solid rod of index 1.5 until it reaches the back surface, ground on the rod with a radius of curvature of −24 mm. How wide a beam, limited by total internal reflection, can pass through the back surface?

8-4. A meridional ray subtends with the optical axis an angle of $-30°$ and another meridional ray subtends an angle of $-45°$. What are the direction cosines of the two rays?

8-5. A ray of light originates at a point $(2, 4, 0)$ and at another point $(4, 7, 10)$ enters a thick, plane-parallel plate of glass of $n = 1.6$, placed normal to the optical axis. What is the angle of incidence at the plate?

8-6. Continue with Problem 8-5. Determine the angle of refraction, using both the skew ray formula and Snell's law.

8-7. Trace a ray through a spherical surface that has a radius of curvature of $+48$ mm. How high above the axis must the ray be to produce an angle of incidence of $30°$?

8-8. Rays emerging from an arbitrary point in the object have a certain (di-)vergence. The same rays also have certain slope angles. Is the slope, therefore, proportional to the vergence?

SUGGESTIONS FOR FURTHER READING

KINGSLAKE, R. *Optical System Design*. New York: Academic Press, Inc., 1983.

Military Standardization Handbook, *Optical Design*, MIL-HDBK-141. Washington, D.C.: Defense Supply Agency, 1962.

O'SHEA, D. C. *Elements of Modern Optical Design*. New York: John Wiley and Sons, 1985.

SMITH, W. J. *Modern Optical Engineering: The Design of Optical Systems*, 2nd ed. New York: McGraw-Hill Book Company, 1990.

Optical Systems

There are certainly hundreds of optical systems. In some of them the image is formed on the retina of the eye. This is so with most telescopes and microscopes. In others, for example with cameras and video recorders, the image is formed in the system and only then is presented to the viewer.

TELESCOPES

Astronomical Telescope

The purpose of an astronomical telescope, it seems, is to look at the stars. In reality, it is much more than that. An astronomical telescope is a combination of lenses that is used for a wide variety of purposes, from collimators to viewfinders to autorefractors. For use in astronomy we have more sophisticated systems, to be discussed later in this Chapter.

Kepler's astronomical telescope* has two positive lenses, an *objective lens* in front and an *eye lens* next to the eye. With the object at infinity, the light enters

---

* Johannis Kepler (1571–1630), German astronomer and physicist. Kepler's most important contributions were to astronomy. Analyzing Tycho Brahe's data he formulated his three laws of planetary motion; recognized the need for better optics in astronomy; investigated both theoretically and experimentally the formation of images by lenses, mirrors, and the eye; explained the rainbow;

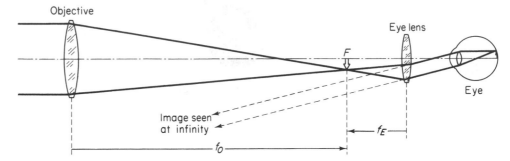

Figure 9-1   Astronomical telescope focused for infinity on both object side and image side. Air image located at *F*.

the objective parallel so that this lens forms in its right-hand focal plane a real image, the *air image* (Figure 9-1). If this focal plane coincides with the left-hand focal plane of the eyelens, rays forming the air image are made parallel again by the eyelens and are still parallel when they enter the eye. Without accommodation, such rays form a real image on the retina, which the observer then sees projected out to infinity. In the afocal mode, therefore, the objective and the eyelens are separated by a distance, *d*, that is equal to the sum of their focal lengths:

$$d = f_O + f_E \qquad \text{[9-1]}$$

But experience shows that often, when a person looks into an optical instrument, he or she involuntarily accommodates, a phenomenon called *instrument myopia*.* If this happens, the light entering the eye is slightly divergent, the same as when reading, the air image formed by the objective has moved *inside* the focal length of the eyepiece, and the image is seen at the distance of most distinct vision, 25 cm in front of the eye (Figure 9-2).

Magnification of a Telescope

The magnification produced by a telescope is a matter of *angular* magnification; it is the ratio of the angular size of the image seen through the telescope to the angular size of the object, seen without the telescope.

---

and was the first, using a pinhole camera, to observe sunspots. In one of his books, *Ad Vitellionem Paralipomena . . . Astronomiae Pars Optica*, 1604, he laid the groundwork for much of today's geometrical optics. In another, *Mathematici Dioptrice*, 1611, he described his telescope. Kepler believed in mysticism, cast horoscopes to finance his studies, and once tried to find the sounds that belong to each planet producing the "music of the spheres."

* Instrument myopia can be avoided by first moving the eyepiece *out* of the telescope beyond the point of best focus, and then moving it in and *stopping* when the image is sharp. Viewing while accommodating is tiring; it also makes the lines of sight converge, which serves no useful purpose.

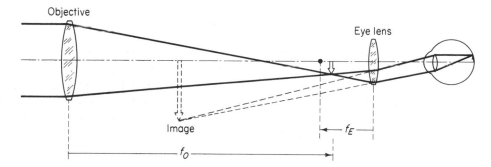

Figure 9-2 Astronomical telescope focused for object at infinity and image at the near-point.

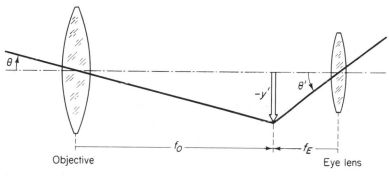

Figure 9-3 Deriving magnification of telescope.

The angular size of a faraway object is simply the angle subtended by rays coming from the top and the bottom of the object. In Figure 9-3 this angle is θ. The same rays entering the eye of the observer subtend a larger angle, θ'. The ratio of these two angles is the magnification provided by the telescope,

$$M_{\text{Telescope}} = \frac{\tan \theta'}{\tan \theta} \qquad [9\text{-}2]$$

From the construction it follows that

$$\tan \theta = \frac{-y'}{f_O} \qquad \text{and} \qquad \tan \theta' = \frac{-y'}{-f_E}$$

Substituting in Equation [9-2] and canceling $-y'$ gives

$$M_{\text{Telescope}} = -\frac{f_O}{f_E} \qquad\qquad [9\text{-}3]$$

where the minus sign means that the image is inverted.*

Now consider a ray, parallel to the optical axis, that enters the objective next to its upper rim [Figure 9-4 (top left)]. The ray goes through F and, with the telescope in the afocal mode, emerges from the eyepiece again parallel to the axis (bottom right). This ray, following our earlier notation, is a *parallel ray* with respect to the objective and a *focal ray* with respect to the eye lens.

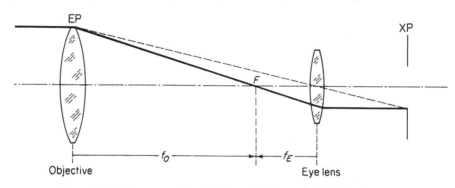

Figure 9-4 Entrance pupil and exit pupil of an astronomical telescope.

The dashed line in Figure 9-4, on the other hand, comes from the upper rim of the objective, goes through the center of the eye lens (without refraction), and at some distance to the right intersects the focal ray. The point of intersection locates the image (of the objective). If, as is usually the case, the aperture of the objective is the *entrance pupil*, EP, the image is the *exit pupil*, XP. The diameters of these two pupils are related through similar triangles:

$$\frac{\text{EP}}{f_O} = \frac{\text{XP}}{-f_E}$$

Consequently, rearranging and substituting in Equation [9-3] shows that the magnification of a telescope is also given by

$$M_{\text{Telescope}} = \frac{\text{diameter of entrance pupil}}{\text{diameter of exit pupil}} \qquad\qquad [9\text{-}4]$$

Based on this definition, there is a simple way of determining the magnification of a telescope. First measure the diameter of the objective lens. Then hold the point of a pencil

---

* This inversion is immaterial when we look at stars. But why do we see *more* stars with a larger telescope? Not because of higher magnification; stars appear as points no matter how much they are magnified. Instead of "magnifying power," it is the *light-gathering power* of the telescope that counts.

or a paperclip next to the objective lens and aim the telescope at the sky or some other diffuse target. Hold a file card to the right of the eyepiece and slowly move the card back and forth until the image of the objective, together with the pencil point or paperclip, is in focus. Mark the diameter of the image (of the objective) on the file card. Divide the diameter of the objective (that is, the entrance pupil) by the diameter of its image on the card (the exit pupil); this gives the telescope's magnification.

If a telescope has the specifications 6 × 30, this means that it has 6× angular magnification and an objective 30 mm in diameter. The inverse ratio, 30/6, is the size of the exit pupil (in millimeters), and the square of that, 25, is the light-gathering power.

The limit of resolution, or *resolving power*, of a telescope is a matter of *Rayleigh's criterion*, which we discuss later.

Matrix Example

An astronomical telescope may have an objective lens of 80 cm focal length and give 8 × angular magnification. It is focused for infinity on both the object and image side. Determine the system matrix.

Solution: First we determine the focal length of the eyepiece. Since

$$M_\theta = \mathbf{b} = 8\times\ = \frac{f_O}{f_E} = \frac{80 \text{ cm}}{f_E}$$

we find that

$$f_E = \frac{80}{8} = 10 \text{ cm}$$

The separation of the two lenses, then, is

$$d = f_O + f_E = 90 \text{ cm}$$

Next we determine the refractive powers,

$$P_O = \frac{1}{0.8} = +1.25 \text{ diopters}$$

$$P_E = \frac{1}{0.1} = +10 \text{ diopters}$$

Inserting these figures in the system matrix gives

$$\mathbf{S} = \mathbf{R_2TR_1} = \begin{bmatrix} 1 & 10 \\ 0 & 1 \end{bmatrix}\begin{bmatrix} 1 & 0 \\ -0.9 & 1 \end{bmatrix}\begin{bmatrix} 1 & 1.25 \\ 0 & 1 \end{bmatrix}$$

$$= \begin{bmatrix} 1 & 10 \\ 0 & 1 \end{bmatrix}\begin{bmatrix} 1 & 1.25 \\ -0.9 & -0.125 \end{bmatrix} = \begin{bmatrix} -8 & 0 \\ -0.9 & -0.125 \end{bmatrix}$$

For an afocal system the equivalent power **a** is zero, the angular magnification **b** = −8×, and the distance between the lenses **d** = (−)0.9 m.

Eyepieces

The field of view of a two-lens telescope is unnecessarily restricted. Some of the light, shown by the dashed line in Figure 9-5 (top), does not even reach the eye at all. Correction is possible by placing an additional lens, called a *field lens*, at, or close to, the air image. This lens has no effect on the size or position of the image; it merely directs all rays passing through into the last lens, the *eye lens*. Consequently, since the field lens is by necessity closer to the entrance pupil than the eye lens, the field of view is larger, without vignetting, than with the eye lens alone.

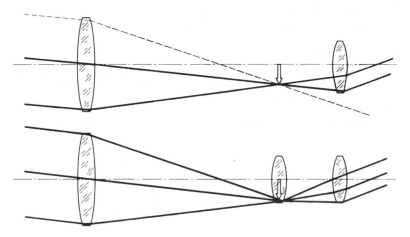

Figure 9-5 Astronomical telescope containing only two lenses (*top*) and with an additional field lens (*bottom*).

Such considerations have led to the development of *eyepieces*. A *Huygens eyepiece* consists of two plano-convex lenses, their convex sides facing the object. The field lens has about twice the focal length of the eye lens, with a separation of about one-half the sum of both. A *Ramsden eyepiece* has two equal, or nearly equal, plano-convex lenses, their convex sides facing each other. Their separation is nearly the same as their focal length. A *Kellner eyepiece* is essentially a Ramsden eyepiece with the eye lens made achromatic.

An *orthoscopic eyepiece* is like a Ramsden eyepiece but has a triplet as the field lens; it has a wide field and good color correction. The *Erfle eyepiece* (Figure 9-6) has two or even three achromatic doublets; it is often used in high-quality astronomical telescopes. A *Barlow lens* is a negative lens placed just ahead of the focal plane of the objective, doubling and sometimes tripling the magnification. A *Gauss eyepiece* contains a reticle, a beamsplitter, and a light bulb to illuminate the reticle (which helps in aiming at a target at night). In a *filar micrometer eyepiece*, a hair line can be moved across the field and the distance of motion read on a drum (which is useful for measuring the image size).

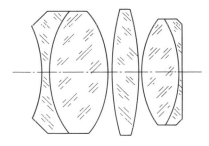

Figure 9-6 Erfle eyepiece.

Terrestrial Telescopes

The drawback of the astronomical telescope is that the image is inverted. That would be awkward for terrestrial observations. Correction is possible by adding a *relay* or *erector lens* (as we will discuss in detail in the example that follows), by using a *prism*, or by designing a *Galilean telescope*.

Prism telescopes, if used for binocular vision, are called *prism binoculars*. Most often they contain a pair of Pechan prisms, also known as Malmros, Schmidt, Thompson, or *Z* prisms (Figure 9-7).

Galileo's Telescope

In the *Galilean telescope*,* as in the astronomical telescope, the objective lens is positive, but in contrast to it, the eyepiece is negative. Still, the two focal points coincide. In Figure 9-8 (top), *F* is both the second (right-hand) focal point of the first lens and the first (left-hand) focal point of the second lens. In Galileo's type (bottom), *F* is still the second focal point of the first lens and again the first (but now the *right*-hand) focal point of the (negative) second lens. Parallel light incident on the system is refracted toward *F* but before it reaches the focus, the light

---

* Named after Galileo Galilei (1564–1642), Italian scientist. While, as a medical student, attending mass in a cathedral in Pisa, Galileo is said to have watched a bronze chandelier swinging in the breeze and discovered that the frequency of oscillation, no matter what the amplitude, was constant. When in the early 1600s Dutch opticians had found that a combination of two lenses showed distinct objects upright as well as magnified, Galileo quickly figured out the lenses necessary, stating in his *Il Saggiatore* (Roma: Appresso Giacomo Mascardi, MDCXXIII) that he had "discovered the same instrument, not by chance, but by the way of pure reasoning." In 1609 he demonstrated his "cannocchiale" in public, looking from the Campanile of San Marco in Venice at the Campanile San Giustinio in Padua (where he taught mathematics), 33 km away. The word *telescope* was coined later by a Greek poet and theologian, Ioannis Demisiani of Cephalonia, combining τῆλη = far and σκοπὸς = viewer. The second of more than 100 telescopes that Galileo made was built into a lead tube 70 cm long and had 3× angular magnification; the fifth, called *Discoverer*, had 30×. With it he resolved the Milky Way into a myriad of stars and discovered the moons of Jupiter and the phases of Venus—which proved its rotation around the sun.

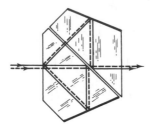

Figure 9-7 Pechan prism.

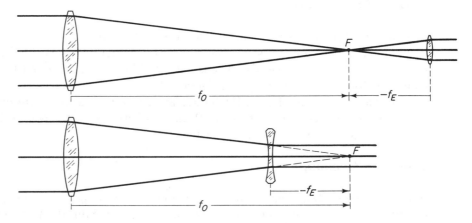

Figure 9-8 Focal points coincide in both the astronomical telescope (*top*) and Galileo's telescope (*bottom*).

is intercepted by the negative lens and the emergent light is parallel again. The image, therefore, is seen at infinity. *With* accommodation, the image is seen at the distance of most distinct vision (not shown).

The magnification, as before, is

$$M_{\text{Telescope}} = -\frac{f_O}{f_E} \qquad [9\text{-}3]$$

but this time it is positive and the image upright. Telescopes of the Galilean type are limited to about 3 × magnification (because the exit pupil lies to the left of the last lens and the eye cannot be brought there) but they are short, and hence are often used as opera glasses and as telescopic spectacles to help people with impaired vision.

Example

Design a telescopic sight, to be used on a hunting rifle.

Solution: With no more information given than this, consider first the diameters of the various pupils and the magnification desired. The diameter of the pupil of the eye varies from 2 to 5 mm, depending on the ambient light. In daylight, the pupil is about 3 mm in size and not much is gained by making the exit pupil of the telescope larger, except that aligning the eye with the eyepiece becomes easier. Let us begin with an exit pupil 3.75 mm in diameter [Figure 9-9(a)].

Since a telescope's magnification is equal to the ratio of the diameter of the objective lens to the diameter of the exit pupil, a higher magnification, for a given exit pupil, requires a larger-diameter objective. For daylight use, reasonable length, light weight, and moderate cost, we may settle for $M_T = 8\times$. The objective lens, then, should have a diameter of

$$D = (8)(3.75) = 30 \text{ mm}$$

Certainly, we must have an upright image. But for reasons explained earlier, an $8\times$ telescope cannot be built as a Galilean type. Instead, we use an astronomical telescope together with an *erector* [Figure 9-9(b)]. We choose, somewhat arbitrarily, 12 cm as the focal length, f_O, of the objective lens. Since $M = f_O/f_E$, this requires an eyepiece of focal length

$$f_E = \frac{f_O}{M} = \frac{120 \text{ mm}}{8} = 15 \text{ mm}$$

With this focal length, the exit pupil is

$$s' = \frac{sf}{s + f} = \frac{-(120 + 15)(15)}{-120 - 15 + 15} = +16.9 \text{ mm}$$

to the right of the eyepiece. But this does not give enough *eye relief* (= distance from the last lens to the exit pupil, which should coincide with the pupil of the eye). The pupil of the eye is about 3.6 mm behind the cornea and the lens is held in a mount; therefore, the actual clearance between the telescope and the eye might be less than 10 mm, not enough for comfort and to avoid the danger of recoil, and certainly not enough for people who wear glasses.

The solution lies in making the erector provide some magnification too. For example, if the eyepiece is given a more reasonable focal length, perhaps 30 mm, the erector can make up for the deficit and provide $2\times$. Assume that the erector has a focal length of 32 mm. Solving $s'/s = -2$ for s' and inserting this in $s' = sf/(s + f)$ gives

$$(-2)(s) = \frac{(s)(32)}{s + 32}$$

and therefore,

$$s = -48 \text{ mm}$$

and

$$s' = 96 \text{ mm}$$

The design so far looks as shown in Figure 9-9(c). The total length is now about 30 cm, acceptable for a riflescope.

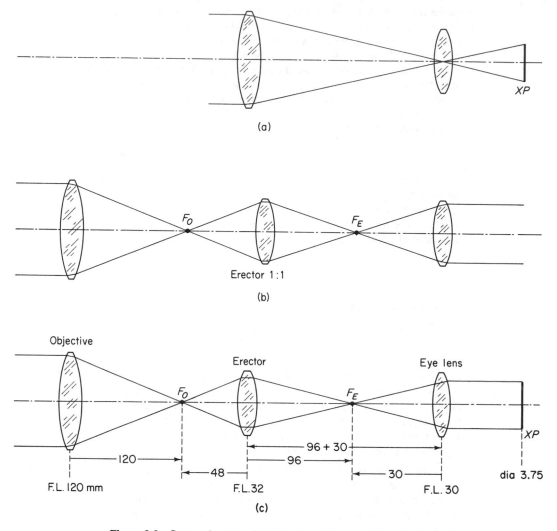

Figure 9-9   Successive steps in designing a telescope. For details see the text.

Finally, we place another field lens at F_E, designing in effect an eyepiece. This lens should have a focal length of

$$f = \frac{(-96)(30)}{-96 - 30} = 22.9 \text{ mm}$$

We have now arrived at the *thin-lens solution*. Next we change to realistic, "thick" lenses, move the two field lenses slightly away from the foci (so that dust accumulating on them will not show), achromatize the system, trace a series of rays including skew rays through it, and correct for other aberrations. Finally, we place cross-hairs at either F_O or F_E. This completes our design.

THE MICROSCOPE

A microscope is an instrument for viewing small objects. It has two positive lenses, an objective lens and an eyepiece. The objective forms a real, inverted, magnified image. This image is further magnified by the eyepiece. Such a system is called a *compound microscope.**

So far, that looks the same as a telescope. What, then, is the difference between the two? Consider a telescope that is first focused at infinity and used without accommodation. The inner foci coincide (as shown earlier in Figure 9-1). As the object is slowly brought closer, the air image moves away, through a distance that in essence is Newton's image distance with respect to the objective. It is this distance, from the right-hand focus of the objective lens, F_O in Figure 9-10, to the left-hand focus of the eyepiece, F_F, that is characteristic of a *micro*scope, in contrast to a *tele*scope. This distance, called the *optical tube length*, T, is set at 160 mm.

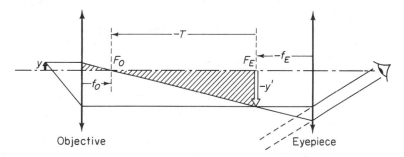

Figure 9-10 Ray trace through a compound microscope.

From the shaded triangles in Figure 9-10 we see that the height of the object, y, and the height of the air image, $-y'$, are related as

$$\frac{y}{-y'} = \frac{f_O}{-T}$$

The inverse ratio, $-y'/y$, by definition, is the magnification provided by the objective lens,

---

* The *simple microscope* is merely a single plus lens of very short focal length; it was used first by Antoni van Leeuwenhoek (1632–1723), Dutch clerk and biologist. In 401 publications, including 375 letters to the Royal Society of London and a book, *Arcana Naturae per Microscopum Detecta* (Secrets of Nature Discovered through the Microscope), van Leeuwenhoek described bacteria and yeast cells, the circulation of blood, red blood cells (which had been misinterpreted by Malpighi as fat cells), spermatozoa (though he fancied seeing tiny human heads inside), and the cross-striation of muscle fibers. He also explained that certain insects do not develop from mud, but from eggs laid there. The word *micro · scope* derives from the Greek μικρός = small and σκοπεῖν = to view. Descartes called these low-power instruments "perspicilia pulicaria," Latin for "viewing glasses for fleas."

$$M_O = \frac{-T}{f_O} \qquad [9\text{-}5]$$

The eyepiece, used without accommodation and following Equation [2-23], gives a magnification

$$M_E = \frac{25}{f_E} \ (f \text{ in cm}) \qquad [9\text{-}6]$$

The total magnification of the microscope is the product of both,

$$\boxed{M_{\text{Microscope}} = -\frac{T}{f_O} \frac{25}{f_E}} \qquad [9\text{-}7]$$

Example

The light follows a different path depending on whether the microscope is used for projection and microphotography or for viewing. For instance, if the objective lens bears the designation "16 mm" and the eyepiece "12.5×," by how much must the tube of the microscope be raised or lowered when changing from visual observation (without accommodation) to photography, assuming that the photographic film is 60 mm away from the eyepiece?

Solution: With the microscope focused for projection, the air image is formed *outside*, ahead of, the first focal length of the eyepiece. The objective forms a real air image, and this air image is relayed by the eyepiece, again as a real image, to the screen or film [Figure 9-11 (left)].

With the microscope used for viewing, the objective lens is brought closer to the specimen. This causes the air image to move farther away, to a position *inside* the focal length of the eyepiece, which now acts as a magnifier (right).

From the data given, we determine first the focal length of the eyepiece,

$$f_E = \frac{25 \text{ cm}}{12.5} = 20 \text{ mm}$$

If the film, and therefore the image, is 60 mm away (from the eyepiece), the object distance (for the eyepiece) is

$$s_E = \frac{s'f}{f - s'} = \frac{(60)(20)}{20 - 60} = -30 \text{ mm}$$

or **10** mm more than for visual observation (−20 mm).

Next, we use the thin-lens equation to find the two object distances (for the objective lens). For visual observation,

$$s_v = \frac{s'f}{f - s'} = \frac{(16 + 160)(16)}{16 - (16 + 160)} = -17.6 \text{ mm}$$

while for photography,

$$s_p = \frac{(16 + 160 - \mathbf{10})(16)}{16 - (16 + 160 - \mathbf{10})} = -17.7 \text{ mm}$$

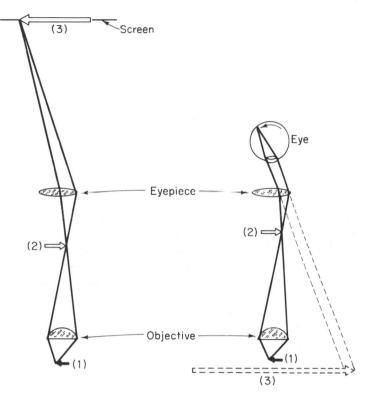

Figure 9-11

Thus, when changing from visual observation to photography, the tube must be *raised*

$$17.7 - 17.6 = \boxed{0.1 \text{ mm}}$$

The *numerical aperture* of a microscope objective is defined the same as that of an optical fiber,

$$NA = n_0 \sin I \qquad [9\text{-}8]$$

The term n_0 is the index of the medium between the cover glass placed on top of the specimen and the front lens of the objective, and I is the angle of total internal reflection at the glass-air boundary of the cover [Figure 9-12 (left)].

With air ($n' = 1.00$) filling the space between cover glass and front lens, the numerical aperture cannot exceed unity. But, if the air is replaced by oil [$n = 1.515$ (right)], the angle of total reflection becomes larger and the cone of light wider, producing a better and brighter image.

As a rule of thumb, the *useful magnification* of a microscope is about 600 times the numerical aperture. Although it is easy to go beyond this limit by using a

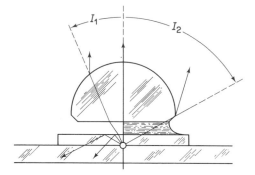

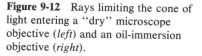

Figure 9-12 Rays limiting the cone of light entering a "dry" microscope objective (*left*) and an oil-immersion objective (*right*).

higher-power eyepiece or by projecting the image on a distant screen, the result is merely a larger image but not the disclosure of more detail. Any magnification beyond the useful limit is an "empty magnification."

The *confocal microscope* uses a point light source, such as the tip of a fiber, on one side of the specimen, and a small aperture in front of a photodetector on the other; the specimen is placed on a moving (scanning) stage. In this way images are formed of only one point at a time. Light from other nearby points can thus be excluded, producing images that are particularly crisp, clear, and free of unwanted scattered light.

THE EYE

At first, the human eye seems to be a simple optical system containing a lens that forms an image on a light-sensitive screen. But it is much more than that. The eye's lens is of the GRIN type and is capable of changing its shape.

On entering the eye, the light passes first through the *cornea*, the first surface of the system, and then through the *anterior chamber* (Figure 9-13). The anterior chamber, filled with the *aqueous humor*, is bounded by an aperture stop, the *iris*, with a central hole in it, the *pupil*. The diameter of the pupil varies as a function of the amount (and wavelength) of the light.

Next comes the *crystalline lens*. An elastic, jellylike composite of long cells, the crystalline lens is particularly rich in proteins (33%, more than any other tissue in the human body) and potassium (25 times the concentration in the surrounding liquids), factors that help maintain its transparency and refractive index. The index is highest in the center (1.41), lowest at the equator (1.37), and intermediate (around 1.39) at the vertices where the lens borders the aqueous in front and the *vitreous humor* in the rear. The curvatures of the surfaces of the lens are controlled by the circumferential *ciliary muscle*. After passing through the vitreous, the light reaches the *retina*, which through the *optical nerve* connects to the brain.

The optical properties of the eye can be represented either by the simpler *reduced eye* or by the more realistic *schematic eye*. The reduced eye is assumed to

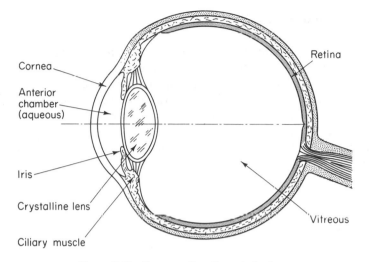

Figure 9-13   Cross-section through the human eye.

have only one surface, with a radius of curvature of $R = +5.7$ mm. The two focal lengths are $f_1 = -16.8$ mm and $f_2 = +22.5$ mm, and the power is $+60$ diopters.

Gullstrand's *schematic eye* has three surfaces, the front surface of the cornea and the front and rear surfaces of the lens. Their dimensions are listed in Table 9-1. The total power is again 60 diopters. Of this, about 43 diopters is contributed by the cornea, the remainder by the lens.

Focusing on a nearby object is brought about by *accommodation*. Accommodation is accomplished by a change of power of the lens, resulting from a change in shape. In its relaxed state, the lens is nearly spherical. But the lens is suspended by ligaments and these ligaments are attached to the circumferential ciliary muscle; thus, with the muscle relaxed, the lens is stretched into a flatter shape. During accommodation, the ciliary muscle contracts, allowing the lens to return to its more spherical shape. This change affects mainly the front surface, whose radius of curvature shrinks, from $+10.0$ mm $+5.3$ mm.

Table 9-1   Schematic eye

Radii of curvature	
Cornea	$+7.80$ mm
Lens	
Anterior surface	$+10.00$ mm
Posterior surface	-6.00 mm
Refractive indices	
Aqueous	1.336
Lens (cumulative)	1.413
Vitreous	1.336

Refractive Anomalies

The two most common refractive errors of the eye are *myopia* (nearsightedness) and *hyperopia* (hypermetropia, farsightedness). In myopia, the image of a distant object is formed in front of the retina and thus is seen out of focus [Figure 9-14(a)]. Only objects that are not farther away than the *far point* will focus on the retina (b). Correction for myopia requires a minus lens. Its second (left-hand) focal point should coincide with the far point (c). Then, parallel light incident on the eye becomes divergent, so that it appears to come from the far point and *will* focus on the retina (d).

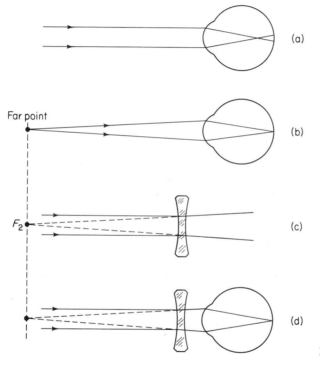

Figure 9-14   Myopia and its correction.

In *hyperopia*, parallel light comes to a focus behind the eye [Figure 9-15(a)]. The far point is to the right of the retina. It is defined as the point to which rays from an object must converge for a sharp image to be formed on the retina (b). For correction, again the second focal point of the correcting lens should coincide with the far point (c). Then rays entering the eye are already converging and will focus on the retina (d).

Intraocular lenses (IOL) are implanted in the eye to replace the natural crystalline lens of the eye that may have been removed because of cataract. A typical IOL is made from polymethyl methacrylate, has a diameter of 5 to 7 mm,

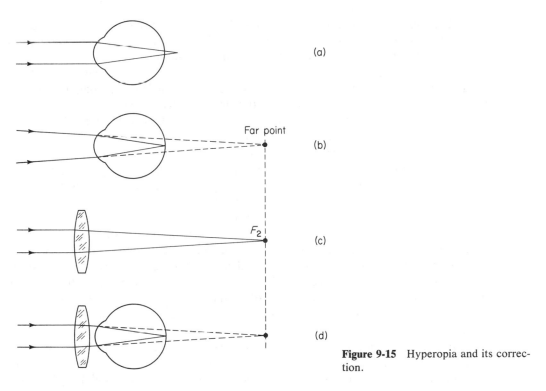

(a)

Far point

(b)

F_2

(c)

(d)

Figure 9-15 Hyperopia and its correction.

and is supported in the eye by flexible loops called *haptics*. Most IOLs are mono-focal but bifocal and even multifocal lenses can be made.

CAMERA LENSES

Now let us study the evolution of camera lenses. They have gradually evolved from very simple to highly complex systems, their progress punctuated from time to time by fundamentally new concepts.

The simplest type of a camera lens, known since the late sixteenth century, is the *biconvex lens*. It suffers from every type of aberration, and is all but useless for any except the most rudimentary form of photography.

Some 250 years later it was found that a *meniscus*, placed with its concave side facing the object, has little astigmatism and coma. However, spherical and chromatic aberration, distortion, and curvature of field are still severe and limit its use. Its aperture can be no better than *f*/16. A slight improvement is possible by using an *achromatic meniscus*, the so-called "landscape lens."

A combination of *two achromatic menisci*, however, their concave sides facing each other, is much better. An example is the *Rapid Rectilinear*, developed in 1866.

The design of camera lenses then took a dramatically different turn, due to the invention of the *Taylor–Cooke triplet.** This is the three-lens system used repeatedly as an illustration to introduce various chapters.

The Taylor triplet consists of two positive lenses, in front and back, both made of crown, and a negative lens, in between, made of flint. The negative lens is the crucial element. Its distance from the first (positive) lens is relatively large. Thus the light, as it reaches the minus lens, has fairly high vergence. Hence, the power of the minus lens can also be kept high to correct for spherical and chromatic aberration and to reduce coma, astigmatism, curvature of field, and distortion, still leaving sufficient power in the system. Because of its good correction, the Taylor triplet could be made with much higher speeds (*f*/6.3 and better) and a wider field than any other camera lens known at that time.

From this concept of the Taylor–Cooke triplet there derived numerous other camera lenses. The most successful of these is the *Zeiss Tessar.*† As shown in Figure 9-16, the rear element became an achromatic cemented doublet. The Tessar has excellent correction for spherical aberration, good achromatism, sufficiently high speed (*f*/3.5 and even *f*/2.8), very little astigmatism, and a field free of curvature and distortion out to about 60°. Today the Tessar and its derivatives are the most widely used high-quality camera lenses. Table 9-2 lists some typical parameters.

Besides such more or less standard lenses, there are certain types of more specialized camera lenses. A *telephoto lens* has a focal length longer than the diagonal across the film negative. A *wide-angle lens* has a focal length much shorter than the diagonal.

A *zoom lens* is a system whose focal length can be varied without moving the image out of focus; a zoom lens, in other words, provides variable magnification.

---

* Designed in 1893 by Harold Dennis Taylor (1862–1943), lens designer at the T. Cooke and Sons Optical Company in Bishophill, York, England. It seems that early camera lenses were often "corrected" for aberrations by stops to eliminate unwanted marginal rays (and making long exposures, up to 40 minutes, a virtual necessity). Taylor broke with tradition. In his invention disclosure he describes his "idea of eliminating the diaphragm corrections and throwing the whole burden of flattening the final image and correcting marginal astigmatism upon a negative lens." H. D. Taylor, *A Simplified Form and Improved Type of Photographic Lens*, Brit. Patent 22,607, 6 Oct. 1894, and *Lens*, U.S. Patent 540,122, May 28, 1895. His triplet was used first as a high-quality portrait lens, but since Cooke, a maker of large telescopes, did not want to go into the production of camera lenses, the design was offered to another company, with the proviso that the new lens, out of respect to Taylor's employer, be called the *Cooke Triplet*.

† Invented by Paul Rudolph (1858–1935), lens designer at the Carl Zeiss optical company in Jena, Germany. In 1895 Rudolph designed the Planar, in 1902 the Tessar. In his invention disclosures, *Sphärisch, chromatisch und astigmatisch korrigiertes Objektiv aus vier, durch die Blende in zwei Gruppen geteilte Linsen*, Dtsch. Reichspatent 142 294, 25 Apr. 1902, and *Photographic Objective*, U.S. Patent 721,240, Feb. 24, 1903, Rudolph describes how with a "comparatively small number of components" he developed a "spherically, chromatically, and astigmatically corrected objective" that is "fruitful in an extraordinary degree." For details on modern Tessars, see R. Kingslake, *Lens Design Fundamentals*, pp. 277–86 (New York: Academic Press, Inc., 1978).

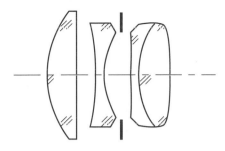

Figure 9-16 *Tessar* camera lens.

Many zoom lenses are derived from the Taylor triplet, plus-minus-plus. The various lenses, or groups of lenses, move in a complex, often nonlinear way. The front lens in Figure 9-17, for example, is stationary, to provide a tight seal. The second lens and the third lens both move, the rates and directions of motion being governed by cams, or slots cut into a rotatable cylinder.

With the minus lens in midposition, the image has a certain size. Moving the minus lens forward makes the image smaller, as with a wide-angle lens; moving it backward makes the image larger, as with a telephoto lens.

But why is this so? Look at the angles subtended by peripheral rays at the image and consider the optical invariant, nyU, the product of refractive index, image size, and slope angle (page 42). Since the product yU must by necessity be constant, a larger angle (the upper example in Figure 9-17) produces a smaller image, as with a wide-angle lens; and a smaller angle (bottom) produces a larger image, as with a telephoto lens.

Table 9-2 Tessar camera lens*

Axial Thickness, d (mm) Refractive Index, n	Radius of Curvature (mm)	Air Space (mm)
Lens 1		———
$d = 3.57$	$R_1 = +16.28$	
$n = 1.6116$	$R_2 = -275.7$	
		1.89
Lens 2		———
$d = 0.81$	$R_3 = -34.57$	
$n = 1.6053$	$R_4 = +15.82$	
		3.25
Lens 3		———
$d = 2.17$	$R_5 = \infty$	
$n = 1.5123$	$R_6 = +19.20$	
Lens 4		
$d = 3.96$	$R_7 = +19.20$	
$n = 1.6116$	$R_8 = -24.00$	

* Focal length: 50.82 mm.

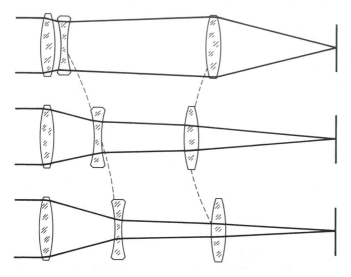

Figure 9-17   Zoom lens showing wide-angle position (*top*), telephoto position (*bottom*).

From simple landscape lenses to highly complex 31-element zoom systems, camera lenses can be built to remarkable specifications. For example, cameras in high-flying aircraft or satellites can resolve details as small as the numbers on an automobile license plate from a height of more than 12 km (40,000 ft).

Example

Telephoto lenses for cameras are often built in the form of a Galilean telescope. Assume that the front lens has +50 mm focal length and the second lens −25 mm focal length. The distance between both, to provide focusing on the film plane, is 30 mm. Determine:

   (a) The focal length.
   (b) The actual physical length of the system.

Solution: (a) The term *focal length* refers, more precisely, to *equivalent* focal length. With the dimensions given, the equivalent focal length of the system is

$$\mathbf{f} = \frac{f_1 f_2}{f_1 + f_2 - d} = \frac{(+50)(-25)}{+50 - 25 - 30} = \boxed{+250 \text{ mm}}$$

This focal length, by definition, is the distance from the second principal plane, $\mathbf{H}_2$, of the system to the film in the camera (Figure 9-18).

(b) In order to find the physical length of the system, we determine first the back vertex power (of the system). We convert the focal lengths into powers, express the distance d between the lenses in meters, and set $n = 1$, because the medium between the lenses is air. Then

$$\mathbf{P}_{BV} = \frac{P_1}{1 - P_1(d/n)} + P_2$$

$$= \frac{+20}{1 - (+20)(0.03)} + (-40) = 10 \text{ diopters}$$

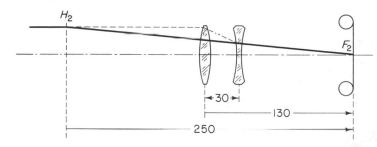

Figure 9-18   Example of telephoto lens. To scale.

The reciprocal is the back vertex focal length, v_2, the distance from the second lens to the plane of the film,

$$v_2 = \frac{1}{10} = 0.1 \text{ m} = 100 \text{ mm}$$

The total length of the system, from the first lens to the film, therefore, is

$$30 + 100 = \boxed{130 \text{ mm}}$$

considerably less than its focal length, 250 mm. This is very convenient when toting around such gear.

REFLECTING SYSTEMS

Conventional telescopes and microscopes are combinations of *lenses*; they are *dioptric systems*. But telescopes and microscopes can also be built as combinations of *mirrors*; these are *catoptric systems*. Combinations of lenses and mirrors are *catadioptric systems*.*

Catoptric Systems

Mirrors have no chromatic aberration and, in addition to visible light, they also work in the UV, the IR, and even in the X-ray region where no refractive materials are known. Reflecting telescopes are widely used in astronomy. In principle, an astronomical reflector is similar to a refractor, except that the objective *lens* is replaced by a *mirror*. In early reflectors, a small plane mirror or a prism was used to deflect the beam out of the tube for viewing; that is the essence of *Newton's telescope* (Figure 9-19).

Gregory's telescope† is the equivalent of the astronomical telescope; it has two curved mirrors, a concave primary (mirror) with a central hole, and a concave

---

* The term *dioptric* comes from the Greek διά = [light passing] through, and catoptric comes from κάτοπτρον = mirror; catadioptric is a combination of both.

† James Gregory (1638–1675), Scottish mathematician and astronomer, suggested the system named after him in 1663, but opticians at that time could not grind the mirrors required.

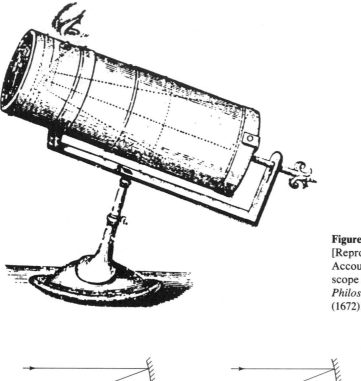

Figure 9-19 Newton's telescope.
[Reproduced from I. Newton, "An
Account of a New Catadioptrical Tele-
scope Invented by Mr. Newton,"
Philos. Trans. Roy. Soc. London **7**
(1672), 4004–10.]

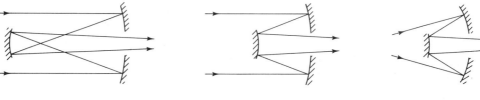

Gregory   Cassegrain   Schwarzschild

Figure 9-20 (*Left to right*) Gregory's telescope, Cassegrain's telescope, Schwarzschild's
microscope.

secondary mounted as shown in Figure 9-20 (left). The distance between the two
mirrors is equal to, or slightly larger than, the sum of their focal lengths.

*Cassegrain's telescope** is the equivalent of Galileo's type, with the secondary being convex [Figure 9-20 (center)]. The secondary contributes most to the
overall magnification because its distance from the primary's focus is much less

---

* Guillaume Cassegrain, French. Little is known about Cassegrain's life; according to some, he
was a professor of physics at the College of Chartres, according to others, a sculptor at the court of
Louis XIV. In 1672, nine years after Gregory, Cassegrain combined a concave and a *convex* mirror,
intercepting the rays before they came to a focus (an idea quickly dismissed by Newton as only a minor

(about one-fourth) than that from the secondary to the image. Cassegrainian telescopes are widely used; they are much shorter than a Gregorian of the same focal length.

The *Schwarzschild type* (right), in principle similar to the Cassegrainian, is used in microscopy.* Its major advantage is that it can be used also in the UV and the IR; in addition, it needs to be focused only in the visible and then is set correctly also for the UV or the IR. The focal length can be derived from the equivalent-power equation,

$$\mathbf{P}_{equivalent} = P_1 + P_2 - P_1 P_2 d$$

If we substitute $-2/R$ for the powers of the two mirrors and, because the light changes direction, change the sign in the R_2 term, then

$$\frac{1}{f} = \frac{-2}{R_1} + \frac{2}{R_2} + \frac{4d}{R_1 R_2} \qquad [9\text{-}9]$$

where d is the distance between the two mirrors. If the mirrors have a common center, that is, if they are concentric, and if they have radii with a ratio of

$$\frac{\sqrt{5} + 1}{\sqrt{5} - 1} = 2.618 \qquad [9\text{-}10]$$

the system is free from spherical aberration, coma, and oblique astigmatism.

Example

A microscope objective of the Schwarzschild type has a primary (mirror) of 50 mm and a secondary of 10 mm radius of curvature. The distance between the two mirrors is 40 mm. Where must the object be placed to produce an image 18 cm from the primary?

Solution: While the light is reflected back and forth between the mirrors, it is much easier to work such a problem with the system *unfolded*, as shown in Figure 9-21.

---

modification of Gregory's precedent). The real virtue of Cassegrain's design is that the spherical aberration introduced by one mirror is partially canceled by the other, a fact established by Ramsden a century later. Cassegrain's original system had a paraboloid primary and a hyperboloid secondary; since then the term *Cassegrainian* has broadened to mean any combination of a concave primary and a convex secondary. In the *Ritchey–Chrétien* type, for example, both mirrors are hyperboloid, completely eliminating spherical aberration and coma.

* Karl Schwarzschild (1873–1916), German astronomer and physicist. At age 16, Schwarzschild published a paper on the theory of celestial orbits, at 28 became professor and director of the observatory at the University of Göttingen and eight years later director of the Astrophysical Observatory at Potsdam. Schwarzschild used diffraction gratings for the separation of double stars, discovered that stellar motion follows ellipsoidal paths, and contributed to geometrical optics, to the quantum theory of molecular spectra, and to the relativistic deformation of space near celestial bodies of sufficiently large mass, which makes such bodies invisible. K. Schwarzschild, *Theorie der Spiegelteleskope*, Ges. Wiss. Göttingen, Math.-Phys. Classe, Neue Folge **IV**, 2 (1905).

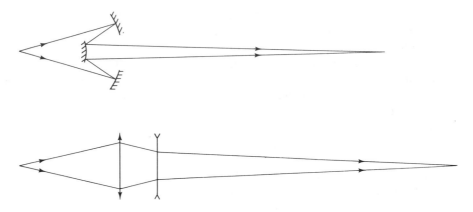

Figure 9-21   Schwarzschild-type microscope objective in its actual configuration (*top*) and unfolded (*bottom*).

First determine the powers of the two mirrors:

$$P_1 = \frac{-2}{R_1} = \frac{-2}{-0.05} = +40 \text{ diopters}$$

$$P_2 = \frac{-2}{+0.01} = -200 \text{ diopters}$$

With the image 18 cm from the primary, or $18 + 4 = 22$ cm from the secondary, the exit vergence at the secondary is

$$\mathbf{V}_2' = \frac{1}{+0.22} = +4.545 \text{ diopters}$$

and the entrance vergence (at the secondary) is

$$\mathbf{V}_2 = \mathbf{V}_2' - P_2 = +4.545 - (-200) = +204.545 \text{ diopters}$$

Without the secondary in place, light of such vergence would come to a focus $1/204.545 = 4.89$ mm to the right of the secondary (considering the *unfolded* system), or 44.9 mm to the right of the primary. The exit vergence at the primary, therefore, is

$$\mathbf{V}_1' = \frac{1}{+0.0449} = +22.27 \text{ diopters}$$

and the entrance vergence is

$$\mathbf{V}_1 = 22.27 - (+40) = -17.73 \text{ diopters}$$

This means that the object distance is

$$s = \frac{1}{-17.73} = -56.4 \text{ mm}$$

and the object must be placed

$$-56.4 + 40 = \boxed{-16.4 \text{ mm}}$$

in front of the secondary.

Catadioptric Systems

The two-mirror astronomical systems described so far still have certain off-axis aberrations and rather limited fields. A significant improvement is possible by adding an aspherical *Schmidt corrector*, mounted at the center of curvature of the primary (mirror) as shown in Figure 9-22.*

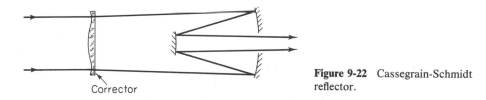

Corrector

Figure 9-22 Cassegrain-Schmidt reflector.

Schmidt's concept of a corrector plate revolutionized the design of large-aperture, wide-field reflectors. But these correctors are difficult to design and produce. This difficulty was overcome by *Maksutov*† who used only spherical surfaces, a meniscus and a concave primary, shown in Figure 9-23. The three surfaces have the same center of curvature. The meniscus has radii given by

$$R_1 - R_2 = \left(\frac{n^2 - 1}{n^2}\right) d \qquad [9\text{-}11]$$

with *d* being its thickness.

---

* Bernhard Voldemar Schmidt (1879–1935). Estonian optician and instrument maker. A rather independent, self-taught, highly skilled craftsman, despite the loss of his right forearm. Schmidt set up shop in Mittweida near Jena, Germany, and later joined the Bergedorf astronomical observatory near Hamburg, where in 1930 he designed and built the reflectors named after him. His first reflector had a speed of *f*/1.75, unheard of at that time, and gave photographs of superb quality. B. Schmidt, *Mitt. Hamburg Sternw. Bergedorf* **7**, 15 (1932).

† Dmitry Dmitrievich Maksutov (1896–1964), Russian physicist and astronomer. After building his first reflector at age 13, Maksutov at 15 was elected a member of the Russian Astronomical Society, in 1914 graduated from a military engineering school, in 1944 became professor and head of Instrument Construction at the Pulkovo Observatory. Maksutov wrote several books, among them *Astronomical Optics*, Moscow 1946, and *Manufacture and Testing of Astronomical Optics*, Moscow 1948. His most significant contribution is "Novye katadioptricheskie meniskovye sistemy," *Dokl. Akad. Nauk SSR* **37**, 147–52 (1942), and *Zh. Tekh. Fiz.* **13**, 87–108 (1943), translated as "New Catadioptric Meniscus Systems," *J. Opt. Soc. Am.* **34** (1944), 270–84. In his original design, Maksutov used the following parameters (in millimeters): primary R_1 −823.2; meniscus diameter 100, thickness 10.0, radius R_a −152.8, R_b −158.6, $n = 1.5163$, $\nu = 64.1$; meniscus-mirror distance 539.1; *f*-stop number *f*/4.

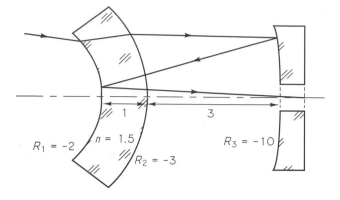

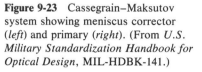

Figure 9-23 Cassegrain–Maksutov system showing meniscus corrector (*left*) and primary (*right*). (From *U.S. Military Standardization Handbook for Optical Design*, MIL-HDBK-141.)

The Cassegrain telescope is short, sturdy, and of light weight; it has a large aperture and a wide field and is completely sealed, making it impervious to air turbulence and contamination. It is a very popular design, widely used in astronomical observatories and also for telephoto camera lenses.

The largest catadioptric telescopes in existence are the 5-m (200-inch) Hale telescope on Mount Palomar, a Russian 6-m telescope in the Caucasus, and the recently built 10-m Keck telescope, actually a mosaic of 36 hexagonal *multiple mirrors*, each 1.8 m across, on the top of Mauna Kea on the island of Hawaii. Such multiple mirrors are mounted in a common frame; they are easier to protect from thermal and other distortion. Their combined light-gathering power is equivalent to that of a mirror 10 m in diameter, at a fraction of the cost.

Besides multiple mirrors, the trend is also toward other new designs such as *deformable mirrors*. These mirrors have a thin, flexible faceplate supported on a rigid backplate by a series of thick piezoelectric crystals. A complex sensor monitors distortions of the incoming wavefronts (distorted because of air turbulence). An elaborate real-time computer compensates for these distortions, eliminating the poor ''seeing'' that so often afflicts astronomical observations. This is called *adaptive optics*.

SUMMARY OF EQUATIONS

Telescope,
 separation of lenses:

$$d = f_O + f_E \qquad\qquad [9\text{-}1]$$

 magnification:

$$M_{\text{Telescope}} = -\frac{f_O}{f_E} \qquad\qquad [9\text{-}3]$$

$$M_{\text{Telescope}} = \frac{\text{diameter of entrance pupil}}{\text{diameter of exit pupil}} \qquad \text{[9-4]}$$

Microscope,
magnification:

$$M_{Microscope} = -\frac{T}{f_O}\frac{25}{f_E} \qquad \text{[9-7]}$$

numerical aperture:

$$NA = n_0 \sin I \qquad \text{[9-8]}$$

PROBLEMS

9-1. An astronomical telescope consists of two lenses separated by a distance of 80 cm. If the angular magnification, when focusing at infinity, is 5×, what are the powers of the two lenses?

9-2. A telescope has an objective of +60 cm focal length and an eyepiece of +5 cm focal length. With the telescope focused at a target 3 m away and used without accommodation, what is:
(a) The distance between the two lenses?
(b) The magnification?

9-3. The objective lens of a telescope has a diameter of 40 mm and a focal length of 32 cm. When focused at infinity, the exit pupil is 2.5 mm in diameter. What is:
(a) The magnification of the telescope?
(b) The focal length of the eyepiece?

9-4. A 10× telescope is pointed at the moon and focused to give an image at infinity. If the focal length of the objective is 60 cm, how far and in which direction must the eyepiece be moved to project an image of the moon on a screen 30 cm away from the eyepiece?

9-5. An astronomical telescope is focused for an object at infinity, without accommodation. How must the distance between the telescope objective and the eyepiece be changed if the telescope is used:
(a) For taking photographs with a camera focused for infinity?
(b) For taking photographs using only the camera body, without the camera lens?
(c) By an observer who does accommodate?

9-6. What is the magnification of a telescope that has an objective of 30 cm focal length and an eyepiece of 3.5 cm focal length and that is used for looking, *with* accommodation, at an object 13.12 m away?

9-7. A 2.5× Galilean telescope has a −25.00-diopter eye lens. When focused for infinity and used without accommodation, what is the distance between the lenses?

9-8. A Galilean telescope, containing a +50 cm objective and a −8 cm eye lens, is focused at an object 6.25 m away and used without accommodation. How far are the two lenses apart from each other? Solve by the vergence method.

9-9. A low-vision aid can be realized by wearing +20.00-diopter spectacle lenses over −32.00-diopter contact lenses. When focusing at infinity, what is:
 (a) The angular magnification?
 (b) The distance between lenses?

9-10. A binocular magnifying loupe has objectives of +12 cm focal length, an axial distance between lenses of 10 cm, and is designed for a working distance (from the eye) of 40 cm. What is the magnification?

9-11. One of *van Leeuwenhoek*'s early microscopes gave an angular magnification of 125×. What was the focal length of the lens he used?

9-12. If a microscope has an objective of 10 mm focal length, an optical tube length of 16 cm, and an eyepiece marked 5×, what is the total magnification?

9-13. A microscope is made from two equal lenses of +50.00 diopters power each, separated by a distance of 14 cm. If used without accommodation, what is:
 (a) The object distance?
 (b) The magnification?

9-14. An oil-immersion microscope objective is marked 1.6 mm, 1.25 N.A. It is used, at 16 cm optical tube length, together with an eyepiece marked 10×. If the oil has an index of 1.515, is the microscope's total magnification within the useful limit?

9-15. If a nearsighted person cannot see objects clearly that are farther away from the eye than 40 cm, what power lenses are needed to see distant objects in focus?

9-16. When a myopic observer, using an astronomical telescope, takes off her corrective glasses, in which direction must she move the eyepiece of the telescope?

9-17. A patient has 5 diopters of myopia and, when reading, can provide an additional 2 diopters of accommodation.
 (a) Where is the far point, without correction and without accommodation?
 (b) Where is the *near point*, the nearest equivalent point *with* accommodation?

9-18. Continue with Problem 9-17 and assume that now the myopia is *corrected*. Where is:
 (a) The far point?
 (b) The near point?

9-19. A telephoto lens of the Galilean type has two elements, of focal lengths +120 mm and −60 mm, respectively, separated by a distance of 10 cm. What is the focal length of the system?

9-20. A telephoto lens can also be used in reverse, with the minus lens in front. If $f_1 =$ −30 mm, $d = 60$ mm, and $f_2 = +45$ mm:
 (a) What is the focal length of the combination?
 (b) What is its purpose?

9-21. When light comes from infinity, the first element of almost any optical system makes the light converge toward, but perhaps without actually reaching, the *prime focus* of the system. Therefore, where is the prime focus of
 (a) Newton's telescope?
 (b) Gregory's telescope?
 (c) Cassegrain's telescope?

9-22. A Cassegrainian telescope of $R_1 = -40$ cm, $R_2 = -8$ cm, and $d = 16$ cm is aimed at infinity. What is the image distance?

SUGGESTIONS FOR FURTHER READING

BEGUNOV, B. N., N. P. ZAKAZNOV, S. I. KIRYUSHIN, and V. I. KUZICHEV. *Optical Instrumentation, Theory and Design*. Moscow: Mir Publishers, 1988.

INOUÉ S. *Video Microscopy*. New York: Plenum Press, 1986.

KINGSLAKE, R. *A History of the Photographic Lens*. Boston: Academic Press, Inc., 1989.

MAXWELL, J. *Catadioptric Imaging Systems*. New York: American Elsevier Publishing Company, Inc., 1972.

RAY, S. F. *Applied Photographic Optics*. London: Focal Press, 1988.

SCHROEDER, D. J. *Astronomical Optics*. New York: Academic Press, Inc., 1987.

TYSON, R. K. *Principles of Adaptive Optics*. San Diego: Academic Press, Inc., 1991.

10

Systems Evaluation

With the design of the system completed, we now turn to testing and evaluation. The modern way of evaluating an optical system is by the use of *transfer functions. Transfer functions are a measure of performance.* They let us predict theoretically, as well as confirm or disprove experimentally, how the individual lenses that constitute the system perform and how the system as a whole measures up to specifications and expectations. Transfer functions can also be used to evaluate peripheral components such as photographic film, video equipment, the eye, and even the atmosphere through which the light passes on its way to the system.

TRANSFER FUNCTIONS

Contrast

In earlier times, optical systems were often tested by using as the object a target like that shown in Figure 10-1. Such targets have *high* contrast: the lines are a deep black on a white background.

But realistic objects are rarely ever black and white. Most often they are in color or in shades of gray. A gray object in a gray fog has low contrast and, surely, two lines of low contrast cannot be resolved as well as two lines of the same

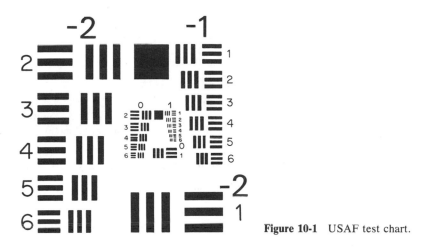

Figure 10-1 USAF test chart.

spacing and high contrast. Any evaluation of an optical system, therefore, must take into account the *contrast*.

More specifically, if the pattern is repetitive, such as a series of dark bars on a bright background, the *contrast modulation*, γ, is defined as a percentage, the ratio of the difference between the amounts of light, I_{max}, returned by the bright intervals, and the amounts of light, I_{min}, returned by the dark bars, to the sum of such amounts,

$$\gamma = \frac{I_{max} - I_{min}}{I_{max} + I_{min}} \times 100\% \qquad \text{[10-1]}$$

If $I_{max} = I_{min}$, the percentage is zero. A modulation of 80% means high contrast; such objects are well visible. A percentage of 20% means low contrast; such objects are barely visible.

Contrast Transfer

Actually, whenever we test for performance, we are not as much interested in the contrast as such as in how much contrast can be *transferred* from object to image. It is the *contrast modulation transfer* that counts.

Consider an object of repetitive, sinusoidal light distribution. Scanning across the object, we find an alternating sequence of maxima and minima (Figure 10-2). These maxima and minima follow each other at a certain *spatial frequency*.

The term *spatial frequency* refers to the number of lines, or other detail, within a given length. A technical line drawing, for example, may have four lines, and four intervals adjoining these lines, per centimeter. The spatial frequency, then, is 4 lines/cm, but since spatial frequency is customarily written in units of inverse *millimeters*, we have $R = 0.4$ mm^{-1}. If 1000 lines, and the intervals between them, fit into 1 mm, $R = 1000$ mm^{-1}. If one bar alone is 1 cm wide, $R =$

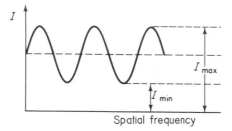

Figure 10-2 Sinusoidal light distribution as a function of spatial frequency.

0.05 mm^{-1}. The higher the spatial frequency that a given optical system can resolve, the better the system.

In the propagation of light, in wave motion in general, and in electric circuits, frequency usually means *time frequency*. In transfer functions and in optical data processing, it means *space frequency*.

We then compare the contrast modulation in the image, M_{image}, to the contrast modulation in the object, M_{object}. The factor by which M_{image} changes is called the *modulation transfer factor*, **T**,

$$M_{\text{image}} = \mathbf{T}M_{\text{object}} \qquad [10\text{-}2]$$

In Figure 10-3, for example, the image modulation is one-half the object modulation, and hence $\mathbf{T} = \frac{1}{2}$.

Is the image contrast always less than the object contrast? Not necessarily. For example, a portrait photograph taken on high-contrast film used for technical reproductions would show harsh, unnatural shadows, of a contrast *higher* than that of the object.

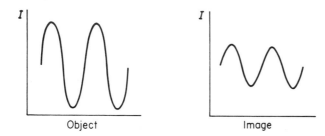

Figure 10-3 Scanning across lines shows high contrast in the object (*left*), low contrast in the image (*right*).

Contrary to what Equation [10-2] seems to suggest, the transfer factor **T** of a given lens does not have a single, unique value. Instead, the factor varies as a function of spatial frequency, R,

$$\mathbf{T} = f(R) \qquad [10\text{-}3]$$

In other words, instead of a single-number transfer factor, **T**, we have a transfer *function*, $\mathbf{T}(R)$, called *modulation transfer function*, **MTF**.

Phase Transfer

Due to aberrations such as coma and distortion, or due to passage through the turbulent atmosphere, some image details may also change with respect to their nominal (conjugate) positions. That is called a change of (transfer of) *phase*. Such phase (shift) ϕ, likewise, is a function of spatial frequency,

$$\phi = f(R) \qquad [10\text{-}4]$$

In addition to the modulation transfer function, therefore, we also have a *phase transfer function*. Both functions are connected with each other, forming the exponential *optical transfer function*, **OTF**:

$$\boxed{\text{OTF} = \text{MTF} \cdot e^{i\phi(R)}} \qquad [10\text{-}5]$$

The optical transfer function, in short, is a spatial-frequency-dependent complex quantity whose modulus, appropriately enough, is the modulation transfer function and whose phase is the phase transfer function.

In the most general terms, the optical transfer function describes the degradation of an image, at different space frequencies. A perfect lens should have a modulation transfer factor **T** of unity, and a phase transfer factor ϕ of zero, *at all space frequencies*. In reality, this can rarely, if ever, be so, because of diffraction and aberrations. At low space frequencies, the transfer factor of a good lens may be close to unity. At higher frequencies, the **T**(R) curve gradually declines, approaching **T** $= 0$ somewhere between 100 and 1000 mm^{-1}. The $\phi(R)$ curve, at low frequencies, will be close to zero. At higher frequencies, it increases. In a way, the $\phi(R)$ curve is a mirror image of the **T**(R) curve, as shown in Figure 10-4.

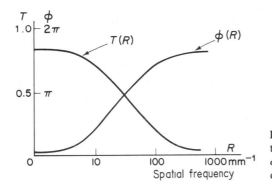

Figure 10-4 Modulation transfer factor, **T**(R), and phase transfer factor, $\phi(R)$, both as functions of spatial frequency.

THE EXPERIMENTAL DETERMINATION OF TRANSFER FUNCTIONS

The essential point in determining the optical transfer function of a lens or lens system is to form and scan an image of a test grid. Often used for this purpose are

Foucault grids (also called Sayce targets); these contain a series of black-and-white, square-wave or sinusoidal, parallel bars of varying spatial frequency, widely spaced at one end and close together at the other. The grid can be used as the object, with a scanning slit in the conjugate image plane; or the slit can be the object, with the grid placed in the image plane. Sometimes the grid is made into a drum, as shown in Figure 10-5.

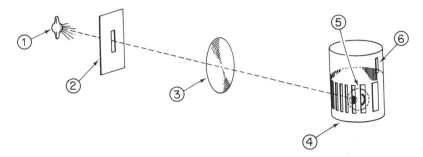

Figure 10-5 Experimental arrangement for measuring transfer functions: 1, light source; 2, slit; 3, lens under test; 4, rotating drum; 5, photodetector; 6, reference mark for determining phase transfer.

At the low spatial frequencies (widely spaced slots), the photocell faithfully registers all bars and intervals. At the high frequencies, depending on the quality of the lens, the individual signals fuse together, the lines are no longer "resolved," and the envelope across the trace goes *down*.

If a *combination of lenses* is tested, the total **MTF** is the product of the MTFs of the individual lenses. But that holds only if any aberrations introduced by one lens are not counteracted by aberrations of another.

Example 1

Photographs are taken from a high-altitude aircraft of a cruise ship, painted white. Assume that the **MTF** of a typical camera lens is that shown in Figure 10-4. What focal length is necessary to obtain enough detail to identify the ship?

Solution: The ship may have a brightness of 5 arbitrary units, and the surrounding ocean may have 2 units. Then its contrast is

$$\gamma = \frac{5 - 2}{5 + 2} \times 100\% = 43\%$$

Initially, the focal length of the system may be chosen so that the ship's image is 0.5 mm wide. This means a spatial frequency of $R = 1$ mm^{-1}. From Figure 10-4 we find that at this frequency the contrast transfer factor **T** is 0.8. At $R = 10$, **T** is 0.7, and at $R = 100$, it is 0.2.

Knowing that the object is of limited contrast (0.43), we now find that for different spatial frequencies the contrast, transferred to the image, changes as follows:

Object contrast   Transfer   Image contrast
                   Space frequency R
                   Transfer factor **T**

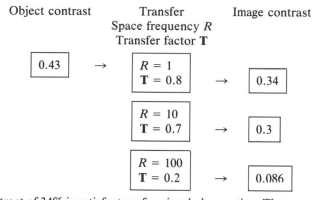

A contrast of 34% is satisfactory for visual observation. That means the ship is easy to see. But fine details, at $R = 100$, cannot be seen. At this frequency the image contrast is so low, less than 9%, that a lens of longer focal length is needed to obtain a larger image. The larger image contains *lower* frequencies, which in turn permit more contrast to be transferred.

Example 2

What effect does atmospheric turbulence have on aerial reconnaissance?

Solution: From Figure 10-6 we conclude that if the details on the ground have high contrast, details of a given spatial frequency can be resolved well both through calm air and through turbulent air. Turbulence merely causes the contrast to become less (arrow A) but since the details have high contrast, they remain well visible. On the other hand, if the details have low contrast, for example if the target is camouflaged, only large details can be identified (arrow B), a result that surely we have anticipated by intuition.

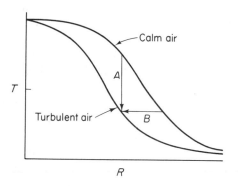

Figure 10-6   Modulation transfer through calm air and turbulent air.

PATENT CONSIDERATIONS

Perhaps the system we have designed and tested looks promising, and we consider whether it might be worth trying to obtain a *patent*. Patents are granted for developments in applied technology; they are not granted for ideas, scientific principles, or for newly discovered laws of nature. Only when the principles are applied to the invention of a "new and useful process, machine, manufacture, or composition of matter, or any new and useful improvements thereof," to quote the United States Patent Act of 1952, may a patent be issued. To be patentable, in short, an invention must be new, useful, and inventive. It must not result from merely exercising mechanical skill or substituting materials, and it must be different from "prior art."

If you think some of your work may lead to an invention, you should keep a written record of the work, best in a notebook that is bound, has numbered pages, and has no pages torn out or left blank. Entries in such a notebook need not be an elaborate literary effort; simply state how the elements are arranged and how they work. Write in ink, and include a sketch. Date and sign what you have written.

Have two witnesses add a statement, "I have read and understood this disclosure," which they also should date and sign. Note that simple reading is not enough; the witnesses should have understood how the invention works, which means they must have familiarity with the subject matter.

Next, things are made easier if you consult a *patent attorney*. Take along the written description. The attorney will quickly understand what the invention is all about and will give further advice as to its merits and chances of patentability.

If you decide to go ahead, the patent attorney will initiate a *patent search*. That search will usually turn up several patents that come close. Carefully read these patents and discuss with the attorney their merits and differences, compared with your invention.

If your disclosure seems clearly superior, a formal *Patent Application*, couched in legal terms, is filed, with drawings made to the Patent Office's specifications. All of this takes time; in general, start the application process no later than six months after the initial notebook entry. Your application must be placed on file within one year of the first public use or disclosure. When the Patent Office receives the application, the patent is "pending."

More often than not, in the first office action the patent examiner will reject some or all the claims made in the application. That will necessitate a further exchange of letters, handled through the patent attorney. If things go well, the patent will be issued within one to three years. Protection of priority, however, begins with the patent "pending." This advantage is sometimes used by makers of certain gadgets, who quickly tool up, manufacture the product, and saturate the market. Such items often become obsolete soon; thus, when the patent is finally denied, the product is no longer marketable anyway. Needless to say, a worthwhile *optical design* does not suffer such an ignominious fate.

This concludes our discussion of *geometrical optics*. Beginning with the next chapter, we turn to *physical optics*.

SUMMARY OF EQUATIONS

Contrast:

$$\gamma = \frac{I_{max} - I_{min}}{I_{max} + I_{min}} \times 100\%$$ [10-1]

Modulation transfer factor:

$$\mathbf{T} = \frac{M_{image}}{M_{object}}$$ [10-2]

PROBLEMS

10-1. A target contains a series of black lines, each line 8 mm wide and separated from the next line by an interval also 8 mm wide. What is the spatial frequency?

10-2. What is the spatial frequency, in mm^{-1}, of the bar pattern in the USAF test chart shown in Figure 10-1:
(a) Near the left-hand lower corner, number 5?
(b) At the right-hand margin, number 4?

10-3. What is the power of a lens that has an angular resolution of 0.05° and resolves details in the focal plane that have a spatial frequency of 7 mm^{-1}?

10-4. If in a periodic test pattern the minima receive two-thirds the amount of light that the maxima receive, what is the contrast?

10-5. If the contrast in a test pattern is 60%, and if the maxima receive 20 units of light, how much do the minima receive?

10-6. Assume that the primary mirror of a reflecting telescope is:
(a) Of perfect paraboloidal shape.
(b) Distorted and not truly paraboloidal.
(c) Of the correct shape but merely ground to a silky finish, rather than polished.
Plot the **MTF**s that correspond to these three conditions.

10-7. An extended object is projected by a converging lens into a real image, at about unit magnification. Assume that a sheet of distorted window glass is placed:
(a) Directly over the object.
(b) Immediately in front of, or closely behind, the lens.
(c) Directly in front of the image.
Neglecting the need for refocusing, how will the image change?

10-8. Continue with Problem 10-7 and assume now that the sheet of glass is precisely plane-parallel but only finely ground, rather than polished. How does this change the image?

SUGGESTIONS FOR FURTHER READING

General Information Concerning Patents. Washington, D.C.: U.S. Department of Commerce, Patent and Trademark Office, 1986.

PERRIN, F. H. "Methods of Appraising Photographic Systems: I. Historical Review." *J. Soc. Motion Pict. Telev. Eng.* **69** (1960), 151–56; II. "Manipulation and Significance of the Sine-Wave Response Function." *Ibid.* **69** (1960), 239–49.

WILLIAMS C. S., and O. A. BECKLUND. *Introduction to the Optical Transfer Function.* New York: Wiley-Interscience, 1989.

11

Interference

Light can be represented at several levels of sophistication. The simplest of these is to represent the light by *rays*. That is the domain of geometrical optics. But, if the wavelength of the light cannot be neglected in comparison to the dimensions of the system as a whole, we need to take into account the wave nature of the light: we need to represent the light by *waves*. That is the domain of physical optics.

YOUNG'S DOUBLE-SLIT EXPERIMENT

One of the phenomena that is easy to explain by the wave nature of light is *interference*. Interference means that under certain conditions light waves can intensify or weaken each other. A good example is *Young's double-slit experiment*. It offered the first proof that light indeed consists of waves. Initially, Young let light pass through two close-by pinholes.* But soon he realized that the result

---

* Thomas Young (1773–1829), British physician and physicist. The son of a Quaker family, Young could read at age 2, at 6 began studying Latin, and at 13 had also mastered Greek, Hebrew, Italian, and French. At 19 he entered medical school, correctly explained the accommodation of the eye, and was elected Fellow of the Royal Society. In 1796, Young graduated from the University of Göttingen Medical School, opened a practice in London, and five years later became Professor of Natural Philosophy (physics) at the Royal Institution of London. That same year, 1801, he read before

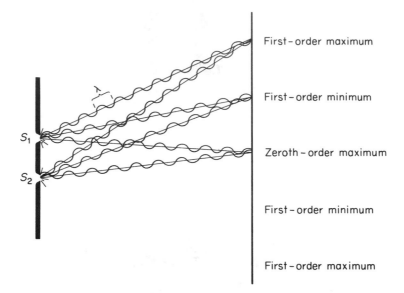

First - order maximum

First - order minimum

Zeroth - order maximum

First - order minimum

First - order maximum

Figure 11-1   Maxima and minima in Young's double-slit experiment. Vertical dimensions exaggerated.

was easier to see when instead of pinholes he used two parallel slits, S_1 and S_2, as shown in Figure 11-1.

In the center of the field where the contributions from the two slits have traveled through equal distances and where the difference between paths, the *path difference* Γ, is zero,

$$\Gamma = 0$$

we have the *zeroth-order maximum*. But maxima will also occur whenever the path difference is one wavelength, λ, or an integral multiple of wavelengths, $m\lambda$,

$$\Gamma = m\lambda \qquad m = 0, 1, 2, \ldots \qquad [11\text{-}1]$$

The integer m is called the *order* of interference.

To calculate the positions of the maxima, we call d the distance between the centers of the two slits, θ the change in direction of the light, and Γ the path

the Royal Society the first of several papers presenting the wave theory of light and the principle of interference, much to the opposition of Newton's followers. Young made noteworthy contributions also to acoustics, atmospheric refraction, elasticity, fluid dynamics, and color vision; later in his life, he helped decipher the Egyptian Rosetta Stone hieroglyphics. Th. Young, "On the Theory of Light and Colours," *Philos. Trans. Roy. Soc. London* **92** (1802), 12–48; and "Experiments and calculations relative to Physical Optics," *ibid.* **94** (1804), 1–16.

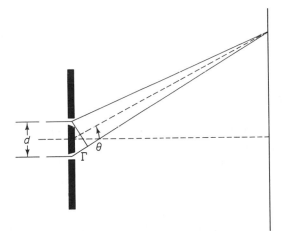

Figure 11-2 Deriving the double-slit equation.

difference between the two contributions. Then from Figure 11-2,

$$\sin \theta = \frac{\Gamma}{d}$$

Combining the last two equations by eliminating Γ gives

$$\boxed{d \sin \theta = m\lambda} \qquad \text{[11-2]}$$

which is *Young's double-slit equation for maxima.* Interference *minima* occur whenever one of the contributions has shifted in phase by $\frac{1}{2}\lambda$, that is, when

$$\boxed{d \sin \theta = (m - \tfrac{1}{2})\lambda} \qquad \text{[11-3]}$$

In a typical double-slit experiment, the angle θ is small and, since θ and d are easy to measure, Young's experiment can be used to determine the wavelength.

Example

Light passes through two slits separated by a distance $d = 0.8$ mm. On a screen 1.6 m away, the distance between the two second-order maxima is 5 mm. What is the wavelength of the light?

Solution: We call x the distance from the double slits to the screen and y the distance of a given maximum, or minimum, from the optical axis. For small angles $\sin \theta$ can be set equal to y/x; thus, when Equation [11-2] is solved for λ,

$$\lambda = \frac{dy}{xm}$$

Since the 5-mm distance given is the distance *between* the two maxima, rather than the distance of one maximum from the zeroth order, we use one-half that distance,

$y = 2.5$ mm, so that

$$\lambda = \frac{(0.8 \text{ mm})(2.5 \text{ mm})}{(1600 \text{ mm})(2)} = \boxed{625 \text{ nm}}$$

LIGHT AS A WAVE PHENOMENON

We now turn to a more general discussion of *light as a wave phenomenon*. Waves are all around us. Some of them we can see, for example the waves on the surface of a lake. Some we can hear, such as musical sounds. Others are very big or submicroscopically small, such as seismic waves produced by an earthquake or the waves that represent atomic particles.

In the most general terms, a motion that repeats itself periodically, in equal intervals of time, is called a *periodic motion*. If the motion follows a sine function, or a cosine function, and the waveform, therefore, is sinusoidal or cosinusoidal, it is a *simple harmonic motion*. These harmonic motions include light and other electromagnetic waves.

A good way of describing simple harmonic motion is by considering it a projection of uniform *circular motion*. Look at the reference circle (Figure 11-3) where a point P_0 travels in a circular path, counterclockwise, at an angular velocity ω. At a given time the point has reached P_1. The projection of P_1 on the y-axis is point Q. Thus as P moves around the circle, Q moves up and down on the y-axis.

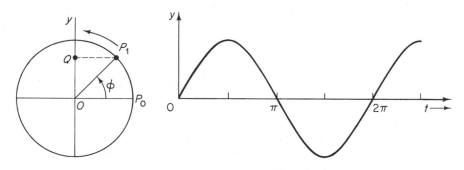

Figure 11-3   Reference circle (*left*) and simple harmonic motion (*right*).

It is customary to plot on the x axis the independent variable, *time*, and on the y axis the dependent variable, the *displacement*. Let point P begin its motion at the initial position P_0, where $t = 0$ and $y = 0$. When P has advanced through $\phi = 45°$, it has reached P_1. Since the total angle around a point is 360° or 2π radians, point P has moved through $2\pi(45°/360°) = \frac{1}{4}\pi$ rad.

After another equal interval of time, P has moved through a total of $\phi = 90°$, or $\frac{1}{2}\pi$ rad. The projection, y, has reached a maximum. After turning through $\phi =$

180° = π rad, P is at a position where again $y = 0$. After $\phi = 270° = \frac{3}{2}\pi$ rad, P is at the lower maximum, and after $\phi = 360° - 2\pi$ rad, it is back at its initial position. Connecting all points gives the sinusoidal curve shown in Figure 11-3 (right). A wave of length λ, therefore, corresponds to a full circumference, 2π, on the reference circle,

$$\lambda = 2\pi \qquad [11\text{-}4]$$

The simple harmonic motion produced by such construction is a *transverse* sinusoidal wave. Consider again the waves on the surface of a body of water. There are two ways of looking at such waves. If we take a snapshot of the waves, their motion is frozen at a time $t = 0$, which means that we consider the shape of the wave as a *function of space*,

$$y = f(x) \qquad [11\text{-}5]$$

But we can also look at the waves through a narrow slot, as in Figure 11-4. A given point (on the water surface) will then move up and down, the x position remaining constant. In contrast to Equation [11-5], we now have

$$y = f(t) \qquad [11\text{-}6]$$

which means that we consider the position of the point as a *function of time*.

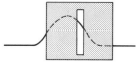

Figure 11-4   Moving wave as seen through slot.

We call the radius of the reference circle **A**, for *amplitude*, and the angle ϕ, for *phase*. Consequently,

$$\sin \phi = \frac{y}{\mathbf{A}}$$

Solving for y and substituting for ϕ the definition of angular velocity, $\omega = \phi/t$, gives

$$\boxed{y = \mathbf{A} \sin (\omega t)} \qquad [11\text{-}7]$$

which is the *equation of motion* (of Q), that is, the *displacement of Q as a function of time*.

We continue with the angular velocity. We call T the *period* of the wave, the reciprocal of the frequency, $1/T = \nu$. For a single oscillation we then have

$$\omega = \frac{2\pi}{T}$$

Substitution gives

$$\omega = 2\pi\nu$$

and substituting this in Equation [11-7]

$$y = \mathbf{A} \sin(2\pi\nu t) \qquad\qquad [11\text{-}8]$$

which again describes the displacement in terms of time.

Now we assume that we have a traveling wave that moves forward, to the right. If v is the velocity of travel of the wave, then after time t the wave has moved through a distance $x = vt$. Therefore, in order to retain the equality in $y = f(x)$, vt must be subtracted from x_0 and

$$y = f(x_0 - vt) \qquad\qquad [11\text{-}9]$$

We then take the relationship

$$v = \lambda\nu = \frac{x}{t}$$

We retain the two right-hand terms and solve for ν:

$$\nu = \frac{x}{\lambda t} \qquad\qquad [11\text{-}10]$$

We substitute Equation [11-10] in [11-8] and cancel t:

$$y = \mathbf{A} \sin\left(2\pi\,\frac{x}{\lambda}\right) \qquad\qquad [11\text{-}11]$$

This describes the *displacement as a function of space.*

Now we are ready to represent the displacement y in terms of both space and time, that is, we make the wave configuration *move*. We combine Equations [11-9] and [11-11]. That gives

$$\boxed{y = \mathbf{A} \sin\left[\frac{2\pi}{\lambda}(x - vt)\right]} \qquad\qquad [11\text{-}12]$$

which shows how the displacement, y, varies in terms of *both space, x, and time, t.*

SUPERPOSITION OF WAVES

Now we come to the essence of interference, which is the *superposition of waves.* We consider two examples.

1. **Superposition of waves of the same frequency and phase.** Assume that two sinusoidal waves of the same frequency are traveling side by side in the same

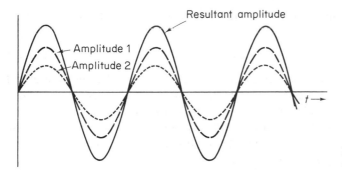

Figure 11-5 Superposition of waves of equal phase and frequency.

direction and that they have the *same phase*. We plot the amplitudes of the waves, as shown in Figure 11-5, and add them, point by point. The resultant amplitude is the sum of the individual amplitudes,

$$\mathbf{A} = \mathbf{A}_1 + \mathbf{A}_2 \qquad [11\text{-}13]$$

2. **Superposition of waves of the same frequency but with a constant phase difference.** Instead we may have two sinusoidal waves that still have the same frequency but now have *different initial phase angles*, ϕ. The resultant wave can be found in different ways, for example by *vector addition*.

One of the waves may follow the equation

$$y = \mathbf{A}_1 \sin(\omega t) \qquad [11\text{-}14]$$

The phasor representing this wave at a given time is shown in Figure 11-6 (left). As before, the phasor rotates counterclockwise and with angular velocity ω. The other wave has the same angular velocity but a different amplitude, $\mathbf{A}_2$; it also lags

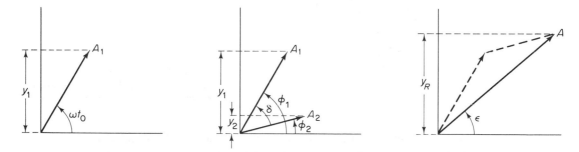

Figure 11-6 (*Left*) Phasor representing wave amplitude $\mathbf{A}_1$; (*center*) two phasors representing waves of different amplitudes and with a constant phase difference δ; (*right*) vector triangle showing resultant phase.

behind the first wave by a (constant) phase difference δ (center). Its equation, therefore, is

$$y = \mathbf{A}_2 \sin(\omega t - \delta) \qquad [11\text{-}15]$$

The resultant of the two phasors can be found by the parallelogram method of vector addition or, more easily, by a *vector triangle* (right). Note that both phasors, and the vector triangle formed by them, *rotate as a unit*.

If we want to add two wave equations algebraically, it is probably best to use *complex numbers* because they are easier to manipulate than sines and cosines. The connection is given by *Euler's formula*,

$$\cos \phi + i \sin \phi = e^{i\phi} \qquad [11\text{-}16]$$

For example, if we have two waves with different initial phase angles, ϕ_1 and ϕ_2, we may write

and

$$\mathbf{A}_1 = A_1 e^{i(\omega t + \phi_1)}$$
$$\mathbf{A}_2 = A_2 e^{i(\omega t + \phi_2)} \qquad [11\text{-}17]$$

where $\mathbf{A}$ is the (complex) amplitudes of the waves, ω the angular frequency, and t the time.

Then if we plot the displacements, y, versus time, t, for the two waves and for the resultant, we arrive at Figure 11-7. It shows that *the resultant of two* (or more) *sinusoidal waves of the same frequency is again a sinusoidal wave*, of the same frequency but with a new amplitude, $\mathbf{A}$, and a new phase angle.

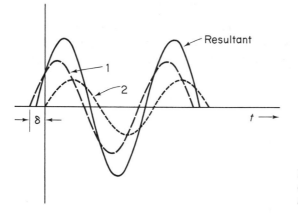

Figure 11-7 Two sine waves, 1 and 2, of constant phase difference, δ, and the *resultant*, also a sine wave.

As was discussed, the radius of the reference circle is called $\mathbf{A}$, for *amplitude*. But actual light, generated, propagating, or received, is measured in terms of *intensity*, $\mathbf{I}$. We note that intensity is proportional to the *square of the amplitude*,

$$I \propto \mathbf{A}^2 \qquad [11\text{-}18]$$

Whenever the phase difference between two waves is zero, $\delta = 0$, the intensity reaches a *maximum*,

$$\mathbf{I}_{max} = I_1 + I_2 + 2\sqrt{I_1 I_2}$$

which, when $I_1 = I_2$, becomes

$$\mathbf{I}_{max} = 4I_1 \qquad\qquad [11\text{-}19]$$

On the other hand, whenever the phase difference $\delta = 180°$, we have a *minimum*,

$$\mathbf{I}_{min} = I_1 + I_2 - 2\sqrt{I_1 I_2}$$

which, when $I_1 = I_2$, becomes

$$\mathbf{I}_{min} = 0 \qquad\qquad [11\text{-}20]$$

At points in between, and assuming again that $I_1 = I_2$, we find that

$$\mathbf{I} = I_1 + I_1 + 2I_1 \cos \delta$$
$$= 2I_1(1 + \cos \delta) \qquad\qquad [11\text{-}21]$$
$$= 4I_1 \cos^2 \tfrac{1}{2}\delta$$

A plot of the resultant sinusoidal wave is shown in Figure 11-8.

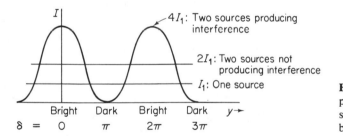

Figure 11-8 Intensity distribution as produced by one source, by two sources not causing interference, and by two sources causing interference.

This figure shows the intensity distribution across a few fringes projected on a screen. If only one beam of light illuminates the screen, the distribution is uniform throughout (lower horizontal line marked I_1). If we have two beams of light of equal intensity, but lacking the property necessary to produce interference, the distribution is again uniform, but twice as much light will reach the screen (upper horizontal line marked $2I_1$). But if the two beams *are capable of producing interference*,* they form alternate maxima and minima. If the initial contributions have equal amplitudes, and therefore equal intensities, then the

---

* The property of light necessary to produce interference is called *coherence*, to be discussed in Chapter 13.

maxima, from Equation [11-19], contain *four times* the intensity of the individual contribution; the minima, from Equation [11-20], contain nothing. The result is the $\cos^2$ curve labeled $4I_1$. Note that the *integrated* values, that is, the areas under the curves labeled $2I_1$ and $4I_1$, *are the same*. Indeed, expressions such as "constructive" and "destructive" interference should not be used; they seem to imply that somehow there could be a "destruction" of light. That is not so. If less light reaches a given point, more light reaches some other point: interference merely causes a *redistribution* of the light.

THE MICHELSON INTERFEROMETER

Look again at Young's double slits. While using two slits makes it easy to understand the meaning of a superposition of waves, and hence the essence of interference, any use of slits significantly restricts the amount of light that can pass through. Moreover, it is next to impossible to design a double-slit mechanism where the two slits could be brought together *down to a separation of zero*. These disadvantages do not exist in the *Michelson interferometer*.

The Michelson interferometer* is very versatile. Using it we can bring two optical planes into coincidence and move them virtually through one another. The Michelson instrument also lets us use an extended light source, which results in much brighter fringes. Michelson's interferometer can also be built using fibers, which makes the alignment, and isolation from vibration, much easier.

Still another difference is that Young's double slits separate the light into two beams that run side by side, a process called *division of wavefronts*. By contrast, Michelson's instrument separates the light along the axis of propagation; that is called *division of amplitude*.

In its most elementary form, the Michelson interferometer consists of a beamsplitter, *A*, and of two mirrors, *C* and *D* (Figure 11-9). The beamsplitter divides the light into two parts, one part being transmitted toward mirror *C*, the other part being reflected toward mirror *D*. The two mirrors, *C* and *D*, return the light to *A*. There the beams recombine and proceed toward *E* where interference is observed.

---

* Albert Abraham Michelson (1852–1931). Born in Strelno, Prussia (now Strzelno, Poland), Michelson was two years old when his parents brought him to the United States. He graduated from, and taught at, the U.S. Naval Academy and later worked at the Case School of Applied Science in Cleveland, at Clark University in Worcester, Massachusetts, and at the University of Chicago. In 1907 he was awarded the Nobel Prize in physics, the first American scientist to be so honored. Michelson is best known for his precise determination of the velocity of light, for inventing the interferometer that bears his name, for the Michelson-Morley aether drift experiment, and for establishing the length of the meter in terms of wavelength of light. He also made noteworthy contributions to astronomy, spectroscopy, and geophysics, was proficient in tennis and other sports, played the violin, and liked to paint landscapes. The interferometer was first described in A. A. Michelson, "Interference Phenomena in a new form of Refractometer," *Am. J. Sci.* (3) **23** (1882), 395–400, and *Philos. Mag.* (5) **13** (1882), 236–42.

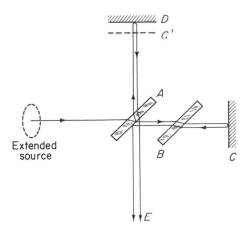

Figure 11-9 The Michelson interferometer.

One of the mirrors is mounted so that it can be moved along the axis. But note that, if reflection at *A* occurs at the rear surface as shown, the light reflected at *D* will pass through *A* three times while the light reflected at *C* will pass through only once. For this reason, a compensating plate, *B*, of the same thickness and inclination as *A*, is inserted into the *A–C* path. If we look into the instrument from *E*, we see mirror *D*, and in addition we see a virtual image, *C'*, of mirror *C*. Depending on the positions of the mirrors, image *C'* may be in front of, or behind, or exactly coincident with mirror *D*.

Fringe Formation

If image *C'* coincides with mirror *D*, the two arms of the interferometer are equal in length. If they are not, the distance between *C'* and *D* is finite, *C'D* = *d*. In addition, the observer may be looking into the system subtending an angle *I* with the optical axis, as shown in Figure 11-10.

Assume that the light comes from a point *S* and is reflected by both *C'* and *D*. The observer will see two virtual images: *S'*, which is due to reflection at *C'*, and *S''*, which is due to reflection at *D*. Sighting along the axis, the two images *S'* and *S''* are a distance 2*d* apart, which is the *path difference* Γ,

$$\Gamma = 2d$$

But to the observer looking into the system at an angle *I*, the apparent distance between *S'* and *S''* is less; in fact,

$$\Gamma = 2d \cos I$$

Replacing Γ by *m*λ, following Equation [11-1], results in

$$\boxed{2d \cos I = m\lambda}$$

[11-22]

which is the *Michelson interferometer equation*.

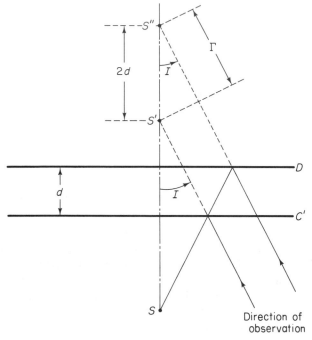

Direction of
observation

Figure 11-10   Looking off-axis into the
Michelson interferometer.

For a given mirror separation d, and a given order m and wavelength λ, angle
I is constant. This means that the fringes are of rotational symmetry: They are
circles concentric around the axis. They are *fringes of equal inclination.*

At the coincidence position where $d = 0$, we expect the two waves to
reinforce each other and to form a maximum. But this is not so. The reason is a
180° or π phase change that occurs whenever the light is reflected at the boundary
between a lower-index material and a higher-index material (such as air and glass).
It does not occur on reflection at a higher-to-lower boundary (such as glass and
air), and it does not occur on transmission or refraction either. Figure 11-11 shows
two examples.

Figure 11-11   A π phase change occurs
on reflection at an air-glass boundary
(*left*) but not at a glass-air boundary
(*right*).

Look again at Figure 11-9 and note: It is only the light that comes from C and
goes to E that is reflected at an air-glass boundary and therefore shows a π phase
change. Because of this disparity between the two beams, there will be a *minimum*
at the coincidence position: the center of the field will be *dark.*

If we now *move* one of the mirrors through a distance $\frac{1}{4}\lambda$, the path length changes by $\frac{1}{2}\lambda$, the two contributions get out of phase by 180°, the phase change compensates for that, and we have a *maximum*. Moving the mirror by another $\frac{1}{4}\lambda$ gives another minimum, moving it another $\frac{1}{4}\lambda$ again a maximum, and so on.

As d is gradually made larger, a given ring—which has a certain order m—increases in size because the product $2d \cos I$ in Equation [11-22] must by necessity remain constant. Consequently, the cosine of I must become smaller, and the angle I and the rings larger, a new ring appearing in the center of the field each time one of the mirrors has moved through $\frac{1}{2}\lambda$. Moreover, as d increases, new rings appear in the center faster than rings already present disappear in the periphery; thus the field will become more crowded with thinner rings (Figure 11-12). Conversely, as d is made smaller, the rings contract and disappear in the center.

Figure 11-12 Appearance of fringes in the Michelson interferometer as the mirrors are moved away from each other. Arrows at the far right indicate motion of the fringes.

As the mirrors are *tilted*, and are no longer perpendicular to each other, the fringes appear to be straight. Actually, they are sections of large hyperbolas. If now distance d is changed, that is, if one of the mirrors is moved, the fringes move across the field and can easily be counted as they pass a reference mark. For each fringe that passes, the mirror has moved through one-half of a wavelength; hence, if m fringes are counted as the mirror is moved through d, the wavelength of the light is

$$\lambda = \frac{2d}{m} \tag{11-23}$$

With care, $\frac{1}{100}$ of the width of a fringe can be measured.

Aligning the Michelson Interferometer

Tape a black mark to a ground glass placed between a light source and a beamsplitter. Make the two paths, from the beamsplitter to the two mirrors, equal to within a millimeter.

Now when you look into the interferometer, you see three reflected images of the alignment mark [Figure 11-13(A)]. A slight turn of the mirror tilt controls will show you which of these images can be moved. Superimpose the movable image exactly on the right-hand image of the stationary pair of images, aligning the

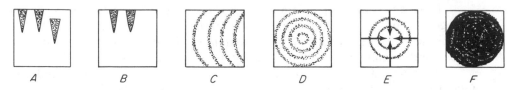

Figure 11-13   Aligning the Michelson interferometer.

movable image first in *height* and then *sideways* (*B*). At this point, interference fringes should become visible (*C*).

Using the tilt controls, bring the center of the fringes into the center of the field of view (*D*). Move the other mirror along the axis so that the fringes move *inward* (*E*). Near the zeroth order, alignment becomes critical. Even placing your hand over and close to the paths—which will raise the temperature of the air and lower its refractive index—will make the fringes wiggle and move in and out of the field. At the zeroth order, the center of the field will be dark (*F*).

APPLICATION AND EVALUATION

The most important application of interferometry is to *precision measurements* of length or thickness and the like. In principle almost any interferometer will do, but for practical purposes some types are better suited than others. We distinguish two groups, depending on whether the instrument is used for *reflecting objects* or *transparent objects*.

The first category includes the *examination of surfaces*, *metrology*, and *alignment*. Here we use the Michelson interferometer, or one of its many variations. For example, if we want to know whether there are scratch marks or other defects in a surface, we use that surface in place of one of the mirrors. The other mirror serves as a reference. Any defect present in the surface shows as a characteristic distortion of the fringe pattern, as we see from the following example.

Example

When examining the surface of a polished workpiece in thallium light (535 nm), some fringes are seen to be distorted by $\frac{4}{10}$ the distance between them (Figure 11-14). How deep is the defect?

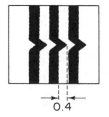

0.4   **Figure 11-14**

Solution: As always in interferometry, consecutive fringes indicate a path difference of one wavelength, $\Gamma = 1\lambda$. But the light is reflected, and goes back and forth through the depth of the defect; therefore, a Γ path difference is produced by a defect only one-half that deep. In our example, the distortion is $\frac{4}{10}$ of the fringe separation and hence the depth of the defect, from Equation [11-22], is

$$d = (0.4)(\tfrac{1}{2})(535 \times 10^{-9}) = \boxed{0.1 \ \mu m}$$

For transparent objects the *Mach–Zehnder interferometer** is preferred, in part because the light passes through the sampling field only once and evaluation is easier. The Mach–Zehnder interferometer, illustrated in Figure 11-15, consists of two mirrors and two beamsplitters. The first beamsplitter divides the incoming light, the mirrors reflect the beams as shown, and the second beamsplitter brings the beams together again. The two paths may be widely separated, which permits testing large objects, as in a windtunnel.

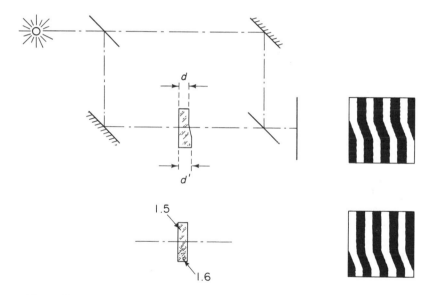

Figure 11-15   Mach–Zehnder interferometer showing fringe shift due to local variation of thickness (*top*) or refractive index (*bottom*).

---

* Ernst Waldfried Joseph Wenzel Mach (1838–1916), Austrian physicist, best known for his investigation of shock waves caused by a projectile as it reaches the speed of sound, the ratio of the speed of the projectile to the speed of sound called the *Mach number*. An ardent experimentalist, Mach did not believe in then unobservable and, to him, rather mystical ideas such as the atomic structure of matter and the theory of relativity. The interferometer is described in L. Mach, ''Modifikation und Anwendung des Jamin Interferenz-Refraktometers,'' *Anz. Akad. Wiss. Wien math. naturwiss.-Klasse* **28** (1891), 223–24, and L. Zehnder, ''Ein neuer Interferenzrefraktor,'' *Z. Instrumentenkd.* **11** (1891), 275–85.

The *Jamin interferometer* is similar to the Mach–Zehnder type, except that the left-hand side beamsplitter and mirror and the right-hand side mirror and beamsplitter are combined into one thick plane-parallel plate each. This makes for mechanical stability and easier alignment.

Interferometers respond to differences in optical path length. Such *path difference*, Γ, is simply the difference between two optical path lengths,

$$\Gamma = S_2 - S_1$$

optical path length being the product of thickness and refractive index,

$$S = Ln$$

Phase difference, δ, on the other hand, is a wave theory concept; it means that a wave of length λ corresponds to a full circumference, 2π, on the reference circle. Path difference Γ and phase difference δ, therefore, are connected by

$$\frac{\Gamma}{\lambda} = \frac{\delta}{2\pi} \qquad [11\text{-}24]$$

The path difference Γ that we see in an interferometer may be due to either a variation in thickness or a variation in refractive index. In the upper example in Figure 11-15 we assume that the refractive index within the sample is constant but the thickness is not; thus we have a path difference of

$$\Gamma = n(L_2 - L_1)$$

In the lower example, the thickness is constant but the index is not; thus we have

$$\Gamma = L(n_2 - n_1)$$

Since $\Gamma = m\lambda$, therefore, a given path difference will cause either

$$m = \frac{1}{\lambda} n(L_2 - L_1) \qquad [11\text{-}25]$$

or

$$m = \frac{1}{\lambda} L(n_2 - n_1) \qquad [11\text{-}26]$$

additional wavelengths to be present within the thicker, or denser, medium, and a fringe shift by m fringes will result. This is a distinction that the two samples in Figure 11-15 show very well.

SUMMARY OF EQUATIONS

Young's double-slit maxima:

$$d \sin \theta = m\lambda \qquad\qquad [11\text{-}2]$$

Equation of motion:

$$y = \mathbf{A} \sin(\omega t) \qquad\qquad [11\text{-}7]$$

Michelson interferometer:

$$2d \cos I = m\lambda \qquad\qquad [11\text{-}22]$$

Path difference and phase difference:

$$\frac{\Gamma}{\lambda} = \frac{\delta}{2\pi} \qquad\qquad [11\text{-}24]$$

PROBLEMS

11-1. Monochromatic light passes through a double slit, producing interference. The distance between the slit centers is 1.2 mm. When on a screen 5 m away from the slits the distance between consecutive fringes (on one side of the zeroth order) is 2 mm, what is the color of the light?

11-2. Two narrow slits, 0.8 mm apart from each other, are used to produce interference. If on a screen 80 cm away the distance between the two second-order maxima is 2 mm, what is the wavelength of the light?

11-3. Light of 600 nm wavelength passes through a double slit and forms interference fringes on a screen 1.2 m away. If the slits are 0.2 mm apart, what is the distance between the zeroth-order maximum and a third-order minimum?

11-4. When two narrow slits are 0.375 mm apart, on a screen 1.2 m away the distance between the two first-order maxima is found to be 4 mm. What is the wavelength of the light?

11-5. When one of the slits in Young's experiment is covered by a film of transparent material, the zeroth order is seen to shift by 2.2 fringes. If the refractive index of the material is 1.4 and the wavelength of the light 500 nm, how thick is the film?

11-6. A satellite circling Earth is transmitting microwaves of 15 cm wavelength. When the satellite is above a ground station which has two antennas 100 m apart and located in the plane of the orbit, a total signal is received by the two antennas that fluctuates with a period of $\frac{1}{10}$ s. If the satellite is known to be at an altitude of 400 km and if we neglect the curvature of Earth, what is the satellite's velocity?

11-7. Assume that point P on the reference circle moves at a constant angular velocity of 4 rad/s. How long does it take the projection, Q, to move from $y = 0$ to a position halfway up to the maximum amplitude?

11-8. What is the wavelength and the velocity of a wave that can be represented by $y = 4\sin(10x - 20t)$?

11-9. Using red cadmium light of 643.8 nm wavelength, Michelson in his original experiment could still see interference fringes after he had moved one of the mirrors 25 cm away from the coincidence position. How many fringes did he count?

11-10. If one arm of a Michelson interferometer contains a tube 2.5 cm long that is first evacuated and then slowly filled with air ($n = 1.0003$), how many fringes will cross the center? Assume light of 600 nm wavelength.

11-11. The *Rayleigh interferometer* is derived from the double-slit design. It has two test chambers placed side by side as shown in Figure 11-16. Assume that the chambers are 30 cm long and that one is completely empty (evacuated) while the other is gradually being filled with a certain gas. If, with light of 500 nm wavelength, 240 fringes are seen to cross the field of view, what is the refractive index of the gas?

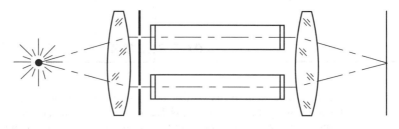

Figure 11-16

11-12. The two tubes of a Rayleigh interferometer are first filled with air ($n = 1.0003$). The air in one tube is then gradually replaced by a gas of unknown refractive index. If the tubes are 10 cm long, the wavelength is 500 nm, and 70 fringes cross the center of the field, what is the refractive index of the gas?

SUGGESTIONS FOR FURTHER READING

BORN, M., and E. WOLF. *Principles of Optics*, 6th ed. Elmsford, N.Y.: Pergamon Press, Inc., 1980.

HARIHARAN, P. *Basics of Interferometry*. Cambridge, Mass.: Academic Press, Inc., 1992.

INGARD, K. U. *Fundamentals of Waves and Oscillations*. New York: Cambridge University Press, 1988.

MAIN, I. G. *Vibrations and Waves in Physics*, 3rd ed. New York: Cambridge University Press, 1993.

STEEL, W. H. *Interferometry*, 2nd ed. New York: Cambridge University Press, 1985.

12

Thin Films

Some of the most lustrous colors in nature are the iridescent colors of insects. These colors are due to interference, in contrast to colors caused by pigments in watercolors or oil paints, which are due to absorption. Some other, more practical results of interference include interference filters, antireflection coatings, and especially the high-resolution Fabry–Perot interferometer, the forerunner of the laser cavity. What these systems have in common is that they are based on multiple-beam interference as it occurs most often in *thin films*.

PLANE-PARALLEL PLATES

Consider two plane surfaces, parallel to one another and a short distance apart. Light incident on the surfaces consists of many rays but for simplification we draw only two of the rays that are incident, and reflected again, as shown in Figure 12-1. This kind of reflection reminds us of the reflection of light at the two surfaces, C' and D, in a Michelson interferometer. Indeed, the same equation that we had derived there also applies here; we only need to include the index of refraction of the medium between the surfaces, replacing the distance d by the optical path length nd:

$$\boxed{2nd \cos I = m\lambda} \qquad m = 1, 2, 3 \qquad [12\text{-}1]$$

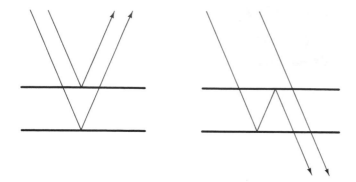

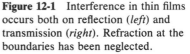

Figure 12-1 Interference in thin films occurs both on reflection (*left*) and transmission (*right*). Refraction at the boundaries has been neglected.

(I use $m = 1$ as the lowest order because $m = 0$ would imply that either n or d is zero or that $I = 90°$, none of which is possible.)

Again we need to consider whether or not there is a phase change. A π phase change occurs only on external (lower-to-higher index) reflection. It does not occur on internal (higher-to-lower index) reflection and it does not occur on transmission or refraction. For example, if the light is transmitted through a film in air, both reflections are internal (high-to-low) and neither will cause a π change. In that case, Equation [12-1] describes the condition for *maxima*. Conversely, *minima* will occur whenever

$$2nd \cos I = (m - \tfrac{1}{2})\lambda \qquad [12\text{-}2]$$

But if the light is *reflected* by the two surfaces of a thin film in air, the light reflected at the first surface *will* undergo a π phase change (because that is now an external, low-to-high reflection), but the light reflected at the second surface is internal and *will not*. Essentially the same happens if the light is reflected by an air film between plates: again there will be one π change (at the second surface but not at the first). Then the two equations change places: Equation [12-1] refers to minima and Equation [12-2] to maxima. In addition, at least in the case of transmission, the amount of light transmitted depends on the reflection coefficient.

Example

A soap bubble, made with soapy water of $n = 1.35$ and its wall known to be 300 nm thick, is seen at normal incidence in reflected white light. In what color will it appear?

Solution: While the soapy film, both outside and inside the bubble, borders on air, a π phase change occurs only at the outside surface. To find a reflection *maximum*, therefore, we use Equation [12-2],

$$2nd \cos I = (m - \tfrac{1}{2})\lambda$$

For light at normal incidence, $\cos I = 1$, and thus the wavelength of the reflected light is

$$\lambda = \frac{2nd}{m - \frac{1}{2}} = \frac{(2)(1.35)(300)}{0.5} = 1620 \text{ nm}$$

But that is in the infrared, and cannot be seen. We then go to the next higher order, setting $m = 2$, so that

$$\lambda = \frac{(2)(1.35)(300)}{2 - \frac{1}{2}} = 540 \text{ nm}$$

which means the bubble's reflection is $\boxed{\text{green}}$.

The next higher order, where $m = 3$, would be at 324 nm, in the UV, and that again is not visible.

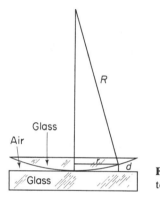

Figure 12-2 Notation used with Newton's rings.

Newton's Rings

Newton's rings* are a special case of thin-film interference. They occur when a convex surface is placed in contact with a plane surface. Light passing through the

---

* Sir Isaac Newton (1642–1727), British scientist, professor of mathematics at Cambridge University, long-time president of the Royal Society. After graduating from Trinity College, the bubonic plague then raging forced Newton into seclusion for two years. During that time he invented differential and integral calculus, discovered the composition of white light, formulated his three laws of motion, and conceived the idea of universal gravitation and the dynamics of the solar system. Around 1670 at Cambridge he gave a series of lectures on optics, five years later sent a copy of his notes to the secretary of the Royal Society, and in 1704 published them under the title *Opticks: or, a Treatise of the Reflexions, Refractions, Inflexions and Colours of Light*. A Latin version came out two years later. The English edition begins with these words: "My Design in this Book is not to explain the Properties of Light by Hypotheses, but to propose and prove them by Reason and Experiments." There is no question of the extraordinary significance of Newton's work. He often found the solution to problems that had escaped others, adding the mathematical reasoning later. Opinions of Newton the man vary. Some consider him the last of the magicians; others adore him unqualifyingly as English Poet Alexander Pope did when he wrote: "Nature and Nature's laws lay hid in night: God said, *Let Newton be!* and all was light."

two surfaces forms a system of circular fringes around the point of contact. As shown in Figure 12-2, we call R the radius of curvature of the (convex) surface, r the radius of a given Newton ring, and d the separation of the surfaces, the thickness of the air film, at this ring. Then we take the vertex depth formula,

$$d \approx \frac{r^2}{2R}$$

and insert it into the expression for minima in reflected light, Equation [12-1]. That gives

$$r^2 \approx m\lambda R \qquad\qquad [12\text{-}3]$$

Newton's rings are widely used for the testing of surfaces. For example, Equation [12-3] tells us that, if we know the wavelength, we can determine the radius of curvature of the surface by measuring the diameter of just one ring. Essentially the same result is obtained with transmitted light; the two patterns are simply complementary.

THE FABRY–PEROT INTERFEROMETER

The *Fabry–Perot interferometer* is probably the most precise interferometer that exists. It is widely used for precision wavelength measurements, for the analysis of hyperfine spectral structure, for the determination of refractive indices of gases, and, in its most recent evolution, as a laser cavity.*

As shown in Figure 12-3, it consists of two parallel plates of glass, their inner surfaces polished to a flatness better than $\lambda/50$ and coated with a highly reflective layer such as silver or aluminum. The outer surfaces are ground to a small angle (making the plates slightly wedge shaped) so that they do not cause any reflections.

The separation of the two surfaces, ranging from a few millimeters to centimeters, is either fixed or variable. When fixed, the plates are separated by a hollow cylinder; that is called an *etalon*. Variable spacing is possible using a spacer of barium titanate or some other piezoelectric material.

The light, as shown in the figure, comes from an extended source and is focused on a screen on the right. Because of repeated back-and-forth reflections of

---

* Marie Paul Auguste Charles Fabry (1867–1945), French physicist. While at the University of Marseilles, Fabry taught medical students. Later he became director of the French Institut d'Optique and professor of physics at the École Polytechnique (Sorbonne) in Paris and is remembered as a brilliant lecturer of sometimes caustic wit. Besides his work on electricity, spectroscopy, atmospheric ozone, and astrophysics, Fabry today is best known for the instrument he developed together with Jean Baptiste Gaspard Gustave Alfred Perot (1863–1925), which he referred to as "this interferometer whose name I have the honor to bear myself." Ch. Fabry and A. Perot, "Théorie et applications d'une nouvelle méthode de spectroscopie interférentielle," *Ann. Chim. Phys.* (7) **16** (1899), 115–44.

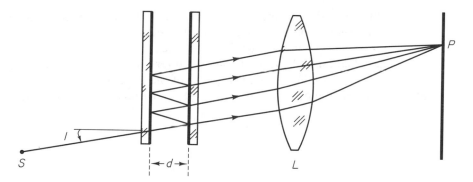

Figure 12-3 Fabry–Perot interferometer. Point S is part of an extended source, forming point P, part of a maximum on the right.

the light, this results in *multiple-beam interference*; the fringes therefore are exceedingly narrow, much more so than in any other type of interferometer.

The medium between the two surfaces is air. Thus π phase changes occur on both of these (air-to-glass) surfaces and

$$2d \cos I = m\lambda \qquad [12\text{-}4]$$

is the equation for *maxima*. Lens L may be a separate lens, as shown, or it may simply be the eye of the observer, in which case the retina serves as the screen. As in Michelson's interferometer, the maxima and minima, for given values of I and m, appear in the form of rings, concentric around the optical axis. Maxima closer to the axis have higher values of m. Maxima farther away have lower values because for these maxima the angle of incidence must by necessity be larger, and its cosine be less.

We call $\Delta\lambda$ the difference between two wavelengths and ΔI the angular separation between maxima formed by these wavelengths. The problem then becomes one of expressing $\Delta\lambda$ in terms of ΔI. To do so, we solve Equation [12-4] for λ and differentiate with respect to I,

$$\frac{d\lambda}{dI} = -\frac{2d}{m} \sin I$$

For paraxial rays, and for small changes in wavelength and angle,

$$\frac{d\lambda}{dI} = \frac{\Delta\lambda}{\Delta I}$$

so that

$$\Delta\lambda = -\frac{2d}{m} I \, \Delta I \qquad [12\text{-}5]$$

which expresses the change of wavelength as a function of the change in angle. (The minus sign is of no consequence because it is immaterial whether $\Delta\lambda$ is measured upward or downward from the mean wavelength.)

The advantage of the Fabry–Perot interferometer, compared with the Michelson instrument, lies in the sharpness of the fringes. At least for paraxial rays the angle of incidence is small, cos $I \approx 1$, and Equation [12-4] reduces to

$$2d \approx m\lambda \qquad\qquad [12\text{-}6]$$

If for a given ring

$$2d \cos I = m\lambda$$

then for the next larger ring,

$$2d \cos I' = (m' - 1)\lambda$$

Substituting the approximation cos $I \approx 1 - I^2/2$ for cos I yields

$$2d\left(1 - \frac{I^2}{2}\right) = (m' - 1)\lambda \qquad\qquad [12\text{-}7]$$

Subtracting Equation [12-6] from [12-7] gives

$$I^2 d = \lambda$$

and dividing this by Equation [12-5] yields

$$\frac{\lambda}{\Delta\lambda} = \frac{m}{2}\frac{I}{\Delta I} \qquad\qquad [12\text{-}8]$$

The quantity $\lambda/\Delta\lambda$ is called the *resolvance* (resolving power) of the interferometer.

When the second wavelength is longer than the first, the λ_2 light forms rings *smaller* than those formed by λ_1, and, at some wavelength difference, an *m*th-order λ_2 ring coincides with an $(m + 1)$th-order λ_1 ring,

$$m\lambda_2 = (m + 1)\lambda_1$$

The wavelength difference at which this occurs is

$$\delta\lambda = \frac{\lambda_1}{m}$$

or, substituting Equation [12-6] and assuming normal incidence,

$$\delta\lambda = \frac{\lambda^2}{2d} \qquad\qquad [12\text{-}9]$$

The interval $\delta\lambda$ is called the *free spectral range*; it is the change in wavelength necessary to shift the fringes by one order, that is, by one fringe. When λ and d are known, in short, the wavelength difference $\delta\lambda$ can be found.

INTERFERENCE FILTERS

Color filters are based either on absorption or on interference. Both types are characterized by three parameters: the *peak wavelength*, λ_{max}, in nm; the peak transmittance or *efficiency*, T_{max}, as a percentage of the total transmittance possible at this wavelength; and the *halfwidth*, HW, the width of the passband at the level of one-half the peak transmittance, also called *full width at half maximum*, FWHM, in nm (Figure 12-4).

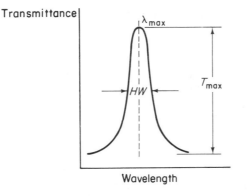

Figure 12-4 Characteristics of a color filter: peak wavelength, λ_{max}; peak transmittance, T_{max}; and halfwidth, HW. The halfwidth is measured at 1/2 of T_{max}.

Transmission-type *interference filters* consist of a spacer layer, for example a thin transparent layer of cryolite, Na_3AlF_6, sandwiched between two reflective coatings. (In older bandpass filters these coatings were metallic; in modern filters they are *dielectric*, that is, nonmetallic; they do not conduct electricity. Dielectric coatings are more efficient.) The coatings and the spacer are vacuum-deposited on a plate of glass, and then another plate is cemented on top for protection (Figure 12-5). Usually, the two reflective coatings have refractive indices higher than that

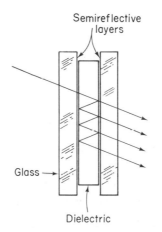

Figure 12-5 Interference filter.

of the spacer between them; thus, there are two π phase changes and, for normal incidence, it is Equation [12-1],

$$\lambda = \frac{2nd}{m} \qquad [12\text{-}10]$$

that determines the wavelength of maximum transmission. Interference filters can be made for any wavelength desired and to specifications much more stringent than those met by other filters. Table 12-1 shows a comparison.

Table 12-1   Typical values for two types of color filters

	Absorption Filter	Interference Filter
T_{max}	30%	90%
HW	50 nm	1 Å
Cost	$10	$250

Example

(a) What is the least thickness required of a layer of cryolite ($n = 1.35$) in an interference filter designed to isolate light of 594 nm wavelength?

(b) How will the peak transmittance change if the filter is tilted by 10°?

Solution: (a) From Equation [12-1] we find that

$$d = \frac{m\lambda}{2n \cos I} = \frac{(1)(594)}{(2)(1.35)(1)} = \boxed{220 \text{ nm}}$$

(b) The angle of tilt is in effect the angle of incidence, I. But we need the angle *inside* the spacer, that is, the angle of refraction, I'. From Snell's law

$$n \sin I = n' \sin I'$$

$$I' = \arcsin \left(\frac{1.00}{1.35} \sin 10° \right) = 7.39°$$

Substituting this figure for I in Equation [12-1] and solving for λ gives

$$\lambda = (2)(1.35)(220 \times 10^{-9})(\cos 7.39°) = 589 \text{ nm}$$

so that

$$\Delta\lambda = 594 - 589 = \boxed{5 \text{ nm}}$$

Note that the transmission of an interference filter, when tilted, always changes toward the *shorter* wavelengths.

Interference filters can also be made for reflection. Such filters are called *dichroic mirrors*; they reflect certain wavelengths and transmit others. Heat reflectors or *hot mirrors* reflect infrared (which on absorption converts into heat) and transmit the visible; *cold mirrors*, used in movie projectors and Klieg lights, transmit infrared and reflect the visible. *Wedge-type interference filters* vary in spacing across the filter. This causes a whole spectrum to be transmitted, from blue at one end of the filter to red at the other. Such filters, combined with a slit for wavelength selection, are a "poor man's monochromator." *Rejection* or *minus filters* eliminate given wavelengths from the spectrum; they are used, for example, in laser safety goggles.

The color by which a filter is named customarily refers to the color of the light that is *transmitted*. A green filter, for example, transmits predominantly green and more or less *blocks* all other colors. Likewise, an UV-filter should mean a filter that *transmits* ultraviolet and blocks other parts of the spectrum. If the filter were to block the ultraviolet, it should be called an UV-rejection or UV-blocking filter.

ANTIREFLECTION COATINGS

Part of the light passing through a boundary is lost due to reflection. Quantitatively, when the two media forming the boundary have refractive indices n_1 and n_2, the *reflectivity*, R, is

$$R = \left(\frac{n_2 - n_1}{n_2 + n_1}\right)^2 \qquad [12\text{-}11]$$

For example, at a boundary between air and glass ($n = 1.5$),

$$R = \left(\frac{1.5 - 1.0}{1.5 + 1.0}\right)^2 = 0.04$$

which means that 4% of the light is reflected. While for a single surface this may not be very much, for a system that contains several lenses the loss is considerable.

A good way of eliminating any reflections is by interference, applying a suitable coating to the surface.* Assume that the light is incident on a glass surface coated with a suitable thin film (Figure 12-6). Both sides of the film will reflect some of the light. But these reflections will cancel if the two reflected waves are out of phase by 180°, and if they have the same amplitude, $A_1 = A_2$.

The phase condition is met if the two contributions are out of phase by 180°. Both of these contributions occur at low-to-high boundaries (air-to-coating first and coating-to-glass second), and both show a π change. Thus, for minimum

---

* This discovery was made by Alexander Smakula (1900–1983), German-born physicist, at that time head of the research department at the Carl Zeiss Optical Company in Jena. Smakula later came to the United States to work at Ft. Belvoir and later at MIT. Fa. Carl Zeiss, *Verfahren zur Erhöhung der Lichtdurchlässigkeit optischer Teile durch Erniedrigung des Brechungsexponenten an den Grenzflächen dieser optischen Teile*, Dtsch. Reichspatent 685 767, 1 Nov. 1935.

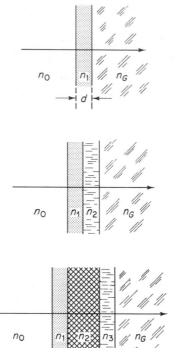

Figure 12-6 Antireflection coatings comprising one layer (*top*), two layers (*center*), and three layers (*bottom*).

reflection to occur, we use Equation [12-2], which, for normal incidence and $m = 1$, becomes

$$2n_{\text{coating}}d = \tfrac{1}{2}\lambda$$

and therefore

$$\boxed{n_{\text{coating}}d = \tfrac{1}{4}\lambda} \qquad\qquad [12\text{-}12]$$

which is the optical thickness of a *quarter-wave coating*.

The amplitude condition, $A_1 = A_2$, is met if

$$\left(\frac{n_1 - n_0}{n_1 + n_0}\right)^2 = \left(\frac{n_g - n_1}{n_g + n_1}\right)^2 \qquad\qquad [12\text{-}13]$$

where n_0 is the outside refractive index, n_1 the index of the coating, and n_g the index of the glass. For $n_0 = 1$,

$$\frac{n_1^2 - 2n_1 + 1}{n_1^2 + 2n_1 + 1} = \frac{n_g^2 - 2n_1 n_g + n_1^2}{n_g^2 + 2n_1 n_g + n_1^2}$$

which reduces to

$$4n_1^3 n_g + 4n_1 n_g = 4n_1^3 + 4n_1 n_g^2$$

Dividing both sides by $4n_1$ and rearranging gives

$$n_1^2 - n_1^2 n_g + n_g^2 - n_g = 0$$

from which

$$\boxed{n_{\text{coating}} \approx \sqrt{n_{\text{glass}}}}$$ [12-14]

With the coating transparent and no light lost to absorption or scattering, the law of conservation of energy tells us that all of the light must go through. Clearly, the coating should also be insoluble and resistant to wear and tear. The best material known today is magnesium fluoride, MgF_2; its refractive index is 1.38.

Multilayer Antireflection Coatings

A single-layer coating works well only for a narrow range of wavelengths. The wavelengths chosen are usually near the center of the visible spectrum, causing part of the blue and red to be reflected, which makes the coating appear purple. A much wider coverage is possible with multiple coatings, called *multilayers*. In a *two-layer antireflection (AR) coating*, for example, each layer is made $\frac{1}{4}\lambda$ thick. This is called a *quarter-quarter coating* [Figure 12-6 (center)]. No reflection will occur if

$$\frac{n_1^2 n_3}{n_2^2} = n_0$$ [12-15]

In many *three-layer AR coatings* the center, or "absentee," layer is made $\frac{1}{2}\lambda$ thick, the other two layers $\frac{1}{4}\lambda$ each. This is called a *quarter-half-quarter coating* [Figure 12-6 (bottom)]. Such coatings are widely used; they are effective over most of the visible spectrum. The first, outside layer is often a $\lambda/4$ coating of magnesium fluoride, the next is $\lambda/2$ zirconium dioxide (ZrO_2, index 2.10), and the layer next to the substrate is $\lambda/4$ cerium trifluoride (CeF_3, index 1.65) or aluminum oxide (Al_2O_3, index 1.76). High-performance interference filters, as well as antireflection coatings, can have up to 200 layers of alternating high- and low-index materials. Some others are of the gradient-index type.

Antireflection coatings for microwaves are of historic as well as current interest. During World War II, the Germans developed AR coatings for the towers of submarines to avoid detection by enemy radar. Such coatings need to have magnetic permeabilities equal in magnitude to their relative dielectric constants. If in addition the coating is absorptive for microwaves, that is, if it is ferromagnetic, the target cannot be detected by radar. But, with a single layer, the same as in light optics, this can be done only for a single radar frequency. However, with a multiple-layer coating an object can be made "invisible" throughout a certain range of frequencies, an approach presumably taken in the design of the *Stealth* aircraft.

SUMMARY OF EQUATIONS

Interference in plane-parallel plates:

$$2nd \cos I = m\lambda \qquad\qquad \text{[12-1]}$$

Newton's rings:

$$r^2 \approx m\lambda R \qquad\qquad \text{[12-3]}$$

Fabry–Perot, free spectral range:

$$\delta\lambda = \frac{\lambda^2}{2d} \qquad\qquad \text{[12-9]}$$

Reflectivity:

$$R = \left(\frac{n_2 - n_1}{n_2 + n_1}\right)^2 \qquad\qquad \text{[12-11]}$$

Antireflection coating,
   thickness of coating:

$$n_{\text{coating}}d = \tfrac{1}{4}\lambda \qquad\qquad \text{[12-12]}$$

   index of coating:

$$n_{\text{coating}} \approx \sqrt{n_{\text{glass}}} \qquad\qquad \text{[12-14]}$$

PROBLEMS

12-1. An oil film, 0.1 μm thick and of index 1.52, rests on a body of water.
   (a) How many π phase changes occur on reflection?
   (b) What is the wavelength reflected by the oil?

12-2. A soap bubble, seen in white light, shows a particularly strong reflection of first-order red (630 nm). If the light is incident normally and the soapy water has a refractive index of 1.4, how thick is the wall of the bubble?

12-3. How many π phase changes occur when light is reflected by, or transmitted through:
   (a) A thin film (in air)?
   (b) An air film (between two plates of glass)?

12-4. A soap bubble ($n = 1.4$) has a wall 0.36 μm thick. If seen at normal incidence in reflected white light, what color is it?

12-5. Two glass plates are in contact at one end and separated by a hair at the other end. If the wedge-like space between the plates is filled with water ($n = \tfrac{4}{3}$) and if with light of 546 nm wavelength 98 fringes are counted across the length of the plates, how thick is the hair?

12-6. A wedge-shaped space between plane plates is filled with water ($n = \frac{4}{3}$) in such a way that a few bubbles of air are trapped between the plates. If 18 fringes are counted within a given distance inside an air bubble, how many fringes, within the same distance, are seen in the water?

12-7. In an experiment on Newton's rings, the diameter of the tenth dark ring formed by yellow sodium light (589 nm) and seen in reflection is 3.6 mm. What is the radius of curvature of the lens surface?

12-8. The diameter of the fourth bright Newton ring is 10 mm. When an unknown liquid is poured into the gap between lens and support, the diameter of this ring shrinks to 8.45 mm. Calculate the liquid's index.

12-9. If used at 600 nm, how deep must a Fabry–Perot etalon be to provide a free spectral range of 0.1 nm (1 Ångström, 1 Å)?

12-10. Plot the transmittance as a function of wavelength of:
   (a) A *hot mirror* of the dichroic type.
   (b) A *cold mirror* of the same type.

12-11. What percentage of the incident light is reflected at the surface of a material of index $n = \frac{5}{3}$?

12-12. The antireflection coating on a lens is made of magnesium fluoride ($n = 1.38$). How thick a coating is needed to produce minimum reflection at 552 nm?

12-13. Light strikes the plane surface of a block of glass ($n = 1.53$) covered with water ($n = \frac{4}{3}$). How much light is reflected at the water-glass boundary?

12-14. When light strikes a water surface ($n = \frac{4}{3}$) covered with a film of oil ($n = 1.65$), how much light is lost due to reflection on both sides of the oil?

12-15. If the first layer of a two-layer antireflection coating is magnesium fluoride ($n = 1.38$) and the substrate is ophthalmic crown ($n = 1.523$), what index should the second layer have?

SUGGESTIONS FOR FURTHER READING

MacLeod, H. A. *Thin-Film Optical Filters*. New York: Macmillan Publishing Company, 1986.

Pulker, H. K. *Coatings on Glass*. Amsterdam: Elsevier, 1984.

Vaughan, J. M. *The Fabry–Perot Interferometer: History, Theory, Practice, and Applications*. New York: American Institute of Physics, 1989.

Coherence

Ordinary light is disorganized, not capable of producing interference; such light is called *incoherent*. By contrast, the light emitted by a laser is highly organized, easily producing interference; such light is called *coherent*. Sometimes, light is thought to be *either* coherent *or* incoherent. Coherence and incoherence, though, are idealized states and, actually, no light is completely coherent, and none is completely incoherent.

SPATIAL COHERENCE

Although the distinction is sometimes blurred, we distinguish two classes of coherence, *spatial coherence* and *temporal coherence*. Spatial coherence refers to the phase relationship between waves traveling side by side, at the same time. Temporal coherence refers to the constancy, and predictability, of phase as a function of time, in essence the same as *monochromaticity*.

Consider first the case of *transverse spatial coherence*. Let the light come from an extended source (placed to the left in Figure 13-1), pass through a double slit (center), and then reach a screen (right). On the screen we see a series of interference fringes; they consist of maxima and minima that are equidistant as well as parallel to each other and to the slits.

The light, emerging from an extended source, comes from an assembly of groups of atoms that all emit independently from one another. Each group pro-

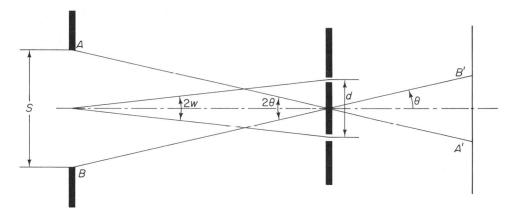

Figure 13-1 Young's double-slit experiment and spatial coherence. Vertical dimensions exaggerated.

duces maxima whose angular distance θ from the axis is given by the double-slit equation,

$$d \sin \theta = m\lambda \qquad [13\text{-}1]$$

If the angles are small (so that $\sin \theta \approx \theta$, in radians), consecutive maxima subtend angular distances

$$\theta = \frac{\lambda}{d}$$

Now consider a slit S that limits the source. As seen from the double slit, the limiting edges of the slit, A and B, subtend an angle 2θ. The double slit, in turn, subtends an angle $2w$ at the slit S. Individually, both A and B produce double-slit interference but the two patterns, as they arrive at the screen, are shifted by 2θ: the center of one pattern lies at A', the other at B'.

The two patterns will cancel whenever a maximum of one falls on a minimum of the other. Since in double-slit interference the minima are halfway between the maxima, that happens when

$$2\theta = \frac{1}{2}\frac{\lambda}{d} \qquad [13\text{-}2]$$

In that case, no fringes are seen on the screen.

As slit S is gradually opened further, and points A and B move apart from each other, patterns A' and B' will move apart too; maxima of one pattern fall on the *maxima* of the other, and the fringes will reappear. That happens when

$$2\theta = \frac{\lambda}{d}$$

As *A* and *B* are moved still farther away from each other, the fringes disappear again, then reappear, and so on, through several cycles.

From the construction in Figure 13-2 it follows that the path difference, Γ, between the two contributions is

$$\Gamma = S \sin w \qquad \text{[13-3]}$$

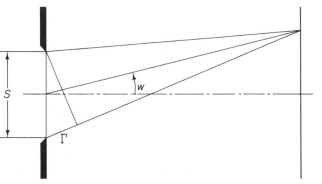

Figure 13-2 Deriving the coherence condition.

Fringes of high contrast will result if the path difference is small; in fact, the path difference becomes negligible only if it is less than one-half of a wavelength:

$$\Gamma \ll \tfrac{1}{2}\lambda \qquad \text{[13-4]}$$

Combining Equations [13-3] and [13-4] gives

$$S \sin w \ll \tfrac{1}{2}\lambda \qquad \text{[13-5]}$$

which is the *coherence condition*. It determines the diameter *S* of a source that, within an angle 2*w*, is sufficiently *spatially coherent* to produce fringes of satisfactory contrast.

TEMPORAL COHERENCE

Temporal coherence is essentially the same as *longitudinal* spatial coherence. Some electromagnetic radiation such as microwaves and radiowaves, and also sound waves, water waves, and other mechanical waves, can be generated as an all but infinite number of waves, one wave after another, but light waves cannot. Light waves come in *wavetrains*. The wavetrains are of finite length, each train containing only a limited number of waves (Figure 13-3). The length of a wavetrain, Δs, is called the *coherence length*. It is the product of the number of waves, *N*, contained in the train and of their wavelength, λ:

$$\Delta s = N\lambda \qquad \text{[13-6]}$$

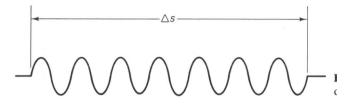

Figure 13-3 Schematic representation of a wavetrain.

Since velocity is the distance traveled per unit of time, it takes a wavetrain of length Δs a certain length of time, Δt, to pass a given point, and therefore

$$\Delta t = \frac{\Delta s}{c} \qquad [13\text{-}7]$$

where c is the velocity of light. The length of time Δt is called the *coherence time*.

To measure the temporal coherence, we could use Young's double slits. We only need to *delay* one of the beams, that is, place a sheet of transparent material over one of the slits. If at a given thickness (of the material) no interference fringes are seen, the path difference introduced by the material exceeds the temporal coherence.

Instead, it is more convenient to use a Michelson interferometer. To obtain high contrast, the two arms of the interferometer are made as equal in length as possible. That is necessary in particular if the wavetrains are fairly short; in that case, the contrast is inversely proportional to the path difference.

Some wavetrains, however, are rather long. Light from the green line of mercury, for example, has a coherence length of about a millimeter. Light from the orange krypton line has wavetrains about 80 cm long, and light from a laser can have a coherence length of many kilometers. Consequently, with the highly coherent light from a laser we can observe fringes even if one arm of the interferometer is considerably longer than the other arm (as in Figure 13-4).*

PARTIAL COHERENCE

So far, we have always assumed that two wavetrains, each of finite length Δs, overlap to their full extent. Such complete overlap will produce fringes of the highest contrast. But even if the wavetrains overlap only in part, as in Figure 13-5, interference is possible; it is only that the degree of contrast becomes less. The

---

* R. B. Herrick and J. R. Meyer-Arendt, "Interferometry through the Turbulent Atmosphere at an Optical Path Difference of 354 m," *Appl. Opt.* **5** (1966), 981–83. Three years later, an even longer path difference was reached: V. V. Pokasov and S. S. Khmelevtsov, "Interferometry with a Path Difference of up to 500 m through the Turbulent Atmosphere" (in Russian); *Izv. Vyssh. Zaved. Fiz.* (*Tomsk U.*) **8** (1969), 139–41.

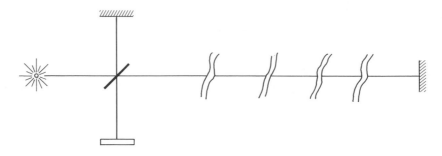

Figure 13-4 Unequal-arm Michelson interferometer.

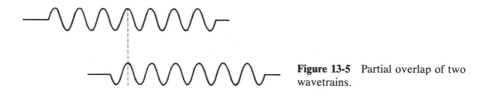

Figure 13-5 Partial overlap of two wavetrains.

question, then, is not how much the wavetrains must overlap to produce interference; the question is how much contrast we *need* to see any fringes.

Contrast, as we have shown earlier, can be defined as

$$\gamma = \frac{I_{max} - I_{min}}{I_{max} + I_{min}} \times 100\% \qquad [13\text{-}8]$$

but how does contrast relate to coherence? Assume that two points on a screen are illuminated by two bundles of light that fall with equal power, or better: with equal *illuminance*, E_0, on the screen. Each of these bundles consists of two parts, A and B. Parts A may be "completely coherent" and cause an illuminance

$$E_A = CE_0 \qquad [13\text{-}9]$$

Parts B may be "completely incoherent," both to themselves and with respect to A, and cause an illuminance

$$E_B = (1 - C)E_0 \qquad [13\text{-}10]$$

The quantity C is called the *degree of coherence*.*

If interference occurs, it is because of parts A. These parts form fringes whose maxima, from Equation [11-19], have an illuminance four times as high as the individual contributions. The maximum illuminance, thus, is $4CE_0$. The minimum illuminance, if both bundles are equal, is zero. On this pattern a uniform

---

* Following P. H. van Cittert, "Degree of Coherence," *Physica* **24** (1958), 505–507.

distribution is superimposed that, because it comes from two sources, has an illuminance twice that of Equation [13-10],

$$E_B = 2(1 - C)E_0$$

As a result, the illuminance in the maxima is

$$E_{max} = 4CE_0 + 2(1 - C)E_0 = 2(1 + C)E_0 \qquad [13\text{-}11]$$

and in the minima it is

$$E_{min} = 2(1 - C)E_0 \qquad [13\text{-}12]$$

If Equations [13-11] and [13-12] are substituted in [13-8], we find that

$$\gamma = \frac{2(1 + C)E_0 - 2(1 - C)E_0}{2(1 + C)E_0 + 2(1 - C)E_0} = \frac{4CE_0}{4E_0} = C \qquad [13\text{-}13]$$

which shows that the *degree of contrast* of the fringes produced by interference of two waves *is equal to the degree of coherence* between these two waves.

The highest contrast will result when, following Equation [13-8], the illuminance in the minima is zero. Both the contrast, and the degree of coherence, will then be *unity*. Although conceivable in theory, this figure cannot be attained in practice because scattering and diffraction prevent the minima from receiving no light at all. *Complete* coherence, in short, is merely a theoretical limit.

But why can't we have complete incoherence? Because diffraction will cause each image *point* to spread out into a more diffuse image *patch*. The patches from image point to image point are incoherent from patch to patch but, since each patch comes from one group of atoms, are coherent *within* each patch. Since the patches overlap, the image will have a certain degree of coherence and *complete* incoherence is not possible either.

APPLICATIONS

Stellar Interferometry

Let us consider some practical applications. Ordinarily, the separation d of the two slits in Young's experiment is a few millimeters at most. We could move the slits farther apart by placing a converging lens next to them. That lens would deflect the light passing through toward the axis, but now the wavefronts in the two contributions subtend a larger angle and the fringes become very closely spaced, requiring a magnifying glass to see them. The lens and the magnifier together then form a *telescope*. Indeed, placing two slits in front of a telescope is a method well known in astronomy for determining the angular separation of binaries (double stars) or the diameter of fixed stars, angular dimensions too small to measure by direct (noninterferometric) observation.

An extension of this principle is found in *Michelson's stellar interferometer.** Here the two slits are separated even farther, exceeding the diameter of the telescope's aperture. This is done by four mirrors mounted in front of the objective as shown in Figure 13-6. The two inner mirrors are fixed, but the two outer mirrors can be moved apart up to several meters; these two mirrors take the place of the double slits.

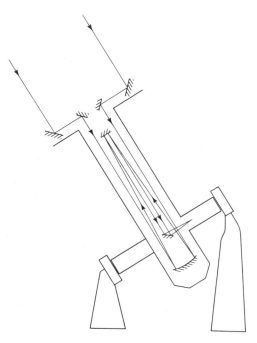

Figure 13-6 Michelson's stellar interferometer.

Interference is not seriously affected by atmospheric turbulence. The reason is that the two bundles that produce interference are small in cross section compared to the air cells causing turbulence, and the resulting fringes, though in motion, remain distinct. A conventional image, on the other hand, is produced by light integrated over the telescope's whole aperture, and in turbulent weather may be so blurred as to be worthless.

---

* A. A. Michelson and F. G. Pease, "Measurement of the Diameter of α Orionis with the Interferometer," *Astrophys. J.* **53** (1921), 249–59. The fringes seen disappeared when the outer mirrors were 307 cm apart. Using Rayleigh's criterion (see Chapter 14) and assuming a wavelength of 575 nm, that gives an angular diameter

$$\theta = 1.22 \frac{\lambda}{d} = 1.22 \frac{575 \times 10^{-9}}{3.07} = 2.29 \times 10^{-7} \text{ rad} = 0.000013°$$

and a linear diameter of 3.8×10^8 km, more than the diameter of the orbit of Earth (3×10^8 km).

Lab Experiment. Take two fiber bundles and let the light emerging from them be reflected at a small metal ball. These reflections represent our double star. First we determine, by interferometry, the *distance between the two stars*. At a distance $L = 1.5$ m from the stars, the light passes through a double slit. Individually, either of the reflections, using both slits, produces double-slit interference. We vary the separation of the two slits, d, such that the fringes disappear. Following Equation [13-2] that happens when

$$2\theta = \frac{1}{2}\frac{\lambda}{d}$$

Assume that the wavelength is $\lambda = 600$ nm and the fringes disappear when $d = 2.5$ mm. Then the angular distance between the two reflections is 2θ and the linear distance between them is

$$D = 2\theta L - \frac{1}{2}\frac{\lambda L}{d} = \frac{1}{2}\frac{(600 \times 10^{-9})(1.5)}{0.0025} = \boxed{0.18 \text{ mm}}$$

Next we determine the *width of a single source*. The source has in front of it a slit of variable width. With the slit very narrow, the light passing through, and then passing through a double slit, will produce typical interference. But as the slit is gradually opened wider, the fringes disappear.

Assume that the dimensions are the same as before. Then, to find the width of the slit, S, we take the upper bound of the coherence condition,

$$S \sin w = \tfrac{1}{2}\lambda \qquad\qquad [13\text{-}14]$$

solve for S and, from Figure 13-1, substitute

$$\sin w = \frac{1}{2}\frac{d}{L}$$

That gives

$$S = \frac{\lambda L}{d} = \frac{(600 \times 10^{-9})(1.5)}{0.0025} = \boxed{0.36 \text{ mm}}$$

twice the separation of the double star. (The factor 1.22, as it occurs in Michelson's calculations, applies only to a circular source, not to a slit as we use it here.)

Retinal Acuity

Both interference and coherence play an important role in some acuity measurements. That is especially true in cases of *cataract*, where, because of the turbidity of the crystalline lens, conventional acuity measurements are not possible. Indeed, even with a cataractous lens blocking the way, fringes can be seen, as shown schematically in Figure 13-7.

But, how can light go through a diffusing medium, and still form distinct fringes? Surely, no distinct *image* could form this way. The answer is that the lines seen by the patient do not pass as ready fringes through the eye; they are *formed on the retina*. Moreover, refractive errors, such as myopia, have no effect

Figure 13-7 Fringes of different spatial frequency and orientation as seen by the patient.

on the sharpness of the fringes, because the fringes are not *focused* on the retina; they fill all space in front of, and behind, the retina.

In practice, a beam of light is split into two and projected into the pupil of the eye. The two beams pass through the lens separately, side by side, but despite the turbidity of the lens they retain most of their spatial coherence and still can form fringes that are (nearly) as distinct as without the cataract.

Intensity Interferometry

In most any interferometer we add *amplitudes*. In the intensity interferometer we add *intensities*. That is of interest primarily in astronomy, and applies to both visible light and radio frequencies. In the light-optical version it made it possible to measure, for the first time, the angular diameter of Sirius. Two searchlight mirrors, each 1.56 m in diameter and with a separation that could be increased up to 14 m, were used to form two images. As shown in Figure 13-8, the images are received by two photomultipliers, amplified, and multiplied with each other. Ordinarily, the amount of current produced by a photodetector is proportional to the

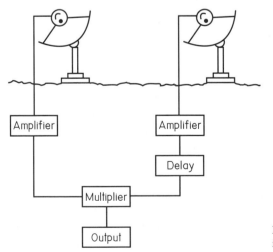

Figure 13-8 Principle of the intensity interferometer.

amount of light received. But if the steady (dc) component of the current is filtered out, the remaining fluctuations are found to be correlated depending on the degree of (spatial) coherence. By gradually increasing the mirrors' separation, the angular diameter of Sirius could be determined as 0.0068 arc sec. But even smaller stars, down to 0.0005 arc sec, could be measured, about 1/100 of what can be done with Michelson's stellar interferometer.

In the radio application, the largest interferometer in operation today is a series of 10 radio antennas extending 8000 km across the United States, providing a resolution comparable to that of a single antenna nearly as big as the diameter of Earth.

SUMMARY OF EQUATIONS

Coherence condition:

$$S \sin w \ll \tfrac{1}{2}\lambda \qquad\qquad [13\text{-}5]$$

Coherence length:

$$\Delta s = N\lambda \qquad\qquad [13\text{-}6]$$

Coherence time:

$$\Delta t = \frac{\Delta s}{c} \qquad\qquad [13\text{-}7]$$

PROBLEMS

13-1. Red light of 600 nm wavelength is used to produce double-slit interference. If the double slits are separated by 1.5 mm, and if they are 2 m from the source, how narrow a slit should be placed in front of the source to obtain fringes of satisfactory contrast?

13-2. The angular size of the sun, as seen from Earth, is approximately 30 arc min. Considering light of 550 nm coming from opposite points on the circumference of the sun, what should be the separation of two slits whose interference patterns just *cancel*?

13-3. If light of $\lambda = 660$ nm has wavetrains 20λ long, what is its:
   (a) Coherence length?
   (b) Coherence time?

13-4. Determine the number of waves per wavetrain in light from:
   (a) The green mercury line (546 nm).
   (b) The orange krypton line (606 nm).
   (c) A helium-neon laser (633 nm), assuming that its coherence length is 20 km.

13-5. If the contrast in an interference pattern is 50%, and if the maxima receive 15 units of light, how much do the minima receive?

13-6. Two wavetrains overlap to 29% of their length. If the maxima in the resulting interference pattern receive 20 units of light, how much do the minima receive?

13-7. If we put a red filter (with a peak transmission at 600 nm) over a slit that is 1 mm wide and mounted in front of a light source, how close to the source could we place a double slit (with a slit separation of 1.2 mm) to observe fringes of satisfactory contrast?

13-8. Continue with Problem 13-7 and assume that instead of a red filter we now have a *blue* filter (480 nm).

13-9. Atomic hydrogen, with its emission at 21 cm wavelength, is the subject of much of *radio astronomy*. What should be the least distance between two dish antennas, for example, if we wish to measure a radio star with an angular diameter believed to be 10 arc min?

13-10. When determining by interference the acuity of a patient who is highly *myopic*, the fringes will not be recorded at the same distance from the eye's pupil as if the patient had normal vision. How will that affect the result?

SUGGESTIONS FOR FURTHER READING

GOODMAN, J. W. *Statistical Optics*. New York: Wiley-Interscience, 1985.

HANBURY BROWN, R. *The Intensity Interferometer, Its Application to Astronomy*. New York: John Wiley and Sons, Inc., 1974.

MARATHAY, A. S. *Elements of Optical Coherence Theory*. New York: John Wiley and Sons, Inc., 1982.

14

Diffraction

When a point source is casting a shadow on a screen, we expect the shadow to be well defined. But careful inspection shows that it is not. The edges of the shadow are blurred: a certain amount of the light is deflected into the shadow, and another part of the light is deflected out of it forming brighter and darker fringes (Figure 14-1). Such departure from the predictions of geometrical optics is called *diffraction*.

HUYGENS' PRINCIPLE

The concept of diffraction is not restricted to light. It also occurs with other waves such as sound waves, X rays, radio waves, and even with water waves, whenever part of the wave's path is blocked by an obstacle. But how can such waves simply deviate from their initial direction and bend around an obstacle? That can be explained by *Huygens' principle.**

---

* Christiaan Huygens (1629–1695), Dutch mathematician, physicist, and astronomer. Huygens made noteworthy contributions to many fields, invented the pendulum clock, formulated the laws governing conservation of momentum, centrifugal force, and the moment of inertia. He learned to grind lenses, as many of his contemporaries did, built telescopes of superior quality, found that the "arms" of Saturn were actually a ring, discovered Saturn's sixth moon, Titan, and built a planetarium that still stands today in Leiden. In 1678 he presented the concept of what is now known as *Huygens' principle* to the French Academy of Sciences: C. Huygens, *Traité de la lumière* (Leiden: Van der Aa, 1690).

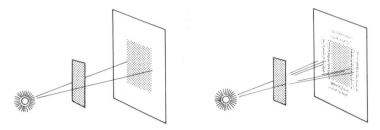

Figure 14-1 Shadow in geometrical optics (*left*) and, more correctly, in wave optics (*right*).

Let a wave be represented by a series of *wavefronts*. A wavefront is a hypothetical surface connecting points of equal phase. On the wavefront next to the edge of the obstacle, for example, I have drawn several points (Figure 14-2). Each of these points is the origin of a secondary wave, a "wavelet." Then an envelope is drawn over these wavelets: that envelope is the next wavefront. Points on this next wavefront are the source of still other wavelets (not shown), these form another wavefront, and so, in this repetitive manner, the wave moves on to the right.

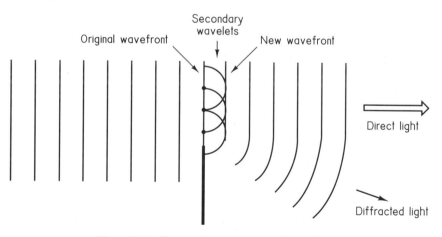

Figure 14-2 Huygens' construction of wavefronts.

Some distance away from the edge, the wavelets have circular symmetry and the wave goes on straight ahead. But close to the edge, the wavelet that originates next to it finds no counterpart coming from below; hence, the wavefront normal will turn (clockwise) and the wave enter a region that by geometrical construction would be in the shadow.

Depending on the distances involved, we distinguish two types of diffraction, *Fraunhofer diffraction* and *Fresnel diffraction*. Fraunhofer diffraction occurs when the source and the screen are far apart and the light is essentially parallel; it is a matter of *far-field diffraction*. Fresnel diffraction occurs when the source or the screen, or both of them, are close-by; it is a matter of *near-field diffraction*. Fresnel diffraction is more general; it includes Fraunhofer diffraction as a special case. But Fraunhofer diffraction is so much easier to discuss that it is customarily presented first.

FRAUNHOFER DIFFRACTION

We begin with Fraunhofer diffraction on a *single slit*.* The slit may have a width s. We divide the slit into several narrow strips of equal width Δs. Light passing through each of these strips has an amplitude A_i, and each can be represented by a short phasor (Figure 14-3). With the light parallel and in phase, these phasors have the same length and direction.

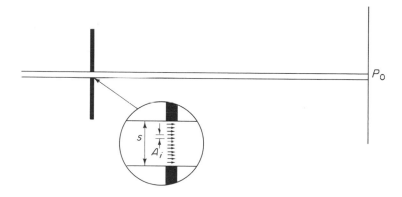

Figure 14-3 Central maximum in Fraunhofer diffraction. Insert shows phasors inside aperture.

At point P_0, the elemental waves arrive in phase and the resultant amplitude, **A**, has its *maximum*, the arithmetic sum of the phasors (Figure 14-4):

$$\mathbf{A} = \sum \mathbf{A}_i \qquad [14\text{-}1]$$

---

* Joseph Fraunhofer (1787–1826), German. After working for a while as a lens grinder and apprentice optician, Fraunhofer became a partner in an optical company that made precision theodolites, was a professor at the University of Munich, and was knighted by King Maximilian of Bavaria. In his short life (he died of tuberculosis at age 39), Fraunhofer produced large-aperture telescope lenses, exceptionally well corrected for spherical and chromatic aberration, ruled precision gratings and discovered their use for spectroscopy, and found that the spectrum of the sun is crossed by dark lines since named *Fraunhofer lines*. His theory of diffraction appeared first in J. Fraunhofer, "Kurzer Bericht von den Resultaten neuerer Versuche über die Gesetze des Lichtes, und die Theorie derselben," *Ann. Phys.* (3) **14** (1823), 377–78.

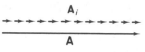

Figure 14-4   Vector representation of amplitudes in the zeroth-order maximum.

At other points P on the screen, above and below the zeroth order point P_0, however, the individual phasors will subtend constant angles with one another, due to the phase angle differences between them (Figure 14-5).

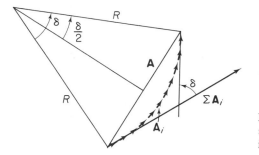

Figure 14-5   Vector construction shows *vibration curve*, the sum of the individual phasors, $\Sigma\mathbf{A}_i$.

We call δ the total phase angle difference, that is, the angle subtended by the extensions of the first and the last phasor. Then

$$\sin \tfrac{1}{2}\delta = \frac{\mathbf{A}/2}{R}$$

$$\mathbf{A} = 2R \sin \tfrac{1}{2}\delta$$

and

$$\delta = \frac{\sum \mathbf{A}_i}{R}$$

Solving the last equation for R and inserting it in the preceding equation gives

$$\mathbf{A} = 2 \frac{\sum \mathbf{A}_i}{\delta} \sin \frac{\delta}{2} = \sum \mathbf{A}_i \frac{\sin \delta/2}{\delta/2} \qquad \text{[14-2]}$$

Consequently, since intensity is proportional to the square of the amplitude, the intensity at any arbitrary point P_θ, at an angular distance θ from the axis, is

$$\boxed{I_\theta = I_0 \left(\frac{\sin \delta/2}{\delta/2}\right)^2} \qquad \text{[14-3]}$$

where I_0 is the intensity in the zeroth-order maximum. This maximum is the central peak in Figure 14-6.

The intensities in the maxima can be calculated to a good approximation by determining $(\sin^2\delta/2)/(\delta/2)$ at the halfway positions, that is, where

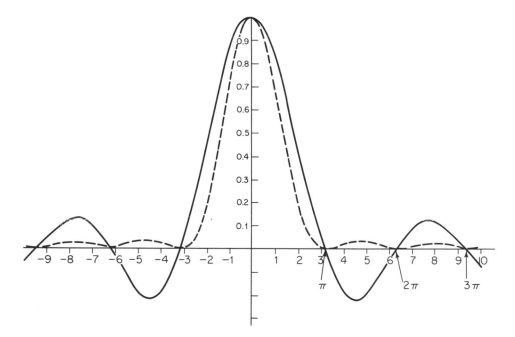

Figure 14-6 Amplitude distribution (*solid line*) and intensity distribution (*dashed line*) in Fraunhofer diffraction by a single slit.

$$\frac{\delta}{2} = \frac{3\pi}{2}, \frac{5\pi}{2}, \frac{7\pi}{2}, \cdot \cdot \cdot$$

[14-4]

This gives for the first maximum $4/(9\pi^2)$ or approximately 4.5%, for the second maximum $4/(25\pi^2)$ or 1.6%, for the third maximum $4/(49\pi^2)$ or 0.8%, and so on, of the intensity in the zeroth-order maximum.

Next consider the *minima*. Assume, as shown in Figure 14-7, that the light passing through the slit contains three rays, 1-2-3. For a minimum to occur, the light must subtend with the axis an angle θ, of a size such that rays 1 and 2 are $\lambda/2$ out of phase, canceling each other.

The first minimum, therefore, occurs when the optical path difference, Γ, between rays 1 and 2 is

$$\Gamma_{12} = \tfrac{1}{2}\lambda$$

or, between rays 1 and 3,

$$\Gamma_{13} = \lambda$$

From the construction it follows that

$$\sin \theta = \frac{\Gamma_{13}}{s}$$

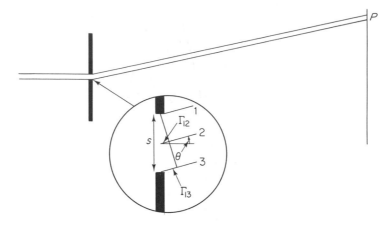

Figure 14-7 Fraunhofer diffraction, first minimum. Insert shows path differences between rays.

Combining both equations gives

$$s \sin \theta = \lambda$$

But there will be other minima (whenever Γ_{13} is an integral multiple of λ); thus

$$\boxed{s \sin \theta = m\lambda} \qquad m = 1, 2, 3, \ldots \qquad [14\text{-}5]$$

which is the equation for minima in Fraunhofer diffraction on a single slit.

The minima, in short, are equidistant, at least for small angles. The maxima, however, do not fall exactly between the minima; they are slightly displaced toward the center. The first maximum, for instance, is located at $y = 1.4303\pi$, the second maximum at 2.4590π, the third at 3.4707π, and so on, with the higher orders even more nearly halfway between the minima.

Keeping the right-hand side of Equation [14-5] constant, we note that, if the slit is made narrower, the angle θ becomes larger and the light spreads out wider. We also note that, as the wavelength increases, θ increases too; *red light is diffracted more than blue light*, the opposite of what occurs in *re*fraction.

If the opening is *rectangular*, rather than a narrow slit, the light is diffracted in two directions, orthogonal to each other. An example is shown in Figure 14-8.

Circular Aperture

Fraunhofer diffraction by a circular aperture is of great practical interest, simply because most lenses and stops are round. The result is again a series of maxima

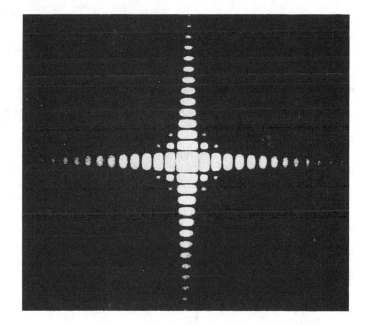

Figure 14-8 Diffraction pattern produced by a rectangular aperture.

and minima, but behind a circular aperture these maxima and minima take the form of concentric rings. The bright central maximum is known as the *Airy disk*.*

   The mathematical analysis of diffraction behind a circular aperture is considerably more complex than diffraction behind a slit. Whereas the positions of the minima behind a slit were given by the simple relationship $s \sin \theta = m\lambda$, we now need to replace m by another factor, J, that derives from *first-order Bessel functions*. These functions also oscillate between maxima and minima but they de-

Table 14-1 Positions of minima in Fraunhofer diffraction

Minimum	Single Slit, m	Circular Aperture, J
First-order	1	1.220
Second-order	2	2.233
Third-order	3	3.238

   * Sir George Biddell Airy (1801–1892), British mathematician, Astronomer Royal, and director of the Greenwich Observatory. Airy is perhaps best known for the disk just mentioned. He described it in G. B. Airy, ''On the Diffraction of an Object-glass with Circular Aperture,'' *Trans. Cambridge Philos. Soc.* **5** (1835), 283–91.

crease in amplitude with increasing distance from the central axis. For the first maximum behind a circular aperture, for example, we have $J = 1.635$, for the second maximum $J = 2.679$, and for the first three minima we use the values listed in Table 14-1.

Example 1

Diffraction behind a circular *aperture* is nearly the same as diffraction behind a circular *obstacle*. The reason is that diffraction is an edge effect. Even better to see is diffraction at *multiple* apertures or obstacles. For example, looking at the sun through the hazy atmosphere, with its many droplets or ice crystals suspended therein, we see colored rings around the sun, red on the outside and blue on the inside, as shown in Figure 14-9 (left). That is called a *corona*, in contrast to a *halo*, which is due to refraction (right). If the observer is standing so that the sun is blocked by the corner of a house, the rings, if caused by turbidity in the atmosphere, will persist or, if caused within the eye, will disappear.

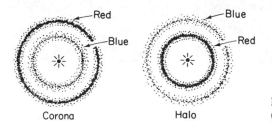

Corona   Halo

Figure 14-9 Corona (*left*) and halo (*right*).

Example 2

If we have an assemblage of particles of uniform size (a *monodisperse* system), the resulting diffraction pattern is fairly distinct. With a mixture of particles of different sizes (a *polydisperse* system), the pattern is more diffuse (Figure 14-10).

We can then construct an *annular mask*, a ring-shaped slot in an opaque screen to match the size of a certain maximum. This will permit only light to pass through that is diffracted by particles of a preselected size. Light diffracted by particles of another size will not pass through. A mask of this kind will discriminate between particles of different sizes.

Rayleigh's Criterion

Diffraction at a circular aperture sets the limit of resolution for virtually any optical system. Consider a telescope that is aimed at two stars next to each other that have about equal magnitude (equal "brightness"). Only when the diffraction patterns of the two stars, seen in the focal plane of a lens or telescope, are separate will the stars appear separate. When the central maxima fuse, the two stars appear as one. When the central maximum of one star coincides with the first minimum of the other, resolution is marginal, a condition called *Rayleigh's criterion* (Figure 14-11).

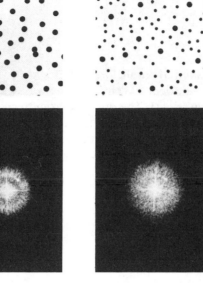

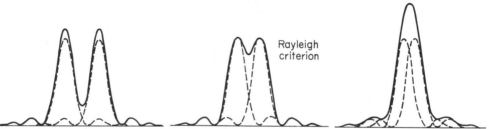

Figure 14-10 Diffraction pattern obtained from a monodisperse system of many dots of equal size (*left*) and from a polydisperse system containing a mixture of large and small dots (*right*).

Rayleigh criterion

Figure 14-11 Rayleigh's criterion. Note the separation of the maxima in the left-hand plot and the close overlap on the right.

Rayleigh's criterion has been criticized for several reasons. Some people claim that it is overgenerous; under favorable conditions details can be resolved that are even smaller in size. Others believe in Rayleigh's criterion as if it were written in tablets of gold by the angels. Rayleigh himself considered it to be no more than an approximation.

From Table 14-1 we see that, if a lens has a diameter D and the light a wavelength λ, the *minimum angle of resolution*, in radians, is

$$\theta_{\min} \approx 1.22 \frac{\lambda}{D} \qquad [14\text{-}6]$$

The lens, in other words, even if it could be fully corrected for all aberrations, is still *diffraction-limited*.

The same condition as with the stars and a telescope applies to a microscope. To a first approximation, the resolution of a microscope is about equal to the wavelength of the light used. That shows why radiation of shorter wavelength such as UV, X rays, and even electrons gives better and higher resolution.

FRESNEL DIFFRACTION

Fresnel diffraction* is not limited to parallel light. In Figure 14-12, for example, a slit source on the left emits cylindrical wavefronts. The wavefronts pass through a slit (center) and reach a screen (right). The distance from the source to the center of the slit is *a* and the distance from there to the screen is *b*. As before in Fraunhofer diffraction, we divide the wavefront (inside the slit) into a series of narrow strips, each of them parallel to the slit.

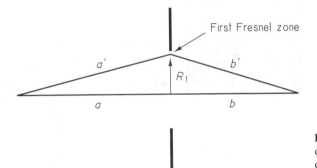

First Fresnel zone

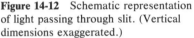

Figure 14-12   Schematic representation of light passing through slit. (Vertical dimensions exaggerated.)

We choose the width of the strips such that they increase from strip to strip by one-half of a wavelength, $\frac{1}{2}\lambda$. Ordinarily, the distance along the axis, from the source to the screen, is $a + b$. Measured through the next higher wavefront element (inside the slit), however, the distance is

$$a' + b' = a + b + \tfrac{1}{2}\lambda \qquad\qquad [14\text{-}7]$$

Through the next higher element the distance is $a + b + 2(\frac{1}{2}\lambda)$, through the third element $a + b + 3(\frac{1}{2}\lambda)$, and so on. The wavefront elements defined this way are known as *Fresnel half-period zones.*

Now we consider the disturbance, dy, caused by a wavefront element dW and acting on the *midpoint* on the screen. From our earlier Equation [11-8], this

---

* Augustin Jean Fresnel (1788–1827), French physicist. Fresnel studied mathematics, then civil engineering, at the French École des Ponts et Chaussées (School of Bridges and Roads); he went into optics later. In his dissertation, "Sur la Diffraction de la lumière, où l'on examine particulièrement le phénomène des franges colorées que présentent les ombres des corps éclairés par un point lumineux," *Ann. Chim. Phys.* (2) **1** (1816), 239–81, Fresnel presented the first rigorous treatment of diffraction. He also worked on more mundane problems and invented flat, weight-saving lenses with circular prismatic grooves, called *Fresnel lenses*; they are used in lighthouses, automobile headlights, and overhead projectors.

disturbance is

$$dy = \mathbf{A}\,\sin(2\pi\nu t)\,dW \qquad [14\text{-}8]$$

For other elements, farther away from the center (inside the slit), the distance to the midpoint on the screen is of course longer. Likewise, for points on the screen farther away from the midpoint the distances also become longer. These longer distances go hand in hand with greater phase differences between the individual wave contributions. Adding all these contributions, to find the total disturbance at any one point on the screen, requires integration. Details on how to do this, I feel, go beyond an *Introduction to Optics*.

At least, I list the result. It takes the form of two integrals, known as *Fresnel's integrals*,

$$x = \int \cos(\tfrac{1}{2}\pi v^2)\,dv$$
$$\qquad\qquad\qquad\qquad\qquad [14\text{-}9]$$
$$y = \int \sin(\tfrac{1}{2}\pi v^2)\,dv$$

The term v is a variable. It is the length of the vibration curve. Each short length on the curve, dv, represents the amplitude vector (phasor) of an individual wavefront element. Such a phasor subtends an angle δ with the x-axis which, from the integrals, is

$$\delta = \tfrac{1}{2}\pi v^2 \qquad [14\text{-}10]$$

As shown by Figure 14-13, the direction of dv is given by the tangent the angle makes with the $+x$-axis,

$$\tan \delta = \frac{y}{x}$$

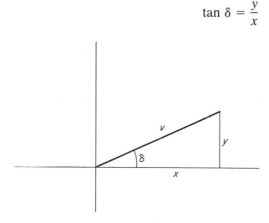

Figure 14-13 Phasor and the angle it subtends with the *x*-axis.

Cornu Spiral

We solve Fresnel's integrals between a lower limit, $v_1 = 0$, and an upper limit, $v_2 = \infty$. The results, x and y, assume the values listed in Table 14-2. If we plot x

Table 14-2 Fresnel integrals

v	x	y	v	x	y
0.00	0.0000	0.0000	2.20	0.6363	0.4557
0.10	0.1000	0.0005	2.30	0.6266	0.5531
0.20	0.1999	0.0042	2.40	0.5550	0.6197
0.30	0.2994	0.0141	2.50	0.4574	0.6192
0.40	0.3975	0.0334	2.60	0.3890	0.5500
0.50	0.4923	0.0647	2.70	0.3925	0.4529
0.60	0.5811	0.1105	2.80	0.4675	0.3915
0.70	0.6597	0.1721	2.90	0.5624	0.4101
0.80	0.7230	0.2493	3.00	0.6058	0.4963
0.90	0.7648	0.3398	3.10	0.5616	0.5818
1.00	0.7799	0.4383	3.20	0.4664	0.5933
1.10	0.7638	0.5365	3.30	0.4058	0.5192
1.20	0.7154	0.6234	3.40	0.4385	0.4296
1.30	0.6386	0.6863	3.50	0.5326	0.4152
1.40	0.5431	0.7135	3.60	0.5880	0.4923
1.50	0.4453	0.6975	3.70	0.5420	0.5750
1.60	0.3655	0.6389	3.80	0.4481	0.5656
1.70	0.3238	0.5492	3.90	0.4223	0.4752
1.80	0.3336	0.4508	4.00	0.4984	0.4204
1.90	0.3944	0.3734	4.10	0.5738	0.4758
2.00	0.4882	0.3434	4.20	0.5418	0.5633
2.10	0.5815	0.3743	4.30	0.4494	0.5540

versus y, we obtain a curve known as *Cornu's spiral** (Figure 14-14). This spiral provides us with a most elegant means for the graphical solution of diffraction problems.

The length of the vibration curve, v, is the length measured *along the spiral*. Because of angle δ, as we move away from the midpoint of the spiral and v *increases*, the *steepness of the curve increases even faster*—which accounts for the spiral shape. The actual, total amplitude, however, corresponds to the shortest distance *between* points on the spiral; it corresponds to the *chord*. The square of this chord represents the light intensity.

Of particular interest are the two endpoints, the *eyes* of the spiral. At these points $v \rightarrow \pm\infty$. Using the identity

$$\int_0^\infty \sin ax^2 \, dx = \int_0^\infty \cos ax^2 \, dx = \frac{1}{2} \sqrt{\frac{\frac{1}{2}\pi}{a}}$$

---

* Marie Alfred Cornu (1841–1902), French physicist, professor of experimental physics at the École Polytechnique in Paris. While working earlier at the École des Mines, Cornu became interested in optics, read Félix Billet's book *Traité d'optique*, and repeated all the experiments described therein. He determined the speed of light by Fizeau's method and made precise measurements of the wavelengths of certain lines in the spectrum of hydrogen. He is probably best known for the spiral he described in A. Cornu, "Méthode nouvelle pour la discussion des problèmes de diffraction dans le cas d'une onde cylindrique," *J. Phys.* **3** (1874), 5–15, 44–52.

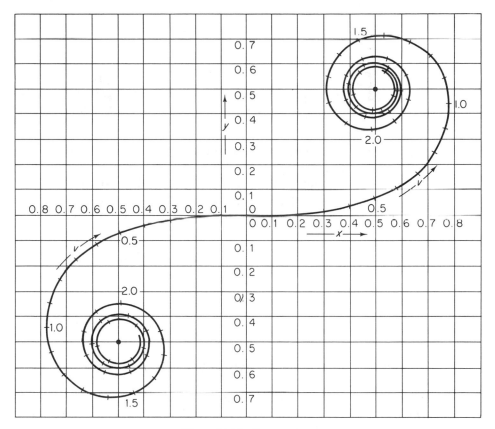

Figure 14-14 Cornu spiral.

we substitute $\frac{1}{2}\pi$ for a, and v for x, and obtain

$$\int_0^\infty \sin \tfrac{1}{2}\pi v^2 \, dv = \int_0^\infty \cos \tfrac{1}{2}\pi v^2 \, dv = \frac{1}{2} \sqrt{\frac{\frac{1}{2}\pi}{\frac{1}{2}\pi}} = \frac{1}{2}$$

which means that the upper eye of the spiral has the coordinates

$$x = y = (+0.5, +0.5) \qquad\qquad [14\text{-}11]$$

and the lower eye,

$$x = y = (-0.5, -0.5) \qquad\qquad [14\text{-}12]$$

Applications

First we apply Cornu's spiral to diffraction at a *knife edge*. Consider point P_0, located at the geometrical limit of the shadow (Figure 14-15). The amplitude at P_0

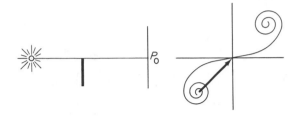

Figure 14-15 Fresnel diffraction at a knife edge (knife edge is heavy line in left-hand diagram); Cornu spiral (*right*) shows amplitude of light at point P_0.

is proportional to the length of the chord which extends from the lower eye of the spiral to the origin, as shown.

As point *P* moves *into* the shadow, the chord on the spiral retracts (Figure 14-16). The lower end of the chord remains anchored to the lower eye, but the upper end retraces the path of the spiral downward, becoming gradually shorter.

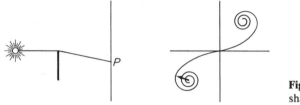

Figure 14-16 Amplitude vector *inside* shadow.

Conversely, as *P* moves *outside* the shadow, the upper end of the chord moves into the upper branch of the spiral (Figure 14-17), while its lower end remains anchored to the lower eye. But the *length of the chord undergoes periodic variations*, passing through a series of maxima and minima, rather than changing monotonically as before.

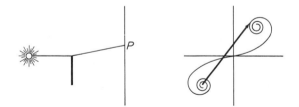

Figure 14-17 Amplitude vector *outside* shadow.

At certain points the amplitude is even higher than the amplitude with no obstacle in the path. In short: *Inside* the shadow the intensity drops off gradually, and *outside* the shadow there are fringes (Figure 14-18). At the edge of the shadow, only one-half of the wave contributes; thus the amplitude has fallen to one-half, and the intensity to one-fourth, of the unobstructed wave.

Example

Whereas Cornu's spiral illustrates the principle very well, higher accuracy is obtained when the tabulated values of Fresnel's integrals are used directly. For example, calculate the relative intensity at a point:

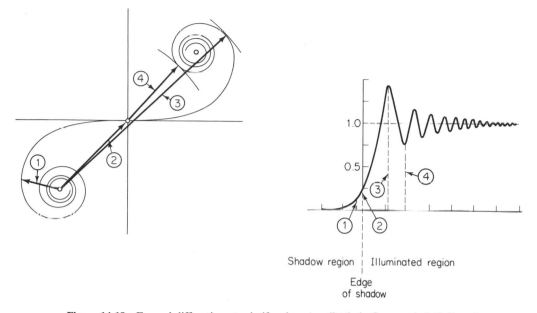

Figure 14-18 Fresnel diffraction at a knife edge. Amplitude in Cornu spiral (*left*) and intensity distribution on screen (*right*). Note corresponding numbers on left and right.

(a) Where $v = -1.00$, which is *inside* the shadow.
(b) Where $v = +1.00$, which is *outside* the shadow.

Solution: From Table 14-2,

$$v = 1.00, \qquad x = 0.7799, \qquad y = 0.4383$$

(a) *Inside* the shadow, the top of the phasor lies in the same (lower left-hand) quadrant as the lower eye of the spiral to which it is anchored. Thus the phasor extends from the coordinates $(-0.5; -0.5)$ to $(-0.7799; -0.4383)$ and the absolute values of the x and y components, respectively, must be *subtracted*:

$$\Delta x = 0.5 - 0.7799 = |0.2799|$$

and

$$\Delta y = 0.5 - 0.4383 = 0.0617$$

We need not solve for the actual length of the phasor (which has to be squared anyway to obtain the intensity). However, we recall that Cornu's construction gives a value of two for the unobstructed wave, and hence the relative intensity at the specified point is

$$I = \tfrac{1}{2}(0.2799^2 + 0.0617^2)I_0 = \boxed{0.041I_0}$$

(b) *Outside* the shadow, the phasor extends from the lower eye to the first (upper right-hand) quadrant; thus the x and y components must be *added*:

$$\Delta x = 0.5 + 0.7799 = 1.2799$$

and

$$\Delta y = 0.5 + 0.4383 = 0.9383$$

The relative intensity then is

$$I = \tfrac{1}{2}(1.2799^2 + 0.9383^2)I_0 = \boxed{1.26I_0}$$

ZONE PLATES

A *zone plate* is a system of concentric rings with a particular spacing: wider near the center and more narrow in the periphery.* The center portion of such a system of rings is shown in Figure 14-19.

Figure 14-19   Center portion of a zone plate.

Light passing through a zone plate is diffracted, the same as light passing by near the edge of an obstacle. A zone plate, therefore, works because of the zones' edges; *they* diffract the light, rather than the zones as such. Consequently, the light, after having passed through the zone plate, is directed not only *toward* the axis but also *away* from it; a zone plate, in other words, acts both as a positive and a negative lens.

A zone plate, furthermore, has not merely one, but several foci. This is because of the higher orders. The most intense, first-order focus is farthest away. The higher-order foci are closer; they are located at distances 1/3, 1/5, 1/7, . . . of that of the first order.

If a zone plate has completely opaque and completely clear zones, it is called a *Fresnel zone plate*. These are used for image formation in regions of the spectrum (X rays, microwaves) where no conventional lenses exist. If the zones,

---

* A zone plate, or *zone lens* because of its focusing properties, is the only image-forming element for which there is no prototype in nature. The first to make a zone plate, in 1871, was Lord Rayleigh. Four years later, J. L. Soret described its properties, "Ueber die durch Kreisgitter erzeugten Diffractionsphänomene," *Ann. Phys.* (6) **6** (1875), 99–113.

instead of opaque and clear, have a sinusoidal transmissivity distribution, it is a *Gabor zone plate*; oddly enough, these have only first-order foci.

To derive the numerical relationships, we consider one of the boundaries between zones, that is, we consider again a *circular aperture*. Assume that the light comes from infinity. The wavefronts that pass through the aperture in Figure 14-20, therefore, are *plane*. We divide the aperture into half-period zones, which from zone to zone are $\frac{1}{2}\lambda$ farther away from point P_0. If b is the distance from the center of the aperture to P_0, then $b + \frac{1}{2}\lambda$ corresponds to a (first) ring around the center of the aperture, $b + 2(\frac{1}{2}\lambda)$ to a second, wider ring, $b + 3(\frac{1}{2}\lambda)$ to the third ring, and $b + N(\frac{1}{2}\lambda)$ to the Nth ring.

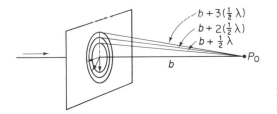

Figure 14-20 Half-period zones in a zone plate.

From Pythagoras' theorem we find that the radius of the first boundary is

$$R_1 = \sqrt{(b + \tfrac{1}{2}\lambda)^2 - b^2} \qquad [14\text{-}13]$$

and the radius of the Nth boundary

$$R_N = \sqrt{(b + \tfrac{1}{2}N\lambda)^2 - b^2} \qquad [14\text{-}14]$$

Furthermore, the *area* of the first *zone* is

$$A_1 = \pi R_1^2 = \pi[(b + \tfrac{1}{2}\lambda)^2 - b^2] = \pi b\lambda + \tfrac{1}{4}\pi\lambda^2$$

or, since λ is small compared with b,

$$A_1 \approx \pi b\lambda$$

The area of the second zone is

$$A_2 = \pi[(b + \lambda)^2 - b^2] - \pi b\lambda \approx 2\pi b\lambda - \pi b\lambda = \pi b\lambda$$

and so on, which means that the individual zones all have approximately the same area. The contributions from these zones, therefore, are very nearly the same but since the boundaries' radii differ by one-half of a wavelength, the resultant at P_0 is zero.

If we now block out all odd-numbered zones, then only zones 2, 4, 6, . . . will contribute to the light reaching P_0, and P_0 will be bright; it will become the *focus of the zone plate*.

In order to find the focal length, we square Equations [14-14] and [14-13] and divide one by the other:

$$\frac{R_N^2}{R_1^2} = \frac{(b + \frac{1}{2}N\lambda)^2 - b^2}{(b + \frac{1}{2}\lambda)^2 - b^2}$$

Canceling *b* and λ gives

$$R_N^2 + \tfrac{1}{2}R_N^2\lambda = N(R_1^2 b + \tfrac{1}{2}R_1^2\lambda)$$

and, since $\lambda \ll R$,

$$R_N = R_1\sqrt{N} \qquad\qquad [14\text{-}15]$$

If we again square Equation [14-14] and rearrange, we obtain

$$b^2 = (b + \tfrac{1}{2}N\lambda)^2 - R^2 = b^2 + bN\lambda + (\tfrac{1}{2}N\lambda)^2 - R^2$$

and, since $\lambda \ll b$,

$$bN\lambda \approx R^2$$

The *focal length* of the zone plate, therefore, is

$$f = b \approx \frac{R^2}{N\lambda} \qquad\qquad [14\text{-}16]$$

Note the wavelength λ in this equation. It shows that a zone plate has much chromatic aberration. (A conventional lens also has chromatic aberration because its refractive index changes with wavelength. This variation, however, is small because *n* varies very little.)

The thin-lens equation, $1/s + 1/f = 1/s'$, applies as well to a zone plate. Rayleigh's criterion, $\theta_{\min} \approx 1.22\,\lambda/D$, also applies. If we replace θ by d/f and *f* by Equation [14-16], then

$$d_{\min} \approx 1.22\,\frac{R^2}{ND}$$

But *D* is twice the radius, and thus

$$d_{\min} \approx 1.22\,\frac{R}{2N} \qquad\qquad [14\text{-}17]$$

Interestingly, λ has dropped out: The wavelength, although it causes much chromatic aberration, has no effect on the resolution. Finally, we see from Equation [14-17] that to make a better zone plate (which will resolve smaller details), (a) the zone plate must be as small as possible and (b) it must have as many zones as possible, two requirements that call for high-resolution photographic material when making such zone plates.

Drawing a Zone Plate by Computer

We start out with Equation [14–15], $R_N = R_1\sqrt{N}$, written in *BASIC*:

$$RN = R1 * SQR(N)$$

Using these radii, RN, we draw a set of concentric circles:

$$CIRCLE\ (160,100),\ RN,\ 2,,,\ .86$$

The last number, .86, is the *aspect*, which, in general, describes the shape of an ellipse; here, this figure has been determined empirically to make the circles as round as possible.

The space between alternate concentric circles is then filled using the **PAINT** command. The coordinates specified with PAINT refer to a pixel (*any* pixel) inside this space. The next number is the color of the paint; that color must match the color of the circle. In Program 14-1, ZONEPA, we make the innermost radius so large, and use so many rings, that they fill the screen and are still sufficiently separate from one another; the printout is then reduced photographically.

Program 14-1 Drawing a Zone Plate

```
10    REM       Program ZONEPA
20    PRINT
30    REM       Drawing a Zone Plate
40    PRINT
50    REM       Radius of the Innermost Boundary = R1
60    REM       Radius of the Nth Boundary = RN
70    REM       Total Number of Boundaries = TOTAL
80    PRINT
90    REM       Determining the Radii of the
                Boundaries
100   CLS
110   R1 = 28
120   TOTAL = 13
130   FOR N = 1 TO TOTAL
140   RN = R1 * SQR (N)
150   PRINT
160   REM       Drawing the Boundaries
170   SCREEN 1
180   CIRCLE (160,100), RN, 2,,, .86
190   NEXT N
```

```
200   PRINT
210   REM       Filling in the Space Between
                Boundaries
220   PAINT (161,100), 2
230   FOR L = 2 TO (TOTAL-1) STEP 2
240   RING2 = R1 * SQR (L) + 160
250   RING3 = R1 * SQR (L+1) + 160
260   MEDIAN = (RING2 + RING3) / 2
270   PAINT (MEDIAN, 100), 2
280   NEXT L
290   END
```

Today zone plates are often made holographically, rather than from a large drawing that is then reduced photographically. This also allows us to make cylindrical zone plates (which act like cylinder lenses) and even more complex systems, rather than just circular ones.

The practical use of zone plates is in *micro-* and *integrated optics*. They often have focal lengths of the order of millimeters and diameters of no more than perhaps 1 mm. They are often used together with a light source such as a diode laser.

Concluding Phenomena

There exist many more diffraction phenomena. We mention only a few. If instead of an (open) aperture we have a (solid) circular obstacle, Fresnel's theory and experimental observation show that in the center of the shadow there is a bright point of light, called *Poisson's spot** (Figure 14-21). The reason is that in diffraction by a circular disk or circular aperture, there are only two rays (from diametrically opposite points) diffracted to any given off-axis point. But there is an infinite number of rays (from the whole circumference) that converge on the axis.

If the disk is illuminated by two point sources, two Poisson spots appear. If there is an array of many point sources, there are as many Poisson spots. And if the many point sources are replaced by a complex object, such as a photographic

---

* Named after Siméon Denis Poisson (1781–1840), French physicist. As a member of the committee which was to judge Fresnel's dissertation, Poisson concluded that, if Fresnel were right, a central bright spot would appear in the shadow of a circular obstacle, a result which he offered as a *reductio ad absurdum*, that is, evidence that Fresnel's theory is absurd. He did not know that the bright spot in question had been discovered by Maraldi and Deslisle some 100 years earlier. After the objection raised by Poisson, Dominique François Arago, also a member of the committee, immediately tried the experiment using a disk 2 mm in diameter and rediscovered Maraldi's spot. But, either despite his objection, or because of it, it was Poisson's name that became immortal, and the phenomenon is since called *Poisson's spot*.

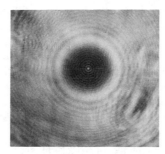

Figure 14-21 Poisson's spot, the bright spot in the center of the shadow.

slide illuminated from behind, a real image of the slide results, having used as the "objective lens" nothing but a solid metal sphere or disk.

Two diffracting objects are said to be complementary if one is the photographic negative of the other, such as a slit and a wire. The (Fraunhofer) diffraction patterns generated by two such objects, except for the zeroth order, are the same, an effect known as *Babinet's principle*.

SUMMARY OF EQUATIONS

Fraunhofer diffraction minima (behind a slit):

$$s \sin \theta = m\lambda \qquad [14\text{-}5]$$

Rayleigh's criterion (using circular aperture):

$$\theta_{min} \approx 1.22 \frac{\lambda}{D} \qquad [14\text{-}6]$$

Zone plate, radius of zone:

$$R_N = R_1 \sqrt{N} \qquad [14\text{-}15]$$

focal length:

$$f \approx \frac{R^2}{N\lambda} \qquad [14\text{-}16]$$

PROBLEMS

14-1. How many wavelengths wide must a single slit be if the first Fraunhofer diffraction minimum occurs at an angular distance of 30° from the optical axis?

14-2. A slit 0.14 mm wide is illuminated by monochromatic light. If on a screen 2 m away from the slit, the two second-order minima are 3 cm apart from each other, what is the wavelength of the light?

14-3. When you are driving at night, the headlight of a motorcycle following you, seen through the foggy rear window of your car, is surrounded by colored rings. Which color is inside, which outside?

14-4. Looking at the sun with the sky slightly overcast, you see a colored ring around the sun that is red outside, and therefore due to diffraction. How can you decide whether the diffraction takes place in the atmosphere or in your eyes?

14-5. A beam of light passes through an aperture that has the shape of a regular triangle (all three sides equal). Another aperture is a regular hexagon (six sides equal). Compare the diffraction patterns produced by the two apertures.

14-6. Lycopodium seeds, which are of spherical shape and nearly uniform size, are dusted on a glass plate. If with parallel light of 640 nm wavelength the angular radius of the first diffraction maximum is 2°, how large are the seeds?

14-7. If a car's headlights are 122 cm apart, then, assuming pupils 4 mm in diameter and light of 500 nm wavelength, what is the maximum distance at which the eye can resolve them?

14-8. How large a target can a telescope resolve that has an aperture 3 cm in diameter, is aimed at an object 5 km away, and uses light of 600 nm wavelength?

14-9. What should be the radius of the twentieth boundary of a zone plate that, with light of 500 nm wavelength, has a focal length of 160 cm?

14-10. If a zone plate with 100 boundaries, when used with light of 500 nm wavelength, has 2 m focal length, what is the least distance that it can resolve?

14-11. **(a)** What should be the diameter of a zone plate that has 55 zones and can resolve details 10 μm in size?
   (b) What wavelength is necessary?

14-12. In the lecture hall we show *Poisson's spot* on a screen 2.25 m away from an obstacle 2 mm in diameter. Why can't we see Poisson's spot also during a solar eclipse? (Angular size of moon $\approx \frac{1}{2}°$.)

SUGGESTIONS FOR FURTHER READING

BORN M., and E. WOLF. *Principles of Optics*, 6th ed., pp. 370–458. Oxford: Pergamon Press Ltd., 1980.

HOOVER R. B., and F. S. HARRIS, JR. "Die Beugungserscheinungen: A Tribute to F. M. Schwerd's Monumental Work on Fraunhofer Diffraction," *Appl. Opt.* **8** (1969), 2161–64.

KLEIN M. V., and TH. E. FURTAK. *Optics*, 2nd ed., pp. 337–468. New York: John Wiley and Sons, Inc., 1986.

15

Diffraction Gratings

A diffraction grating is an extension of Young's double slit and as such is based on both diffraction and interference. In fact, the same equations apply to both double-slit interference and to interference produced by a diffraction grating. It is only because gratings and their applications have developed very much into a technology all their own, that diffraction gratings should be presented in a separate chapter.

THE GRATING EQUATION

In its most elementary form, a diffraction grating consists of a great many parallel grooves drawn on a sheet of glass or plastic. The grooves themselves, because of their corrugated shape, do not let much light pass through; instead, the light passes through the open intervals between them. The same as in double-slit interference, we call, as shown in Figure 15-1, d the distance between the centers of any two adjacent intervals, θ the angle through which the light is bent, m the order, and λ the wavelength. Then

$$\boxed{d \sin \theta = m\lambda} \qquad m = 0, 1, 2, \ldots \qquad [15\text{-}1]$$

which is the *grating equation*, the same equation that we derived for double-slit maxima. If the light is incident on the grating at an angle I, then

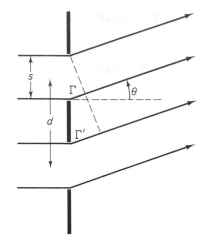

Figure 15-1 Diagram of diffraction grating showing slit width, *s*, and slit separation, *d*.

$$d(\sin \theta - \sin I) = m\lambda \qquad [15\text{-}2]$$

But the slits also have a *finite width*, *s*. That changes the intensity distribution behind the grating. Consider first the light that passes through a *single slit*. The path difference between peripheral rays, touching the edges of the slit, is

$$\Gamma = s \sin \theta$$

We know that path difference and phase difference are related as

$$\frac{\Gamma}{\lambda} = \frac{\delta}{2\pi} \qquad [15\text{-}3]$$

and that the intensity distribution behind a slit, at an angular distance θ from the axis, is

$$I_\theta = I_0 \left(\frac{\sin \delta/2}{\delta/2} \right)^2 \qquad [15\text{-}4]$$

We now substitute Equation [15-3], solved for δ, in Equation [15-4]. That gives

$$I_\theta = I_0 \frac{\sin^2 \left(\dfrac{\pi}{\lambda} s \sin \theta \right)}{\left(\dfrac{\pi}{\lambda} s \sin \theta \right)^2} \qquad [15\text{-}5]$$

which is the *diffraction contribution*. It is due to the *finite width* of the slits.

Next we consider *several, N, consecutive slits*. If the light is incident obliquely, the path difference between rays that pass through (equivalent points in) adjacent slits is

$$\Gamma' = d \sin \theta$$

and the phase difference

$$\delta = \frac{2\pi}{\lambda} d \sin \theta \qquad [15\text{-}6]$$

The intensity distribution behind N slits is found best by adding the complex amplitudes. For one slit, the amplitude of the light passing through is

$$\mathbf{A} = Ae^{i\delta}$$

But since, from slit to slit, there may be a constant phase difference δ', consecutive amplitudes may vary as

$$\mathbf{A}_1 = Ae^{i\delta'}, \qquad \mathbf{A}_2 = Ae^{i2\delta'}, \qquad \mathbf{A}_3 = Ae^{i3\delta'}, \ldots$$

The sum of these amplitudes is that of a geometrical series,

$$Ae^{i\delta'} = A[1 + e^{i\delta'} + e^{i2\delta'} + \cdots + e^{i(N-1)\delta'}] = A\frac{1 - e^{iN\delta'}}{1 - e^{i\delta'}}$$

To find the intensity, we multiply the amplitude with its complex conjugate. This gives

$$\mathsf{A}^2 = A^2 \left[\frac{(1 - e^{iN\delta'})(1 - e^{-iN\delta'})}{(1 - e^{i\delta'})(1 - e^{-i\delta'})} \right]$$

But

$$(1 - e^{iN\delta'})(1 - e^{-iN\delta'}) = 1 - \cos N\delta'$$

and likewise,

$$(1 - e^{i\delta'})(1 - e^{-i\delta'}) = 1 - \cos \delta'$$

Using the trigonometric relationship

$$1 - \cos \alpha = 2 \sin^2(\tfrac{1}{2}\alpha)$$

we obtain

$$\mathsf{A}^2 = A^2 \frac{\sin^2(N\tfrac{1}{2}\delta')}{\sin^2(\tfrac{1}{2}\delta')} \qquad [15\text{-}7]$$

and substituting Equation [15-6] in [15-7], we have

$$I_\theta = I_0 \frac{\sin^2\left(N\dfrac{\pi}{\lambda} d \sin \theta\right)}{\sin^2\left(\dfrac{\pi}{\lambda} d \sin \theta\right)} \qquad [15\text{-}8]$$

which is the *interference contribution*. It is due to the *multiplicity* of slits.

The total, actual pattern behind the grating is found by multiplying Equations [15-5] and [15-8]:

$$I_\theta = I_0 \frac{\sin^2 \left(\dfrac{\pi}{\lambda} s \sin \theta\right) \sin^2 \left(N \dfrac{\pi}{\lambda} d \sin \theta\right)}{\left(\dfrac{\pi}{\lambda} s \sin \theta\right)^2 \sin^2 \left(\dfrac{\pi}{\lambda} d \sin \theta\right)}$$

For simplicity, call D the pertinent term $(\pi s \sin \theta)/\lambda$ in the diffraction contribution and I the term $(\pi d \sin \theta)/\lambda$ in the interference contribution. The total pattern produced by the grating is then

$$I_\theta = I_0 \frac{\sin^2 D}{D^2} \frac{\sin^2 NI}{\sin^2 I} \qquad\qquad [15\text{-}9]$$

which means that one contribution is being *modulated* by the other contribution.

The numerator, $\sin^2 NI$, in the interference contribution becomes zero whenever

$$NI = k\pi \qquad k = 1, 2, 3, \ldots \qquad\qquad [15\text{-}10]$$

The denominator, $\sin^2 I$, similarly, becomes zero whenever

$$I = 0, \pi, 2\pi, \ldots$$

Since the ratio 0/0 is indeterminate, Equation [15-10] is the condition for minima for all values of k *except* where $k = 0, N, 2N, 3N, \ldots, mN$. Substituting $k = mN$ in Equation [15-10], and replacing I by $(\pi d \sin \theta)/\lambda$, gives

$$N \frac{\pi d \sin \theta}{\lambda} = mN\pi$$

This reduces to

$$d \sin \theta = m\lambda \qquad\qquad [15\text{-}1]$$

which again is the grating equation. It refers to the *principal maxima*. Between any two adjacent (principal) maxima, there will be $N - 1$ loci of zero intensity (see Figure 15-2). Between these minima there are *secondary maxima*, but their intensities are much less than those in the principal maxima.

Finally, note the $\sin^2 N$ term in Equation [15-9]. It shows that the intensity in the principal maxima is proportional to the *square* of the number of slits. If there are *more* slits, the principal maxima become higher and higher and narrower and the secondary maxima between them get smaller and are suppressed; in the end, these orders are "missing."

The principal use of diffraction gratings is in spectroscopy. A *grating spectrograph*, as we see from Figure 15-3, is similar to a prism spectrograph. The light passes first through a combination of a slit and a collimating lens. It then reaches the grating, set with its rulings parallel to the slit. Another lens, L_2, focuses the light on the screen. Light of different wavelengths, as a consequence of Equation [15-1], is diffracted at different angles, θ, and for each order, m, is drawn out into a spectrum, the blues nearer to the optic axis and the reds farther away from it. The

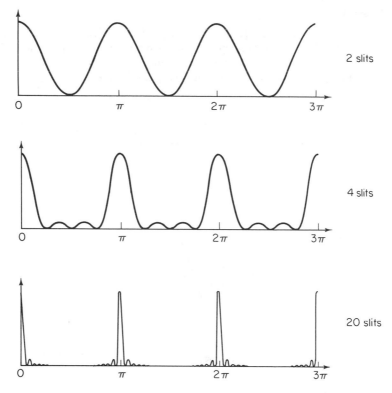

Figure 15-2 Intensity distribution behind a grating with 2 slits (*top*), 4 slits (*center*), and 20 slits (*bottom*).

zeroth order retains the composite color of the source and therefore is easy to identify.

The number of orders possible is limited by the separation of the lines. A coarse grating, with relatively few lines per unit width, will give many orders. A finely ruled grating may give only one or two orders, but their spectra will spread out much wider.

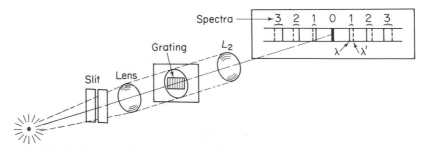

Figure 15-3 Schematic diagram of grating spectrograph. Light source is assumed to emit two wavelengths, $\lambda < \lambda'$.

Example

When looking through a diffraction grating at a hydrogen discharge tube, which emits light of 656 nm wavelength, we see two red lines, to either side of the zeroth order. If at a distance of 90 cm from the grating, the two lines are separated by 62.5 cm, how many lines per millimeter does the grating have?

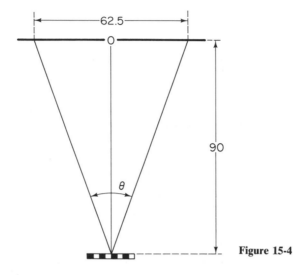

Figure 15-4

Solution: First we determine the angle θ that the lines subtend with the optic axis. From the construction in Figure 15-4 we find that

$$\theta = \arctan \frac{62.5/2}{90} = 19.15°$$

Then we use the grating equation, solve it for d, set $m = 1$, and insert the values given:

$$d = \frac{\lambda}{\sin \theta} = \frac{656 \times 10^{-6} \text{ mm}}{0.328} = 0.002$$

The grating, therefore, has

$$\frac{1}{0.002} = \boxed{500 \text{ lines per millimeter}}$$

Resolvance

The *resolvance* (''resolving power'') of a grating describes its ability to separate closely adjacent spectrum lines. First we take the grating equation,

$$d \sin \theta = m\lambda$$

and differentiate θ with respect to λ. Since $\Delta\lambda \ll \lambda$, this gives

$$\frac{d\theta}{d\lambda} = \frac{m}{d\cos\theta} \qquad \text{[15-11]}$$

The term $d\theta/d\lambda$ is the *angular dispersion* of the grating; it increases as the order, m, increases. Angular dispersion by itself, however, does not make two closely spaced lines more distinct; that is a matter of *resolvance*.

Next consider that the light contains two wavelengths, λ and $(\lambda + \Delta\lambda)$. Whereas for the first wavelength the grating equation holds,

$$d\sin\theta = m\lambda$$

for the second wavelength it becomes

$$d\sin\theta = m(\lambda + \Delta\lambda) \qquad \text{[15-12]}$$

To satisfy Rayleigh's criterion, the second wavelength must be no closer to the first wavelength than the minimum adjacent to λ, in the same order:

$$d\sin\theta = \left(m + \frac{1}{N}\right)\lambda \qquad \text{[15-13]}$$

Setting equation [15-13] equal to [15-12] by eliminating $d\sin\theta$ gives

$$\left(m + \frac{1}{N}\right)\lambda = m(\lambda + \Delta\lambda)$$

and therefore,

$$\boxed{\frac{\lambda}{\Delta\lambda} = Nm} \qquad \text{[15-14]}$$

which is the resolvance of the grating.

We note that the resolvance, $\lambda/\Delta\lambda$, is proportional to the number of lines, N, in the grating, and to the order, m, in which it is used. To obtain higher resolution, we may either use a grating with more lines or go to a higher order. In practice, grating lines are ruled very close to each other in order to keep the aperture of the system to a reasonable size.

Example

The sodium D doublet has wavelengths of 589 nm and 589.6 nm. If only a grating with 400 rulings is available, what is the lowest order possible in which the D lines are resolved?

Solution: From Equation [15-14],

$$m = \frac{\lambda}{\Delta\lambda N} = \frac{589 \times 10^{-9}}{(0.6 \times 10^{-9})(4 \times 10^2)} = 2.45$$

But orders must be *whole numbers* and therefore the lowest possible order in which the D lines are resolved is

$$m = \boxed{3}$$

Types of Gratings

We distinguish *transmission gratings* (which let the light pass through, as in Figures 15-1 and 15-3) from *reflection gratings* (which reflect it, as in Figures 15-5 and 15-6). Either type can be made with more than 1000 lines per millimeter.

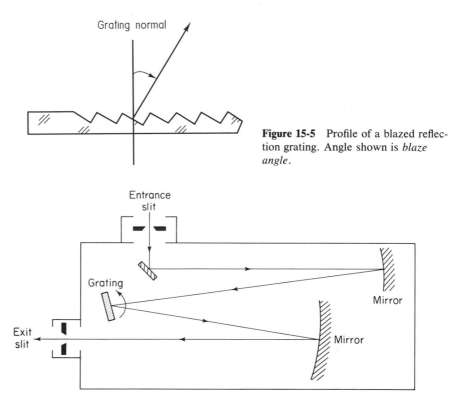

Figure 15-5   Profile of a blazed reflection grating. Angle shown is *blaze angle*.

Figure 15-6   Czerny–Turner mount: diffraction grating and two concave mirrors.

A grating with completely opaque bars and clear intervals is called an *amplitude grating*. If the bars do not block the light but merely retard its phase, we have a *phase grating*. If the light distribution across the bars and intervals is sinusoidal, rather than "square-wave," all the light is diffracted into the two first-order spectra.

Actually, there are no "slits" or "lines" in a grating. Modern gratings are *blazed*. First a plate of low-expansion glass is coated with a suitable metal such as aluminum or gold. Then the coating is ruled with a diamond tool, moved across in repetitive, parallel, equidistant strokes and with just enough pressure to make the metal flow aside. After the diamond has passed, the metal hardens into the desired profile (Figure 15-5). Whereas conventional gratings direct most of the light into the zeroth order, the advantage of blazed gratings is that they direct much more light into some of the higher orders, making more efficient use of the light available.

Gratings can be made with high precision, in sizes up to 50×75 cm^2, using interferometrically and electronically controlled ruling engines. From original *masters*, plastic *replicas* can be produced. But even the best gratings sometimes have periodic errors that give rise to spurious lines in the spectrum, called *ghosts*.

Like prisms, diffraction gratings may be mounted between a collimating lens and a focusing lens. No lens is required with a *concave reflection grating*. But curved gratings are difficult to rule. So, after all, it is better to use a plane grating, often together with two concave mirrors as in the *Czerny–Turner mount* shown in Figure 15-6.

THREE-DIMENSIONAL GRATINGS

The gratings discussed so far diffract the light into a linear spread of wavelengths; they are "one-dimensional" gratings. A two-dimensional or *cross-grating* results if two gratings are laid on top of each other, their rulings oriented at right angles. A zone plate could also be considered a two-dimensional grating, of rotational symmetry (and space-varying frequency). For now, we proceed to *three-dimensional gratings*.

X-ray Diffraction

For a while after Röntgen's discovery of X rays* it was not known whether X rays were particles (like cathode rays) or waves of exceedingly short length. It oc-

---

* X rays were discovered by Wilhelm Conrad Röntgen (1845–1923). On the night of November 8, 1895, Röntgen, at that time head of the Physics Department at the University of Würzburg, Germany, was working with an evacuated *Crookes tube* with two electrodes in it that he had covered with black paper when he saw a faint glow on a sheet of fluorescent paper some 2 m away. Evidently there were some invisible rays coming out of the tube; Röntgen called them X rays. They passed easily through paper and cardboard but were blocked by lead. For the next several weeks Röntgen worked at a feverish pace to find out as much as he could before he submitted a paper, "Über eine neue Art von Strahlen," a model of clarity and completeness, that appeared in *Ann. Physik* **64** (1898), 1–11, and soon was translated and published in English, French, and American journals so that investigators could repeat what he had done, amazing their friends with pictures of their hand bones. In 1901, Röntgen received the first, newly established Nobel prize in physics.

curred to Max von Laue* that a crystal may act on X rays in a way similar to a grating on light. Together with W. Friedrich and P. Knipping, von Laue tried crystals of zinc sulfide and copper sulfate. Both gave fine X-ray diffraction patterns, called *Laue diagrams* (Figure 15-7), which established the wave nature of X rays.

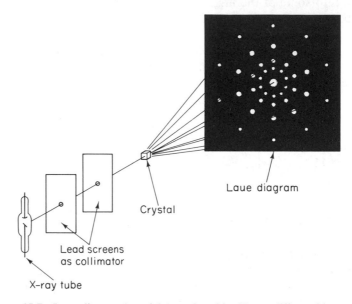

Laue diagram

Crystal

Lead screens
as collimator

X-ray tube

Figure 15-7 Laue diagram (top right) produced by X rays diffracted in a crystal.

A crystal is a three-dimensional lattice of atoms, or groups of atoms, that is built out of repeating fundamental units of structure, called *unit cells*. X-ray diffraction on such units may be visualized as reflections (Figure 15-8). Let a beam of monochromatic X rays incident on a crystal make an angle θ with the surface.† Some of the incident rays are reflected at the top layer, 1, but other rays penetrate to the deeper layers, 2, 3, and so on. The planes have an interplanar spacing d. If the wave reflected at 1 is reinforced by the wave reflected at 2, the path difference Γ between the two rays must be $\Gamma = AB + BC = \lambda$ or it must be a multiple of λ, $\Gamma = m\lambda$.

---

* Max Theodor Felix von Laue (1879–1960), German physicist, professor of theoretical physics and director of the Kaiser Wilhelm Institute, the present Max Planck Institute, in Berlin. W. Friedrich, P. Knipping, and M. von Laue, "Interferenzerscheinungen bei Röntgenstrahlen," *Ann. Phys.* (4) **41** (1913), 971–88. In 1914, von Laue received the Nobel Prize in physics.

† Angles in X-ray crystallography are measured from the surface and not from the surface *normal* as elsewhere in optics.

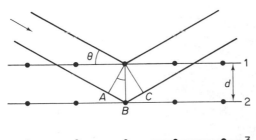

Figure 15-8 Deriving Bragg's law.

Since

$$\sin \theta = \frac{AB}{d}$$

$$AB = BC = d \sin \theta$$

Therefore, in order to produce a maximum, the path difference,

$$\Gamma = 2d \sin \theta$$

must be equal to an integral multiple of a wavelength,

$$\boxed{2d \sin \theta = m\lambda} \qquad m = 0, 1, 2, \ldots \qquad\qquad [15\text{-}15]$$

which is *Bragg's law*.* The same relationship holds if the X rays are transmitted through the crystal, rather than reflected by it.

MOIRÉ FRINGES

When two window screens are placed next to each other, the intersection of their lines produces another sequence of lines called *moiré fringes*. The name comes from the French; it refers to the wavy finish of silk and other fabrics showing shifting patterns of lines.†

---

* Named after (William) Lawrence Bragg (1890–1971), British crystallographer, the son of William (Henry) Bragg (1862–1942). When the older Bragg, professor of mathematics and physics at Adelaide, Australia, and his son saw von Laue's photographs, the younger Bragg, still a student at Cambridge, thought that perhaps the Laue spots could be explained by diffraction, established the law named after him, and thus laid the foundations of X-ray crystallography. He succeeded Rutherford as head of the Cavendish Laboratory and, like his father before him, became president of the Royal Institution. Father and son shared the Nobel Prize in physics in 1915. W. L. Bragg and W. H. Bragg, "The Reflection of X-Rays by Crystals," *Proc. Roy. Soc.* (*London*) **88** (1913), 428–438.

† Moiré fringes were first described by Lord Rayleigh who noticed them when he held the photograph of a diffraction grating next to another such photograph: Lord Rayleigh (J. W. Strutt), "On the Manufacture and Theory of Diffraction-gratings," *Philos. Mag.* (4) **47** (1874), 81–93.

The original grids are called *Ronchi rulings*, after Vasco Ronchi (1897–1988), Italian physicist,

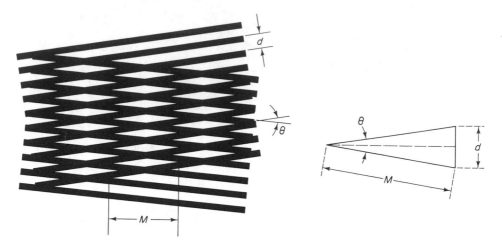

Figure 15-9 Moiré pattern formed by two grids of spacing d, subtending an angle θ. Distance M is spacing of moiré fringes.

There are several ways of producing such fringes. One way is to use two Ronchi rulings held in contact. Assume that the separation of the lines in the two original grids is d, and that one of the grids has been turned through an angle θ with respect to the other. Then from the construction in Figure 15-9 it follows that

$$\sin \frac{\theta}{2} = \frac{d/2}{M}$$

where M is the spacing of the moiré fringes. Solving for M gives

$$M = \frac{d}{2\sin(\theta/2)} \qquad [15\text{-}16]$$

which shows that the more parallel the two grids, the wider the spacing of these fringes.

Moiré fringes can be used for making visible refractive gradients in glass or plastics, or turbulence in air. A microscopic version is illustrated in Figure 15-10.

The Ronchi grid is placed halfway between objective and eyepiece. The exact position is not critical. With the grid closer to the eyepiece, the shadow lines become sharper but the sensitivity is less. With the grid nearer to the objective, the opposite is true.

---

long-time director of the Istituto Nazionale di Ottica in Florence. After graduating from the University of Pisa, Ronchi devoted his life to optics; he wrote 30 books and some 900 papers, most of them on optical testing and interferometry and on the history of science. When asked why he worked so hard, Ronchi replied: "I don't do it for glory or for money or for personal prestige, I do it because I like it. I only regret that I can't work more, to do things which only I can do."

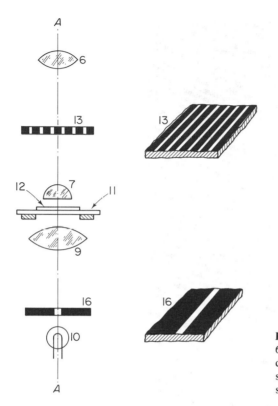

Figure 15-10   Ronchi-grid microscope: 6, eyepiece; 7, objective lens; 9, condenser; 10, light source, 11 plus 12, specimen; 13, Ronchi grid; 16, rotatable slit diaphragm.

With only the Ronchi grid in place, nothing much different is seen than with any other microscope. But then a slit (16 in Figure 15-10) is added in the first (lower) focal plane of the condenser. Turning the slit parallel to the grid, the image seen in the microscope suddenly shows superimposed on it a series of parallel lines, the shadow projection of the grid. If there are no gradients present, these lines are straight. But, if there are gradients, the moiré pattern becomes distorted as shown in Figure 15-11.

Besides transparent matter, we can also analyze *surfaces*. That is called *moiré topography*. Moiré topograms are obtained either by placing a single grid next to a (preferably white) surface and observing or photographing the shadow cast on the surface through the same grid. The shadow, and the grid through which the shadow is seen, superimpose, producing the moiré effect. That is called the *shadow method*.

Or the image of one grid can be projected onto the surface and the image be observed through another grid. That is called the *projection method*. Both versions find use in engineering for measuring stresses and strains, and in goniometry and surface topology. Compared with interferometry, moiré systems are simple to operate; they do not require the high mechanical stability that is essential with interferometry.

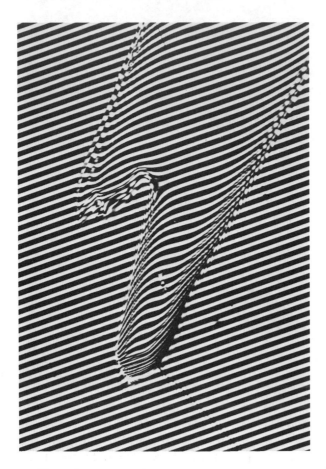

Figure 15-11 Photomicrograph of a thin layer of colorless cement spilled across a glass slide. Thickness gradients near the boundary of the layer cause characteristic distortions.

Example

Fringes obtained by either the shadow method or the projection method are easy enough to see, but how can we use them to draw quantitative conclusions? Assume that the test object is a sphere 16 mm in diameter, its curvature comparable to that of the cornea of the eye, with a small defect in its surface. On a photographic print we see that the defect has caused some of the fringes to become distorted by $\frac{1}{5}$ of their spacing. How deep is the defect?

Solution: First we measure the diameter of the sphere as it appears on the print. If that diameter is 36 mm, the print is magnified $\frac{36}{16}$ times. Then we measure the spacing of the fringes. If the spacing is 2.25 mm, the fringes (on the actual sphere) are separated by

$$2.25 \times \tfrac{16}{36} = 1 \text{ mm}$$

Therefore, the depth of the defect is

$$1 = \tfrac{1}{5} = \boxed{0.2 \text{ mm}}$$

well within the accuracy of the method (estimated to be 0.1 mm). To determine the depth of a defect, in short, we need a mere four measurements, three of them taken on a photograph, without knowing the dimensions of the actual experiment, the magnification, or the rotation of one grid relative to the other.

SUMMARY OF EQUATIONS

Grating equation:

$$d \sin \theta = m\lambda \qquad [15\text{-}1]$$

Resolvance:

$$\frac{\lambda}{\Delta\lambda} = Nm \qquad [15\text{-}14]$$

Bragg's law:

$$2d \sin \theta = m\lambda \qquad [15\text{-}15]$$

PROBLEMS

15-1. A grating, used in the second order, diffracts light of 400 nm wavelength through an angle of 30°. How many lines per millimeter does the grating have?

15-2. A grating has 8000 slits ruled across a width of 4 cm. What is the color of the light whose two fifth-order maxima are 90° apart?

15-3. Red light of 632 nm wavelength is displaced 20 cm from the center of a meter stick mounted 60 cm in front of a grating. Considering the first order only, how many lines per millimeter does the grating have?

15-4. Blue light of 470 nm wavelength is diffracted by a grating ruled with 500 lines per millimeter. Considering that no light can be diffracted more than 90° away from the axis, what is the highest order possible?

15-5. If you compare photographs of a certain line spectrum, one taken with a prism spectrograph and the other with a grating spectrograph, how can you tell which is which?

15-6. Collimated light containing the wavelengths 600 nm and 610 nm is diffracted by a plane grating ruled with 60 lines to the millimeter. If a lens of 2 m focal length is used to focus the light on a screen, what is the linear distance between these two lines in the first order?

15-7. The spectrum of mercury contains, among others, a blue line of 435.8 nm and a green line of 546.1 nm. If these two lines are to have an angular separation of at least 7°, and if the grating available has 500 lines/mm, in which order must the grating be used?

15-8. Red light of 630 nm wavelength, diffracted by a grating in a given order, overlaps blue light of 475 nm wavelength, diffracted by the same grating in the next-higher order. What are the two orders?

15-9. If a diffraction grating is 4 cm wide and if it has 50 lines per millimeter and is used in the first order, what is the least separation of wavelengths that it can resolve near 500 nm?

15-10. How many lines must a grating have, used in the second order and near 550 nm, to resolve two lines 0.1 Å apart?

15-11. X rays of wavelength $\lambda = 1.3$ Å, incident on a crystal, are diffracted at an angle, in the first order, of 22°. What is the interplanar spacing?

15-12. How far apart are the diffracting planes in a NaCl crystal for which X rays of wavelength 1.54 Å in the first order make a glancing angle of 15.9°?

15-13. Two identical grids are laid on top of each other. What angle must one grid subtend with the other so that moiré lines result that have twice the spacing in the original grids?

15-14. Moiré-fringe systems can be *calibrated* by using a small, low-power lens as the object. What will the image look like and how can it be used for calibration?

SUGGESTIONS FOR FURTHER READING

Davis, S. P. *Diffraction Grating Spectrographs.* New York: Holt, Rinehart and Winston, 1970.

Hutley, M. C. *Diffraction Gratings.* New York: Academic Press, Inc., 1982.

Kafri, O., and I. Glatt. *The Physics of Moiré Metrology.* New York: Wiley-Interscience, 1990.

McLachlan, D., Jr., and J. P. Glusker. *Crystallography in North America.* New York: American Crystallographic Association, 1983.

16

Light Scattering

It is now, I believe, generally admitted that the light which we receive from the clear sky is due in one way or another to small suspended particles which divert the light from its regular course.'' With these words John William Strutt, later Lord Rayleigh, began his first paper on *light scattering*. Light scattering is a phenomenon that occurs widely in nature and is of great utility. It accounts for the blue of the sky and the white of the clouds. Interference, as we saw earlier, can be explained by the wave nature of light, regardless of whether the waves are longitudinal or transverse. Scattering requires the light to be transverse. Indeed, light waves consist of transverse, time-varying electric and magnetic "fields." These fields can be described by *Maxwell's equations*.

MAXWELL'S EQUATIONS

Light waves are often compared to mechanical waves. But light is fundamentally different. For example, light requires no medium for propagation, a fact that for a while posed much difficulty when trying to explain the nature of light. The problem was solved by Clerk Maxwell, who showed that light waves are not a mechanical but an electromagnetic phenomenon that can be described by two vectors, the amplitude of the electric field strength, $\mathbf{E}$, and the amplitude of the magnetic field strength, $\mathbf{H}$. These vectors oscillate at right angles to each other and to the direction of propagation. They cannot be separated. They can be expressed in the form of four fundamental equations known as *Maxwell's equations.**

The two vectors go through zero at the same time. By intuition perhaps one could think otherwise, assuming, since one of the fields *induces* the other, that they might differ in phase by 90°. But it seems that this is so only with *standing* waves (Figure 16–1). *Moving* waves proceed without a phase shift.†

Maxwell's first equation can be derived from *Coulomb's law*. This law refers to the force, **F**, between two electric charges *at rest*,

$$\mathbf{F} = \frac{1}{4\pi\varepsilon_0}\frac{q_1 q_2}{R^2}\,\hat{\mathbf{R}}$$ [16-1]

where q_1 and q_2 are the charges separated by a distance R, $\hat{\mathbf{R}}$ is a unit vector indicating the direction from one charge to the other, and ε_0 is the electric permittivity of free space, approximately 8.85×10^{-12}F m^{-1}.

From Coulomb's law we derive *Gauss' law*. It describes the relationship between the electric flux through a closed surface and the charge enclosed by the surface. Expressed in differential form, Gauss' law is in fact *Maxwell's first equation*,

$$\nabla \cdot \mathbf{E} = \frac{\rho}{\varepsilon_0}$$ [16-2]

---

* James Clerk Maxwell (1831–1879), Scottish mathematician and physicist. At age 15, Clerk Maxwell (his last name!) presented a paper "On the Description of Oval Curves and Those Having a Plurality of Foci" before the Royal Society, at 16 became interested in optics when he had the chance of visiting Nicol, who gave him a pair of polarizing prisms. Using these, Clerk Maxwell went to work on photoelasticity, made major contributions also to refraction in nonhomogeneous media, color vision, and heat transfer, and correctly explained the rings of Saturn. In 1864 he read a paper before the Royal Society, published a year later: J. Clerk Maxwell, "A Dynamical Theory of the Electromagnetic Field," *Philos. Trans. Roy. Soc. London* **155** (1865), 459–512. Near the end of part I, on page 466, Clerk Maxwell says that "light itself . . . is an electromagnetic disturbance in the form of waves propagated through the electromagnetic field according to electromagnetic laws."

Some 22 years later, in 1887, Heinrich Rudolf Hertz (1857–1894), German physicist, professor of physics at the Technische Hochschule in Karlsruhe, found the waves that Clerk Maxwell had predicted. During one of his lecture demonstrations on electromagnetic induction, Hertz was surprised to see that the spark produced by the discharge of a Leyden jar could be detected some distance away. The detector he used was a loop of wire with a small gap in it: discharge of the Leyden jar caused a tiny spark to jump across that gap. Since the distance was too large for direct induction, the only alternative seemed to be that the secondary spark was produced by some kind of radiation propagating through space. With a setup as simple as this, Hertz found that the waves were transverse, that their velocity was the same as that of light, and that they could be reflected, refracted, polarized, and brought to interference, just like light waves. In 1901, Guglielmo Marconi (1874–1937), Italian scientist and businessman, transmitted them across the Atlantic Ocean.

† So far I have found only one textbook that shows both standing and moving waves, one type with and the other without a phase shift: A. N. Matveev, *Optics* (Moscow: Mir Publishers, 1988), pp. 31–32 and 47.

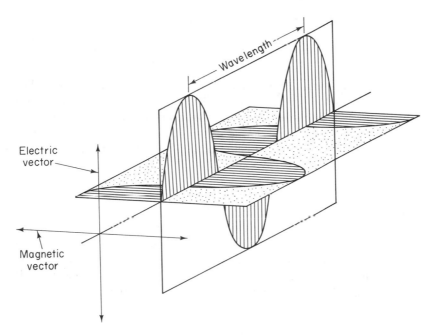

Figure 16-1 Electromagnetic wave showing phase shift between electric and magnetic fields.

where $\nabla \cdot$ is the divergence operator, $\mathbf{E}$ the electric field strength, and ρ the charge density.

Next we consider *moving charges*, that is, we consider a *current*. A current, i, in an element of wire dW, will set up a *magnetic induction field*, $\mathbf{B}$. As before, R is the distance from dW to the test point (Figure 16-2).

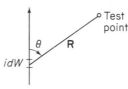

Figure 16-2 Determining the magnetic induction produced by a current.

The moving charge per wire element, $i\, dW$, makes a contribution dB,

$$dB = \mu_0 \frac{1}{4\pi} \frac{1}{R^2} i\, dW \sin\theta$$

or, in vector notation,

$$d\mathbf{B} = \mu_0 \frac{1}{4\pi} \frac{1}{R^2} i\, d\mathbf{W} \times \hat{\mathbf{R}} \qquad [16\text{-}3]$$

The total induction, **B**, is found by integration:

$$\mathbf{B} = \mu_0 \frac{1}{4\pi} i \int \frac{1}{R^2} d\mathbf{W} \times \hat{\mathbf{R}}$$ [16-4]

This is *Biot–Savart's law* where μ_0 is the magnetic permeability of free space, $4\pi \times 10^{-7} NA^{-2}$. The direction of **B** is given by the vector cross product $d\mathbf{W} \times \mathbf{R}$; it points in a direction *normal* to the plane defined by **W** and **R**.

In analogy to the charge density ρ, we replace the current i by the current density, $\mathbf{j} = i/\mathbf{A}$. If **j** is constant as in $\nabla \cdot \mathbf{j} = 0$, then all that is left is $\mathbf{j} \cdot \nabla x(\hat{\mathbf{R}}/R)$, which becomes zero, so that

$$\nabla \cdot \mathbf{B} = 0$$ [16-5]

which is *Maxwell's second equation.*

While Maxwell's first equation deals with a *steady* electric field, and the second equation with a steady magnetic field, we now come to *fields that vary as a function of time.* Assume that a bar magnet is thrust through a loop of wire. This will cause an electromotive force, EMF, to occur in the wire. Its magnitude is proportional to the time rate of change of the magnetic flux, ϕ,

$$\text{EMF} = -\frac{d\phi}{dt}$$ [16-6]

which is *Faraday's law of induced electricity.*

From Faraday's law we derive *Maxwell's third equation,*

$$\nabla \times \mathbf{E} = -\frac{\partial \mathbf{B}}{\partial t}$$ [16-7]

which means, the same as Faraday's law, that whenever a magnetic field changes with time, it generates an electric field.

Finally we come to the reverse of Faraday's law, expecting that a changing electric field generates a magnetic field.

We use again Biot–Savart's law, extending it to fluctuating currents. Consider a capacitor with charge flowing in, which then accumulates on one of the plates. The plate, as shown in Figure 16-3, is surrounded by a closed surface. While the capacitor is being charged, current i is going in and apparently none is coming out but, to assure continuity, Clerk Maxwell reasoned that the same

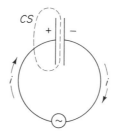

Figure 16-3 Deriving concept of displacement current.

amount of *displacement current* must come out. That leads to Maxwell's fourth equation,

$$\nabla \times \mathbf{B} = \mu_0 \left(\mathbf{j} + \varepsilon_0 \frac{\partial \mathbf{E}}{\partial t} \right)$$ [16-8]

Clearly, displacement current exists only whenever the electric field *changes* ($\partial \mathbf{E}/\partial t$); for steady currents, the last term drops out.

Maxwell's first equation, and Coulomb's law, show that an assembly of charges generates a steady electric field. Maxwell's second equation, and Biot–Savart's law, refer to a steady magnetic field. Maxwell's third equation, and Faraday's law, tell us that a time-varying magnetic field causes an electric field. Maxwell's fourth equation is the final link: it shows that a time-varying electric field causes a magnetic field.

From the vector cross products in Maxwell's third and fourth equations we see that the two fields are normal to each other. They also are normal to the direction of propagation. Most significantly, the electric field arises directly from the (time-varying) magnetic field, and the magnetic field arises directly from the (time-varying) electric field. *No loop of wire, or any other medium, is necessary: Electromagnetic waves by themselves propagate through free space.*

Electromagnetic waves carry energy. That should be obvious from the fact that light, even when it comes from a distant star, still can initiate the process of vision. The *Poynting vector of energy flow*, **S**, is given by the cross product of the electric field **E** and the magnetic field **H**,

$$\mathbf{S} = \mathbf{E} \times \mathbf{H}$$ [16-9]

Presumably, the total energy contained in an electromagnetic wave is equally shared by the two fields; however, the **E** field seems to be more effective than the **H** field. Most phenomena caused by light, from the photoelectric effect to the dissociation of silver halides in photographic film to the response of the retina, are due to the **E** field; the only exception is probably the *pressure* exerted by electromagnetic radiation.

The *velocity of propagation* of electromagnetic energy in free space is

$$c = \frac{1}{\sqrt{\varepsilon_0 \mu_0}}$$ [16-10]

an important relationship that defines the *velocity of light in terms of two electrical constants*.

RAYLEIGH SCATTERING

Now we turn to *light scattering*. Although the term *scattering* as such seems to refer to something that spreads apart, *light* scattering presupposes, first, an *interaction* of the light with matter and only then that the light spreads apart. We distinguish two major classes of light scattering, *Rayleigh scattering* and *Mie scattering*.

It all began when John Tyndall, looking from the side into a beam of white light passing through air or through a variety of gases, noticed a faint blue shine that seemed to emerge from the beam. He thought that this light, of "a colour rivalling that of the purest Italian sky," was due to the presence of small particles.*

At first Rayleigh agreed. Later he recognized that it is the molecules (of the air), rather than other particles, that account for the blue of the sky. Since then, *Rayleigh scattering* has become synonymous with *molecular scattering*. The approximate upper limit of the size of the particles that cause such scattering is *one-tenth of the wavelength of light*.†

Scattering as Dipole Radiation

Consider light that is incident on a single molecule. Without ionization, a molecule contains an equal number of positive and negative charges. Under the influence of the light's electric field, the positive charges are displaced in one direction and the negative charges are displaced in the opposite direction, forming an *electric dipole* (Figure 16-4).

It is convenient to describe the properties of a dipole in terms of its *dipole moment, p*:

$$\mathbf{p} = q\mathbf{d} \qquad\qquad [16\text{-}11]$$

---

* John Tyndall (1820–1893), Irish physicist. Working as a surveyor, railroad engineer, and science teacher, Tyndall obtained a Ph.D. in mathematics, studied chemistry under Bunsen, met Faraday, became professor of physics at the Royal Institution, and, when Faraday retired, succeeded him as superintendent. Tyndall wrote some 30 books, is noted for his work on light scattering, designed foghorns and both emission and absorption spectrophotometers, discovered the effect of penicillin on bacteria, and studied the mechanics of glacier motion, the latter no doubt because he was an accomplished mountain climber who, in 1861, made the first ascent of the Weisshorn, 4512 m, near Zermatt, in the Swiss Alps. J. Tyndall, "On the Blue Colour of the Sky, the Polarization of Skylight, and on the Polarization of Light by Cloudy matter generally," *Philos. Mag.* (4), **37** (1869), 384–94.

† John William Strutt, third Baron Rayleigh (1842–1919), British physicist, professor and later chancellor of Cambridge University, and successor of Clerk Maxwell as director of the Cavendish Laboratory. His first paper on light scattering was J. W. Strutt, "On the Light from the Sky, its Polarization and Colour," *Philos. Mag.* (4), **41** (1871), 107–120, 274–79. Rayleigh also made important contributions to photography, resolving power, zone plates, blackbody radiation, color vision, and acoustics. On his estate at Terling, he had several darkrooms, one of them painted black with a mixture of soot and beer. It was here that he discovered argon, which earned him the 1904 Nobel Prize in physics. In 1899, Rayleigh was admiring the sight of Mt. Everest from a place near Darjeeling, 160 km away. At that distance, the outline of the mountain could barely be seen in the haze. From the degree of visibility and the refractive index of air, Rayleigh calculated the number of molecules per milliliter of air, and found 3×10^{19}, a figure close to today's value.

Figure 16-4 Electric dipole consists of two charges of opposite sign, separated by distance *d*.

where *q* is the charge and *d* the distance between the charges. With an external electric field **E**, the induced dipole moment is

$$\mathbf{p} = \alpha\mathbf{E} \qquad [16\text{-}12]$$

where α is the *polarizability* of the molecule. If the field varies sinusoidally as a function of time, the dipole moment will vary as well:

$$\mathbf{p}_t = \mathbf{p}_0 \sin(\omega t) \qquad [16\text{-}13]$$

The external field, in short, will make the dipole *oscillate*, in synchrony with the impinging field.

The positive charges in a molecule are relatively heavy (they represent the atom's nuclei) and, under the influence of the field, move very little. The negative charges (the electrons) are much lighter and move much farther. Thus the induced oscillations affect mainly the electrons. But electrons that undergo oscillatory accelerations emit electromagnetic waves, just as a dipole antenna emits radio waves; hence, the incident light is almost instantaneously *reradiated*. Light, in other words, does not simply "pass through" transparent matter; instead, it forms dipoles, these dipoles radiate, and this reradiation appears as light.

Assume the incident field is oscillating up and down, along the *y*-axis as shown in Figure 16-5. Consequently, the induced dipoles also oscillate along the *y*-axis. An observer at *S* will see these oscillations as light; in fact, the light is seen from any direction, straight ahead, from the side, and from the rear, *except from a direction in line with the dipole axis*.

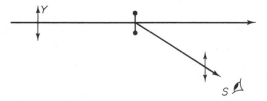

Figure 16-5 Light oscillating up and down, incident from the left, forming dipole. Observer at *S*.

But now let the observer look from a direction *in the plane of the drawing* (the paper plane). Call θ the angle subtended by the forward direction and the direction of scattering. In the forward direction where $\theta = 0$, and in the backward direction where $\theta = 180°$, the scattered light is well visible. But when looking at the dipole from directly above or directly below (where $\theta = 90°$), there will be no light. That is because electromagnetic waves are transverse, rather than longitudinal, the same as with a dipole antenna which radiates in all directions, *except* in the direction of its own length.

In the horizontal plane, in short, the amplitude of the scattered light is the same all around but in the vertical plane it varies as a function of cos θ; the intensity, following $I \propto A^2$, varies as cos$^2 \theta$. Combining the effects in three-dimensional space, a plot of the intensity looks like a doughnut (Figure 16-6).

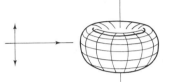

Figure 16-6   Three-dimensional plot of linearly polarized light scattered by a dipole.

Rayleigh Scattering as a Function of Wavelength

Now let a bundle of light pass through a volume of scatterers of thickness Δx. We call I_0 the intensity of the light incident on the volume and I' the intensity passing through. A certain fraction of the light, i, is scattered out of the path and lost:

$$I_0 - I' = i \qquad [16\text{-}14]$$

The magnitude of this loss is proportional to I_0, to the thickness Δx, and to a constant of proportionality called *turbidity*, τ:

$$I' - I_0 = -i = I_0 \tau \, \Delta x \qquad [16\text{-}15]$$

Turbidity, in other words, is the *fractional loss* of light, the ratio i/I_0, per unit thickness; it is a term comparable to *absorptivity* (see Chapter 22) although I hasten to add that in pure scattering there is no absorption, or irretrievable loss of light: all the light removed from the incident beam is reemitted again almost instantaneously in the form of scattered light.

The amplitude of the induced oscillation (of a dipole) increases as the driving frequency (of the incident light) approaches the natural frequency of oscillation of the molecule. Because the natural frequency of a typical molecule is comparable to the frequency of ultraviolet radiation, more light is scattered at shorter wavelengths; this is illustrated in Figure 16-7.

The amplitude of the scattered light is inversely proportional to the *square* of the wavelength and, since $I \propto A^2$, the intensity is inversely proportional to the *fourth power* of the wavelength,

$$\boxed{i \propto \frac{1}{\lambda^4}} \qquad [16\text{-}16]$$

which is the essential point of Rayleigh scattering.

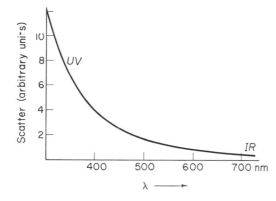

Figure 16-7 Rayleigh scattering: plot of intensity as a function of wavelength.

Example

How much more Rayleigh scattering is produced by light of 528 nm wavelength than by light of 628 nm wavelength?

Solution: From Equation [16-16]

$$i \propto \frac{1}{\lambda^4}$$

we find that

$$\frac{i_1}{i_2} = \left(\frac{\lambda_2}{\lambda_1}\right)^4 = \left(\frac{628}{528}\right)^4 = \boxed{2.0}$$

and, hence, *twice* as much.

Applications

Light scattering can be used to determine the weight of almost any particles, from simple molecules to colloids, polymers, and proteins and other substances of biochemical interest. Conversely, if the composition is known, the number of molecules per unit volume can be found.

Observations are usually made of the intensity of the scattered light as a function of the angle θ of scattering relative to the forward direction. The quantity actually measured is the *Rayleigh ratio*, R_θ, a parameter that is defined as

$$R_\theta = \frac{i_\theta}{I_0} R^2 \qquad \text{[16-17]}$$

and that, as I have indicated by the subscripts, is a function of θ. But, from Equation [16-15], $i/I_0 \propto \tau$, and hence R_θ is also proportional to τ_θ.

Most instruments for measuring light scattering follow the outline shown in Figure 16-8. Light comes from a source (left) and is collimated by a lens, L. After

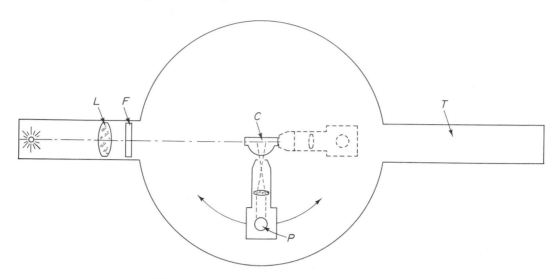

Figure 16-8   Schematic diagram of light-scattering apparatus.

passing through an optional color or polarizing filter, F, the light is incident on a cell, C, containing the scattering medium. Preferably the cell can be quickly exchanged by an opal-glass reference standard, to permit calibration. After passing through the cell, the primary light is absorbed in a light trap, T.

Scattered light from the cell or from the reference standard goes through another slit-lens assembly and perhaps another polarizing filter to a photodetector, P, mounted so that it can be swung around with the cell as the pivot. Its output is plotted versus the angle of rotation, the dashed outline in the figure indicating the forward position where $\theta = 0$.

MIE SCATTERING

Rayleigh Scattering Versus Mie Scattering

When the particles are larger than about one-tenth of a wavelength, the light scattered from one point on the particle may well be out of phase with light scattered from another point. The two contributions then interfere and the scattered-light distribution is no longer symmetrical as in the Rayleigh case. With increasing particle size, in fact, the scattered light becomes more concentrated in the forward direction (Figure 16-9). The difference between forward and back scatter is called *dissymmetry*, which is often defined as the ratio $i_{45°}/i_{135°}$. Dissymmetry can be used to determine the average size of particles, provided that their shape is known. For larger particles, however, the angular dependence of the scattering becomes quite complicated, showing a number of maxima and

Small particles: Rayleigh scattering

Large particles: Mie scattering

Figure 16-9 Angular distribution in Rayleigh and Mie scattering.

minima, and cannot be adequately described by a single parameter such as the dissymmetry.

Mie's theory* takes into account the size of the particles but also their refractive index, and the refractive index of the surrounding medium, as well as the shape, dielectric constant, and absorptivity of the particles. It is based on a formal solution of Clerk Maxwell's equations, leading to a series, in theory an infinite number, of partial waves. The amplitudes of these waves decrease rapidly as the particles become smaller. In Rayleigh scattering, we need to consider only the first few terms, and may neglect all others. Mie scattering, by contrast, requires many more terms; it is more general and includes Rayleigh scattering as a special case.

Mie Scattering Parameters

Assume that light of cross section A is incident on a volume V of thickness Δx, $V = A \Delta x$. If there are N particles per unit volume, the actual volume contains NA Δx particles. If r is the radius of a particle, each particle has a cross section πr^2 and together they have a cross section

$$\sum = NA \, \Delta x \, \pi r^2 \qquad [16\text{-}18]$$

provided that no one particle lies in the shadow of another particle; otherwise, the cross section is less. In most cases, the distances between particles are much

---

* Gustav Adolf Feodor Wilhelm Ludwig Mie (1868–1957), German physicist. Mie began his studies at Rostock and then, following the custom of the time, went to another university, Heidelberg, to major in mathematics and mineralogy. He wrote a thesis on a "Very abstract problem of partial differential equations," became physics assistant at the Technische Hochschule in Karlsruhe and, in 1901, professor of physics at the University of Greifswald where he published the paper on light scattering that made him famous: G. Mie, "Beiträge zur Optik trüber Medien, speziell kolloidaler Metallösungen" (Contributions to the optics of turbid media, in particular colloidal metal solutions), *Ann. Phys.* (4), **25** (1908), 377–445. He also wrote a textbook on electricity and magnetism and worked on theories of matter and energy compatible with his strongly held religious views, of dielectric constants of various materials, and of x-ray crystallography.

larger than their radii, so the actual cross section is about the same as the maximum possible.

But then it turns out that the cross section of the light affected by a particle is not necessarily equal to the cross section of the particle. The reason is the refractive index difference between particles and the medium between them. The ratio of these two cross sections is the *extinction factor, K*. This factor is generally larger than unity, which means that the extinction by a particle is *higher* than what could be expected from its physical size. The ratio i/I_0 in Mie scattering, hence, is

$$\frac{-i}{I_0} = \frac{NA\ \Delta x\ K\pi r^2}{A} \qquad [16\text{-}19]$$

Combining this equation with our earlier Equation [16-15], canceling A and Δx, and replacing τ by the *Mie extinction coefficient, μ*, gives

$$\mu = NK\pi r^2 \qquad [16\text{-}20]$$

which shows that μ is equal to the product of the number of particles times $K\pi r^2$, a term called the *extinction cross section*.

The essential point is that there is *no λ^{-4} relationship*. Indeed, for sufficiently large particles there is no wavelength dependency at all. This is one of the characteristics that distinguish Mie scattering from Rayleigh scattering; it is, of course, the reason why most clouds are white. These clouds, and fog, mist, and aerosol sprays, are composed of droplets at least 10 μm in diameter, 200 times the $\lambda/10$ limit referred to before.

LIGHT SCATTERING AND METEOROLOGY

Light scattering accounts for some of the most beautiful colors that we see in nature. Some examples are the following.

Blue Sky

The blue of the sky is a matter of *Rayleigh scattering*: light of shorter wavelength is scattered more than light of longer wavelength. Indeed, blue is *scattered out* of the light we receive from the sun and therefore the sun appears *yellow*. On the other hand, when we look at the sky in any direction but at the sun, some of the light is *scattered into* the line of sight and the sky appears *blue*. As seen from a high mountain, where there is not much air, or from the moon, where there is none, the sky is black.

The atmosphere may also contain aerosols, colloidal particles both natural and manufactured, that are considerably larger than molecules. Scattering by aerosols does not follow the λ^{-4} law; it is a matter of *Mie scattering*. As seen through polluted air, the sky is not as deeply blue anymore; it looks more gray.

Clouds consist of water droplets, there is no wavelength dependency, and therefore clouds appear *white*.

Airport Lights

Navigational lights seen at an airport are generally green, white, or red. But the lights lining the taxiways are *blue*. Seen from high up in the air, blue is scattered out and all but invisible; thus, the pilot is not misled to land there.

Red Sunset

With the sun high above the horizon and the path through the atmosphere relatively short, only blue is scattered out and the sun appears yellow. With the sun close to the horizon, the light traverses a much longer path, all but red is scattered out and the sun appears *red*.

Green Flash

The green flash is due to a combination of refraction, dispersion, selective absorption, and scattering. Because of refraction in the gradient-index atmosphere, light from the sun reaches the observer on Earth in a curved path. The light is dispersed: blue is refracted more, reaching the observer in a slightly steeper trajectory than red. The blue image of the sun's disk, therefore, appears *above* the red image, separate by about $\frac{1}{60}$ of the sun's diameter. However, the blue and the green segment are so narrow that, with the sun still above the horizon, they are barely visible at all. But as the sun disappears below the horizon, the red disappears too, the orange and yellow are attenuated by absorption in water vapor, oxygen, and ozone, and the blue is lost to Rayleigh scattering. That leaves the green which sometimes, very briefly, is seen as the *green flash*.

SUMMARY OF EQUATIONS

Velocity of electromagnetic radiation:

$$c = \frac{1}{\sqrt{\varepsilon_0 \mu_0}}$$

[16-10]

Rayleigh scattering:
    caused by molecules up to $\lambda/10$ in size,

$$i \propto \frac{1}{\lambda^4}$$

[16-16]

Mie scattering:
    caused by particles larger than $\lambda/10$, no wavelength dependency.

PROBLEMS

16-1. A long straight wire carries a steady current of 6 A. Using Biot–Savart's law, determine the magnetic induction caused by the current within 1 cm length of wire, at a point 20 cm from the wire.

16-2. A point light source is placed:
(a) In the focus of a collimating lens.
(b) In the focus of a paraboloid mirror.
(c) In one focus of a closed ellipsoid cavity, concentrating the light into the other focus.
Does the system, because of radiation pressure, experience any recoil?

16-3. At a wavelength of 546 nm the scattering intensity of pure carbon tetrachloride is found to be 5.9 arbitrary units. What is the intensity at 436 nm?

16-4. Light of 760 nm and 810 nm wavelength is passed through a turbid medium. If 20 percent of the shorter wavelength is lost due to Rayleigh scattering, what percentage is lost of the longer wavelength?

16-5. Infrared radiation of 1.06 μm wavelength, when passing through 10 km of atmosphere, loses (because of Rayleigh scattering) 1.85% of its initial strength. What percentage of red light of 694.3 nm wavelength is lost in the same distance?

16-6. If light of 635 nm wavelength causes a certain amount of Rayleigh scatter, light of what wavelength will give 10 times as much scatter?

16-7. A certain amount of light of 503.3 nm wavelength is scattered out of a beam passing through a volume of clean air at atmospheric pressure. What percentage of that amount is scattered at 632.8 nm wavelength at one-half the pressure?

16-8. Light scattering in pure water, such as in Oregon's Crater Lake, is Rayleigh scattering; hence, the water is very blue. But water that comes out of a glacier is nearly white ("glacier milk"). Discuss the reason for this in terms of both Rayleigh and Mie scattering.

16-9. Automobile foglights are often yellow, presumably because people believe that yellow light can penetrate a dense fog (composed of relatively large droplets) better than white light. Discuss the merits of such a scheme in terms of both Rayleigh and Mie scattering.

16-10. Light scattering is one of several methods that can be used for building a *smoke detector* for the home. If you were given a small light source of low power consumption, such as a light-emitting diode, and a photocell as the receiver, how would you arrange these components?

SUGGESTIONS FOR FURTHER READING

BOHREN C. F., and D. R. HUFFMAN. *Absorption and Scattering of Light by Small Particles*. New York: John Wiley and Sons, Inc., 1983.

LORRAIN, P., D. CORSON, and F. LORRAIN. *Electromagnetic Fields and Waves*, 3rd ed. New York: W. H. Freeman, 1988.

NIETO-VESPERINAS, M. *Scattering and Diffraction in Physical Optics*. New York: Wiley-Interscience, 1991.

VAN DE HULST, H. C. *Light Scattering by Small Particles*. New York: Dover Publications, Inc., 1981.

17

Polarization of Light

The term *polarization* actually has two meanings. It refers to a *property* of light, be it linear, elliptical, circular, or partial polarization. But it also refers to the *process* of producing polarized light, be it by scattering, reflection, selective absorption, or double refraction. These various properties and processes I will discuss in this chapter. I begin with a summary of different *types of polarized light*.

TYPES OF POLARIZED LIGHT

1. Light as it comes from the sun or from most other light sources is *unpolarized*. It is a mixture of light polarized in different directions and to different degrees. These differences result because light is emitted by groups of atoms, each group oscillating independently from another. While within each wavetrain the electric field, the **E** vector, oscillates in the same mode and direction, from train to train the modes and directions of oscillation vary; the result, therefore, is a sequence of oscillations *oriented at random*.

2. If the **E** vector oscillates in a given, constant orientation, the light is said to be *linearly polarized* (Figure 17-1). The term *plane polarized* is deprecated, for reasons that will become apparent soon.

3. If linearly polarized light contains an additional component of natural (unpolarized) light, it is called *partially* (linearly) *polarized*. Such light could come from a perfectly polarizing filter with some holes in it or from a low-quality filter.

Figure 17-1   Linearly polarized light. Projection of wave on a plane intercepting the axis of propagation gives a *line*, hence the term *linear polarization*.

4. In *circular polarization*, the **E** vector no longer oscillates in a plane, as in linear polarization. Instead, the vector retains its magnitude and proceeds in the form of a helix *around* the axis of propagation (Figure 17-2). Within one wavelength, the vector completes one revolution. If the tip of the vector, when seen looking toward the light source, rotates clockwise, the light is said to be right-circularly polarized; if counterclockwise, left.

Figure 17-2   Right-circularly polarized light.

5. *Elliptical polarization* stands between circular and linear polarization. It is the most general type of polarization; linear and circular are the two extremes of elliptical polarization (Figure 17-3). The tip of the vector now proceeds in the form of a *flattened* helix; the vector rotates and, at the same time, it changes in magnitude.

Figure 17-3   Circular, elliptical, and linear polarization.

The term *plane polarization*, as opposed to *linear*, is based on a three-dimensional convention. Circularly polarized light need then be called "helically polarized." To avoid awkward terms such as *circular helical* and *elliptical helical*, I prefer a two-dimensional convention with the terms *linear*, *elliptical*, and *circular*.

6. *Optical rotation* (optical activity) is entirely different. It is a property of matter, rather than a property of light. We will discuss the molecular basis of optical activity later in this chapter.

PRODUCTION OF POLARIZED LIGHT

Light can be polarized in various ways. We distinguish the following.

Polarization by Scattering

Consider a bundle of unpolarized light that passes through an assembly of small particles in suspension (Figure 17-4). As we have seen before, an observer looking

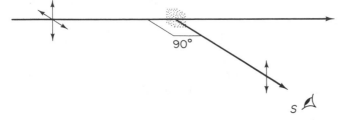

Figure 17-4 Light scattered at right angles to the direction of propagation is linearly polarized.

at the particles from a direction at right angles to the direction of propagation finds the light to be *linearly polarized*. The direction of oscillation of the scattered light is *normal* (perpendicular) to the plane defined by the direction of propagation and the direction of observation.* (Clearly, the light scattered toward S cannot be polarized *parallel* to the incident light because that would require the incident light to oscillate back and forth along the axis, "longitudinally," and there is no such light.)

Polarization by Reflection

Assume that light is reflected at the surface of matter. Actually, the light penetrates a short distance into the matter where it induces molecular oscillations. As in polarization by scattering, the reflected light can oscillate only *normal* to the plane of incidence and hence, the reflected light is *linearly polarized* (Figure 17-5). The plane of incidence, it is important to note, is *not* the surface on which the light is incident; instead, it is the plane defined by the surface normal and the incident, and the reflected, light.

Maximum polarization occurs when the reflected light and the refracted light (not shown) subtend an angle of 90°. If that is the case,

$$I + 90° + I' = 180°$$

$$I' = 90° - I$$

Since

$$\sin(90° - I) = \cos I$$

Snell's law, $n \sin I = n' \sin I'$, becomes

$$\frac{\sin I}{\cos I} = \frac{n'}{n}$$

---

* The polarization of light by scattering was first observed by Lord Rayleigh who used colloidal sulfur as the scattering medium, adding acetic acid to a weak solution of photographic fixer (sodium thiosulfate). A more convenient way to produce a colloidal suspension is by adding to water, and stirring, a few grains of nonfat dry milk.

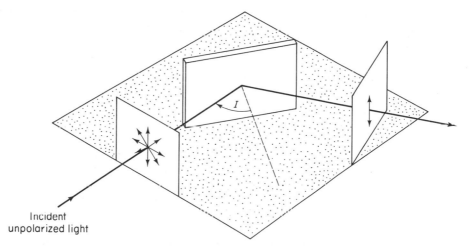

Figure 17-5   Polarization by reflection. The plane of incidence, defined by the incident light and the surface normal, is horizontal.

and, since

$$\frac{\sin I}{\cos I} = \tan I$$

$$\boxed{\tan I = \frac{n'}{n}}$$ [17-1]

which is *Brewster's law.** Light reflected at any other angle but Brewster's is partially linearly polarized.

Polarization by Transmission

Each time the light is reflected at Brewster's angle, the reflected component is completely linearly polarized. Consequently, in the light transmitted through the surface, that component is missing. If the process is repeated often enough, using

---

* Named after Sir David Brewster (1781–1868), Scottish physicist, professor of physics at St. Andrews College. Initially a minister in the Church of Scotland, Brewster became interested in optics, found the angle named after him, contributed also to dichroism, absorption spectra, and stereophotography, invented the kaleidoscope, and wrote a book about it. A prolific writer, Brewster edited several journals, published a 526-page *Treatise on Optics*, the *Life of Sir Isaac Newton*, several more books, and some 315 journal articles. Brewster's law, in his own words, states that "when a ray of light is polarised by reflexion, the reflected ray forms a right angle with the refracted ray." D. Brewster, "On the laws which regulate the polarisation of light by reflexion from transparent bodies," *Philos. Trans. Roy. Soc. London* **105** (1815), 125–59.

a stack of plates, the transmitted light becomes more and more purely linearly polarized. This method of polarization is of interest in the infrared where thin films of selenium take the place of the glass.

Polarization by Selective Absorption

Polarization by selective absorption is an important case which can be best understood by the example of the *wire-grid polarizer*.

A wire-grid polarizer is a grid of parallel wires. In 1888, Heinrich Hertz used such grids as polarizers to test the properties of radiowaves he had discovered the year before. The electric field impinging on the grid may be resolved into two orthogonal components, one oscillating parallel to the wires and the other normal to them. The parallel component drives the conduction electrons in the wires along their lengths; the current encounters resistance, heats up the wire, and thereby loses energy. The normal component, however, has no electrons to drive very far and hence passes through without much loss. In short, it is the *normal* component that goes through (contrary to what one might think).

What the wire-grid polarizer does for longer waves can be accomplished for visible light by a *dichroic crystal*. A good example is *tourmaline*, an aluminoborosilicate containing Al_2O_3, B_2O_3, and SiO_2. On passing through the crystal, the light is split into two components, the horizontal component in Figure 17-6 being attenuated, the vertical component going through. The light emerging from the crystal, therefore, is linearly polarized. (The transmitted light, because of the natural color of tourmaline, is green; in orthogonally polarized light the crystal appears black, hence the term "dichroic" meaning "two colors.")

The same dichroic effect is seen with crystals of quinine sulfo-iodide, called *herapathite*.* These crystals are very small, and single crystals cannot be isolated and used for polarization. However, after a method was found to align the crystals, an efficient and economical polarizing material could be produced, now known as *Polaroid*.†

---

* Named after William Bird Herapath (1820–1868), physician and surgeon at Queen Elizabeth's Hospital in Bristol, England. Herapath worked on a variety of subjects, from chlorophyll and the identification of bloodstains by microspectroscopy to home sanitation. His best known discovery came when his pupil, a Mr. Phelps, dropped iodine into the urine of a dog that had been fed quinine and saw that tiny emerald-green crystals formed there. Herapath, examining them under a microscope, noticed that the crystals were in some places light where they overlapped and in some places they were dark, an observation that he published under the slightly laborious title "On the Optical Properties of a newly-discovered Salt of Quinine, which crystalline substance possesses the power of polarizing a ray of Light, like Tourmaline, and at certain angles of Rotation of depolarizing it, like Selenite," *Philos. Mag.* (4) **3** (1852), 161–73.

† Invented by Edwin Herbert Land (1909–91), American physicist, inventor, businessman. As a 19-year-old undergraduate at Harvard, Land read Brewster's book on the kaleidoscope (which included pieces of polarizing material rather than just colored glass) and set out to develop a better way

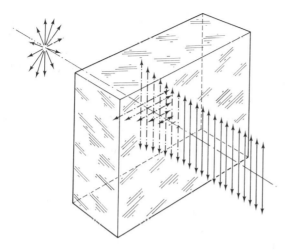

Figure 17-6 Polarization by selective absorption.

The most common type of Polaroid is a sheet of synthetic polyvinyl alcohol, PVA, stretched and then impregnated with iodine. The iodine attaches itself to the long hydrocarbon molecules, letting its conduction electrons move along the molecules as if they were long, thin wires. But because of the resistance that affects a current propagating in a wire, the **E** vector oscillating *parallel* to the iodine is being absorbed and thus weakened; on the other hand, *perpendicular* to the molecules there is little conduction and the vector will go through.

Polarization by Double Refraction

Take a pen and make a dot on a sheet of paper. Place a crystal of calcite or "Iceland spar," calcium carbonate, $CaCO_3$, over the dot and look through. You will see two dots instead of one. Rotate the crystal and observe the dots. One of them will remain in place while the other moves around it. The stationary dot shows the *ordinary ray*, the moving dot the *extraordinary ray*. This was how

---

of producing polarized light. He took a suspension of crystals of herapathite and placed it in a strong magnetic field. It worked; he had the first artificial polarizer for light. Later he used an electric field; he also suspended the crystals in a thermoplastic matrix, extruding the plastic, while soft, through a narrow slot. For details see E. H. Land, "Some Aspects of the Development of Sheet Polarizers," *J. Opt. Soc. Am.* **41** (1951), 12, 957–63.

Land's other major invention is instant photography. When he had the process ready to show to the world, he did not call a news conference but presented a paper—complete with demonstration—at a 1947 meeting of the Optical Society. He wanted his advance to be judged by his peers, but he also had the foresight to invite several prominent newspaper science writers to be present, and an enthusiastic story appeared the next morning in *The New York Times*: the Polaroid Land camera was off to a running start.

Bartholinus* discovered *double refraction*, or *birefringence*. Crystals such as calcite, quartz, mica, and ice that show this effect are called *anisotropic*. In contrast, most noncrystalline solids such as unstressed glass are *isotropic*.

Calcite is a good example of a system of anisotropic crystals called *hexagonal*. To visualize the shape of a hexagonal crystal, think of an orthogonal (right-angled) block of material. Two diagonally opposite corners may be called A and B, as in Figure 17-7. Slightly compress the block along the line A-B: The faces that before were rectangular now become rhombic and corners A and B become blunt; that is, the angles adjacent to them grow larger, the angles away from them acute.

The line A-B is called the *crystal axis*, not to be confused with the optical axis of the system. In the direction specified by this axis the crystal's refractive index, the *ordinary* refractive index, ω, is constant (1.6584 for calcite and sodium light). In other directions, the index changes with the angle subtended with the crystal axis. In a direction 90° from the axis, the index is least (1.4864). This is the *extraordinary* refractive index, ε.

If unpolarized light is incident on the crystal, in a direction different from that of its axis, the light is split into two, the ordinary or *o ray*, and the extraordinary or *e ray*. The two bundles emerging from the crystal are both linearly polarized, at right angles to each other. Hence, such crystals would make good polarizers, if only we could eliminate one of the bundles. This is done in the *Nicol prism*.

Nicol Prism

The Nicol prism† is made out of a crystal of Iceland spar, cut into two and cemented together with a thin layer of Canada balsam in between (Figure 17-8). The two endfaces are polished down from 71° to 68° as shown. The *o* light is refracted more (downward) and is incident on the Canada balsam at an angle slightly larger than the *e* light. But since the refractive index of Canada balsam (1.526) lies between ω and ε for calcite (1.6584 and 1.4864, respectively), the *e* light will pass through but the *o* light will not. The *o* light undergoes total internal reflection and is absorbed in black paint on the sides of the prism, and only the *e* light will emerge from it.

Nicol prisms are good polarizers, but they are expensive and have a limited field of view (28°). This is because, if the light is not nearly parallel, either the *o*

---

* Erasmus Bartholinus (1625–1698), Danish physician and professor of mathematics at the University of Copenhagen. He described his discovery in a 60-page publication, *Experimenta crystalli Islandici disdiaclastici quibus mira et insolita refractio delegitur* (Hafniae, 1670). Bartholinus, however, was not aware of what we now call polarization; he thought his observation to be an unusual case of refraction, *double refraction*.

† Named after William Nicol (1768–1851), Scottish geologist and physicist. A lecturer at the University of Edinburgh, Nicol published his first paper at age 58. His interests were primarily in the fields of crystallography, mineralogy, and paleontology. In 1828 he invented his prism, and described it in the article "On a Method of So Far Increasing the Divergency of the Two Rays in Calcareous Spar That Only One Image May Be Seen at a Time," *Edinburgh New Philos. J.* **6** (1829), 83–84.

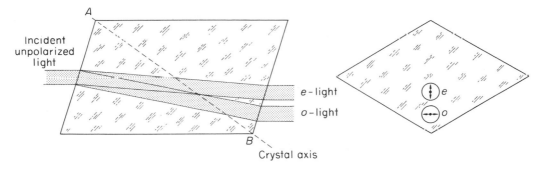

Incident
unpolarized
light

e-light

o-light

Crystal axis

Figure 17-7   Double refraction and polarization of light in calcite.

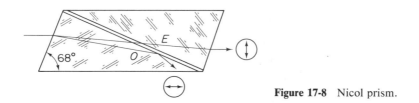

Figure 17-8   Nicol prism.

light will be incident on the balsam at an angle less than the critical angle and go through, or the *e* light will be reflected out of the prism too and be lost.

There are other prisms besides the Nicol type which are useful for special applications. The *Glan–Thompson prism* has a wider angular aperture (40°), but it is wasteful of calcite and hence even more expensive. The *Glan–Foucault prism* has no cement (but a narrow field) and thus is less likely to be damaged at high power densities. In the *Rochon prism* and the *Wollaston prism* the *o* and the *e* rays are transmitted side by side.

Types of Birefringence

The numerical difference between the ε and the ω index of refraction provides us with one definition of birefringence. But since $\varepsilon - \omega$, in effect, is a measure of *path difference*, Γ, this path difference *per unit path length*, L, gives us another definition; thus

$$\text{birefringence} \equiv \varepsilon - \omega = \frac{\Gamma}{L} \qquad [17\text{-}2]$$

If a transparent isotropic substance such as plastic or glass is subject to mechanical stress, it becomes temporarily birefringent. Such birefringence is the basis of *photoelastic stress analysis*, a procedure widely used in the engineering world. In practice, either a model of the test object is made out of transparent plastic (and examined in transmitted light) or a plastic coating is applied to the real structure (and examined in reflected light). In either case, the birefringence reveals the strains induced in the structure.

Malus' Law

Consider a beam of linearly polarized light, coming toward you. The electric vector of the light may oscillate in the vertical direction, as shown in Figure 17-9 (left). The light may then be incident on a polarizer whose *plane of transmission* subtends an angle θ with the direction of oscillation (center). How will the amplitude of the light change?

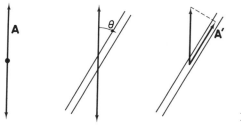

Figure 17-9   Malus' law.

Draw a normal from the tip of vector **A** to the plane of transmission (right). If $\theta = 0$, all the light will go through and $\mathbf{A} = \mathbf{A}'$. If $\theta = 90°$, none will go through and $\mathbf{A}' = 0$. In order to find the amount of light transmitted at intermediate angles, note that $\mathbf{A}'/\mathbf{A} = \cos \theta$. Therefore, since the intensity of light is proportional to the *square* of the amplitude, the intensity transmitted, I', relates to the intensity incident, I_0, as

$$I' = I_0 \cos^2 \theta$$ [17-3]

where θ is the angle through which one of the polarizers has been turned with respect to the other. This relationship is known as *Malus' law.**

---

* Étienne Louis Malus (1775–1812), French army officer and engineer. One evening in 1808 while standing near a window in his home in the Rue d'Enfer in Paris, Malus was looking through a crystal of Iceland spar at the setting sun reflected in the windows of the Palais Luxembourg across the street. As he turned the crystal about the line of sight, the two images of the sun seen through the crystal became alternately darker and brighter, changing every 90° of rotation. After this accidental observation Malus followed it up quickly by more solid experimental work, described his observation in "Sur une propriété de la lumière réfléchie," *Mem. Phys. Chim. Soc. Arcueil, Paris* 2 (1809), 143–58. He concluded that the light, by reflection on the glass, became *polarized* but, by chance, he defined the "plane of polarization" of the reflected light as being the same as the plane of incidence. Today we know that the E vector oscillates *normal* to the plane of incidence; thus Malus' "plane of polarization" is perpendicular to the plane of oscillation.

THE ANALYSIS OF LIGHT OF UNKNOWN POLARIZATION

If two high-quality polarizers are set with their planes of transmission perpendicular to each other (if the polarizers are "crossed"), *no* light will go through: the minimum transmission is zero. But if the light passing through the first polarizer contains some unpolarized light, that is, if the light is partially polarized, at no setting would the light be extinguished completely. The same would happen if circularly polarized light were present.

How, then, can we distinguish between the two? This is possible by using a relatively simple device known as a *quarter-wave plate*. Such a plate is made out of birefringent material, which, as we have seen, makes light of orthogonal polarizations traverse the material at different velocities. One of the wavefronts will be ahead of the other by a distance that depends on the thickness of the plate (Figure 17-10). If one component is ahead by 90°, we have a *quarter*-wave plate. If the thickness is such that the path difference is 180°, we have a *half*-wave plate.

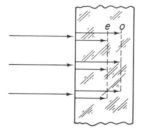

Figure 17-10   Retardation of light in a (positive) birefringent crystal such as quartz.

Example

Assume that we have some birefringent material that has refractive indices $\omega = 1.5000$ and $\varepsilon = 1.5001$. How thick a sheet is needed to make a quarter-wave plate for light of 600 nm?

Solution: First we determine the path difference, using Equation 17-2,

$$\Gamma = L(\varepsilon - \omega)$$

where L is the path length and ε and ω are the indices of refraction for light polarized along the two axes of the material.

For a quarter-wave plate,

$$\Gamma = \frac{\lambda}{4}$$

Combining this with the previous equation by eliminating Γ gives

$$L(\varepsilon - \omega) = \frac{\lambda}{4}$$

$$L = \frac{\lambda}{4(\varepsilon - \omega)}$$

and therefore

$$L = \frac{600 \times 10^{-9}}{(4)(1.5001 - 1.5000)} = \boxed{1.5 \text{ mm}}$$

Let me illustrate how a quarter-wave plate works, using graphical construction. A linearly polarized, sinusoidal wave, A_1 in Figure 17-11 (left), rises up from zero at a time $t = 0$. A second wave of the same amplitude and frequency, A_2, is delayed by one-fourth of a wavelength, 90°. We consider four points evenly spaced in time, 1–2–3–4, and plot the amplitudes of these two waves, at these times, on coordinates labeled A_1 and A_2 (right). At point 1, the A_1 wave has reached a positive maximum and A_2 is zero; at point 2, $A_1 = 0$ and A_2 has reached a positive maximum; at point 3, A_1 is at a negative maximum and $A_2 = 0$; and at point 4, $A_1 = 0$ and A_2 is at a negative maximum. The plot is that of a circle; thus, the resulting light is *circularly polarized*.

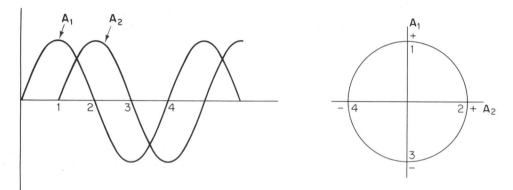

Figure 17-11 Adding two linear polarizations of λ/4 phase difference gives circular polarization.

If there is *no* phase difference between the two contributions, $\delta = 0$, the resultant light is linearly polarized. If the phase difference is $\delta = 90°$ and if in addition the two amplitudes are equal, $A_1 = A_2$, the light is circularly polarized. But this holds true also in the opposite direction: when two circularly polarized waves of the same amplitude but opposite sense of rotation, one right-handed and the other left-handed, interfere with one another, the resultant light is linearly polarized.

Such interference follows *Fresnel–Arago's laws:**

1. Two bundles of light, polarized in the same direction, do interfere.
2. Two bundles, linearly polarized in orthogonal directions, do not interfere.

---

* Dominique François Jean Arago (1786–1853), French physicist. Working at the Paris Observatory, Arago's most noteworthy contributions were to the wave theory of light and to the magnetic effect of electric current. By running a current through a copper wire, Arago in 1820 disproved the then held theory that iron was necessary to produce electromagnetism. When Napoleon made himself emperor in 1852, Arago refused an oath of allegiance and resigned; the new emperor did not accept his

Table 17-1   Analysis of light of unknown polarization

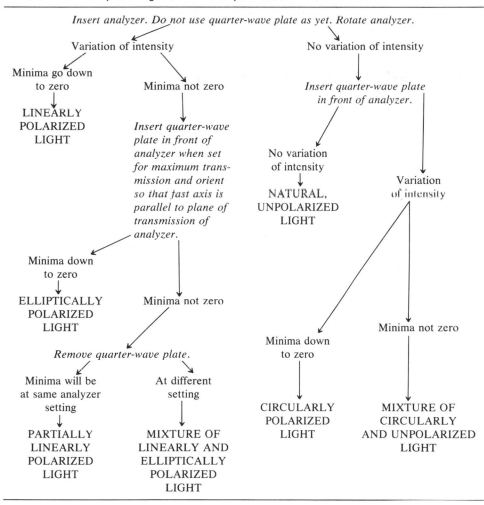

3. Two bundles, derived from orthogonal components of unpolarized light and subsequently brought into the same plane of oscillation, do not interfere.

4. Two bundles, derived from coherent portions of light and subsequently brought into orthogonal planes, do interfere, producing, however, varying degrees of ellipticity.

resignation and let Arago stay on anyway. A. Fresnel and F. Arago, "Sur l'Action que les rayons de lumière polarisés exercent les uns sur les autres," *Ann. Chim. Phys.* (2) **10** (1819), 288–305. (For Fresnel footnote, see page 240.)

Returning to our initial question, by using a polarizer and a quarter-wave plate we can analyze almost any type of light of unknown polarization. The proper steps are listed in Table 17-1.

MATRIX DESCRIPTION OF POLARIZATION

Matrices greatly facilitate the description of polarized light. They also help describe the interaction of polarized light with various types of polarizers and phase retardation plates. There are two approaches which in part overlap and otherwise complement each other. *Mueller matrices* include unpolarized light but require larger (4 × 4) matrices. *Jones matrices* refer to monochromatic polarized light; they are smaller (2 × 2) matrices but the full Jones matrices contain complex numbers.*

Using Mueller matrices, any form of light including partially and elliptically polarized light can be described by four parameters called *Stokes' vectors*. These vectors define the intensity considering light of horizontal polarization, +45° polarization, and right-circular polarization. They are written in the form of a column matrix. The characteristics of a polarizer or a retardation plate can be written as a 4 × 4 matrix and thus a Mueller matrix has 16 elements; fortunately many of them are zero.

Jones matrices are used in either of two forms, depending on whether we need to know the phase of the light. If the phase is needed, the light requires full Jones vectors. If no phase information is needed, linearly polarized light oscillating in the horizontal direction is represented by

$$\begin{bmatrix} 1 \\ 0 \end{bmatrix}$$

Similarly,

$$\begin{bmatrix} 0 \\ 1 \end{bmatrix}$$

represents light linearly polarized in the vertical direction and

$$\begin{bmatrix} \frac{1}{2} \\ \frac{1}{2} \end{bmatrix} = \begin{bmatrix} 1 \\ 1 \end{bmatrix}$$

---

* Named after Robert Clark Jones (1916–), senior physicist and Research Fellow in Physics at the Polaroid Corporation, Fellow of the Optical Society of America and recipient of its Lomb and Ives medals. Jones has also made noteworthy contributions to radiation detection, hydrodynamics, acoustics, photography, and isotope separation, and as a hobby likes to ride railroad locomotives all over the world. R. Clark Jones, "A New Calculus for the Treatment of Optical Systems," *J. Opt. Soc. Am.* **31** (1941), 488–503.

light linearly polarized at an azimuth of 45°. Circular polarization is represented by

$$\begin{bmatrix} 1 \\ -i \end{bmatrix} \quad \text{for right-} \quad \text{and} \quad \begin{bmatrix} 1 \\ i \end{bmatrix} \quad \text{for left-handed rotation}$$

Then, if right- and left-handed circularly polarized light (of equal amplitude) is combined,

$$\begin{bmatrix} 1 \\ -i \end{bmatrix} + \begin{bmatrix} 1 \\ i \end{bmatrix} = \begin{bmatrix} 1+1 \\ -i+i \end{bmatrix} = \begin{bmatrix} 2 \\ 0 \end{bmatrix} = 2\begin{bmatrix} 1 \\ 0 \end{bmatrix} \qquad \text{[17-4]}$$

which means that the result is linearly polarized light of twice the amplitude of either of the initial contributions.

Let the matrix $\begin{bmatrix} A \\ B \end{bmatrix}$ represent the incident light. From Table 17-2 we take the matrix of the element inserted in the path and write it to the left of the AB matrix,

$$\begin{bmatrix} b & a \\ d & c \end{bmatrix}\begin{bmatrix} A \\ B \end{bmatrix} = \begin{bmatrix} A' \\ B' \end{bmatrix} \qquad \text{[17-5]}$$

The resulting matrix, $\begin{bmatrix} A' \\ B' \end{bmatrix}$, describes the emerging light.

Table 17-2   Simplified Jones matrices representing various polarizing elements

Linear polarizer, transmission axis horizontal ⟷		$\begin{bmatrix} 1 & 0 \\ 0 & 0 \end{bmatrix}$
transmission axis vertical ↕		$\begin{bmatrix} 0 & 0 \\ 0 & 1 \end{bmatrix}$
transmission axis at +45° ↗		$\frac{1}{2}\begin{bmatrix} 1 & 1 \\ 1 & 1 \end{bmatrix}$
Circular polarizer, right-handed ↺		$\frac{1}{2}\begin{bmatrix} 1 & i \\ -i & 1 \end{bmatrix}$
left-handed ↻		$\frac{1}{2}\begin{bmatrix} 1 & -i \\ i & 1 \end{bmatrix}$
Quarter-wave plate, fast axis horizontal		$\begin{bmatrix} 1 & 0 \\ 0 & -i \end{bmatrix}$
fast axis vertical		$\begin{bmatrix} 1 & 0 \\ 0 & i \end{bmatrix}$
Half-wave plate, fast axis at +45°		$\begin{bmatrix} 0 & 1 \\ 1 & 0 \end{bmatrix}$

Example 1

Linearly polarized light oscillating in the horizontal direction is incident on a half-wave plate oriented at +45°. What is the state of polarization of the emerging light?

Solution:

$$\begin{bmatrix} 0 & 1 \\ 1 & 0 \end{bmatrix} \begin{bmatrix} 1 \\ 0 \end{bmatrix} = \begin{bmatrix} 0 \\ 1 \end{bmatrix}$$

The emerging light, therefore, is linearly polarized in the vertical direction.

Example 2

A quarter-wave plate, with its fast axis horizontal, is inserted into a beam of light linearly polarized and oscillating at 45°. Determine the polarization of the emerging light.

Solution:

$$\begin{bmatrix} 1 & 0 \\ 1 & -i \end{bmatrix} \begin{bmatrix} 1 \\ 1 \end{bmatrix} = \begin{bmatrix} 1 \\ -i \end{bmatrix}$$

The emerging light is right-circularly polarized.

CRYSTAL OPTICS

It is relatively easy to determine the optical properties of a crystal. Simply add two polarizing filters to a microscope, a *polarizer* below the condensor and an *analyzer* above the objective, or even on top of the eyepiece. That converts a conventional microscope into an elementary type of a polarizing microscope.

There are two ways of using a polarizing microscope, *orthoscopic* and *conoscopic* (Figure 17-12). In orthoscopic observation, the microscope is used like any other microscope; that is, it is focused on the specimen. In the conoscopic mode, an auxiliary lens, called the *Amici–Bertrand lens*,* is inserted about halfway between objective and eyepiece. This turns the microscope into a low-power telescope so that the observer, instead of looking *at* the specimen, looks *through* it.

First we determine whether the sample is *isotropic* or *anisotropic*. If a piece of strain-free glass is placed on the stage between crossed polarizers, the glass remains dark no matter how the stage is turned; such a material is isotropic. Anisotropic matter, in contrast, goes through four positions of maximum extinction, 90° apart.

If the material is isotropic, its refractive index is determined next [Table 17-3, (left-hand column)]. Only a few inorganic substances belong to this group: all

---

* Giovan Battista Amici (1786–1863), Italian physicist and astronomer. Educated as an architect and engineer, Amici became professor of mathematics at the University of Modena, later director of the Astronomical Observatory of the Royal Museum of Physics and Natural History in Florence. He invented the roof prism, and built a large astronomical doublet of 5.33 m focal length, 28.5 cm in diameter. He designed microscope objectives with the front element in the form of a hemisphere; this gave the needed short focal length and high aperture, an ingenious idea still in use today.

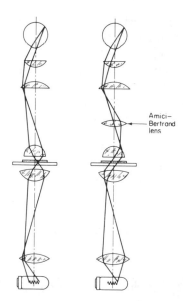

Figure 17-12 In the orthoscopic way of using the polarizing microscope (*left*) the object (the specimen on the stage) is conjugate to the retina of the eye; in the conoscopic way (*right*) it is the light passing parallel through the object that is focused on the retina.

Table 17-3 Outline of steps for identifying a crystal

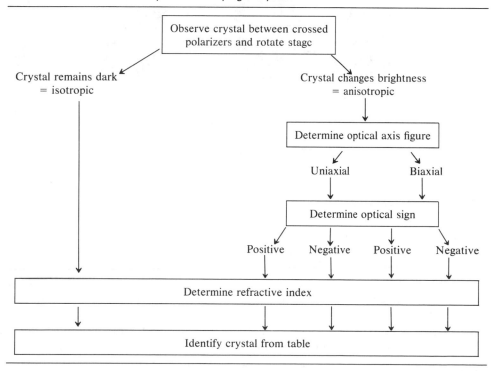

amorphous materials and the crystals of the isometric system. A good way of determining the refractive index is by the *Becke line*, a luminous line seen next to the contour of the crystal. Immerse the crystal in a liquid of known refractive index and observe its contour while slowly raising or lowering the tube of the microscope. When the tube is *raised* and the Becke line is seen to *contract* and to move *into* the crystal, the crystal has a refractive index *higher* than that of the liquid. If the line seems to expand and move out of the crystal, the liquid has the higher index.

On the other hand, if the crystal is anisotropic, look for the *optical-axis figure*. Focus on a group of crystals. Change to the conoscopic configuration and rotate the analyzer to maximum background darkness. Using a high-aperture objective, slowly search the specimen until an optical-axis figure is found. This figure may be either of the *uniaxial* or the *biaxial* type (Figure 17-13). Uniaxial crystals have two principal indices of refraction, ω and ε. Biaxial crystals have three, α, β, and γ.

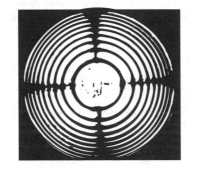 

Figure 17-13 Optical-axis figures of uniaxial and biaxial crystals are distinctly different. Figure on left is that of uniaxial crystal (such as calcite), on the right, biaxial crystal (such as mica). Concentric rings on left are *isochromatic curves*, representing equal path differences 1λ, 2λ, ... Dark cross is formed by *isogyres*, representing directions of oscillation.

OPTICAL ACTIVITY

When linearly polarized light passes through a crystal of quartz, the light remains linearly polarized but its direction of oscillation changes: It *rotates* (Figure 17-14). Such rotation is called *optical activity*. To an observer looking toward the light source, the rotation may be either clockwise or right-handed (as shown) and the substance be *dextrorotatory*, or it may be counterclockwise or left-handed and the substance be *levorotatory*.

The reason for the rotation is the asymmetry of the molecular structure. Even if two molecules contain the same atoms and bonds in the same sequential order, they still may differ in their orientation relative to the coordinate system.

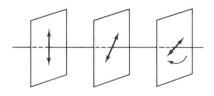

Figure 17-14 Optical rotation.

Such molecules are called *isomers*. If the isomers are exact mirror images of one another, they are called *enantiomers*.*

Optical activity is a matter of *circular birefringence*: an optically active medium has an index of refraction for right-circularly polarized light that is different from the index for left-circularly polarized light. Since linearly polarized light is the resultant of two circular polarizations of the same amplitude but opposite sense of rotation, linearly polarized light passing through the medium will show a rotation of its direction of polarization. The degree of rotation is proportional to the difference between the two indices and to the length of path traversed.

The degree of rotation, for a given wavelength, caused by a 1-mm-thick solid is called *specific rotation*. For liquids, specific rotation is defined as the rotation caused by a solution of 1 g of substrate in 1 ml of solvent through a path 10 cm long. This dependence on concentration is of practical interest, for example when testing syrups or urine for the amount of sugar present. At different wavelengths, the rotation is generally different, an effect called *rotatory dispersion*.

OPTICAL MODULATORS

Optical modulators and light switches based on polarization have become a prominent part of modern optics. Modulators can be made to operate on the amplitude, the phase, the frequency, or the state of polarization of the light, and they can do so at very high rates, up to several GHz (10^9 Hz). A modulator's prime function is to impress information on a light wave by temporarily varying one of its parameters.

A light *switch* changes the direction of the light, or simply turns it on and off. Light switches can have exceedingly short switching times (less than a pico-

---

* The link between optical rotation and asymmetry was found by Louis Pasteur (1822–1895), French chemist. When Pasteur saw that some crystals of sodium ammonium tartrate were mirror images of others, painstakingly, using a pair of tweezers, he separated them into two groups. Checking a solution of each in a polarimeter, he found that one was dextrorotatory and the other levorotatory. But optical activity occurs also in solutions, where no crystals exist; hence, he concluded, the asymmetry must be a property of the molecules, rather than the crystals. Pasteur also disproved the doctrine of spontaneous generation (of life) and found that infectious diseases such as anthrax and rabies were transmitted by microorganisms, opening the way to vaccination.

second, 10^{-12} s), faster than any electronic switch. Both light modulators and light switches can be built out of materials that, when exposed to a *magnetic* field, change their refractive index and become birefringent. Others change when exposed to an *electric* field. We distinguish the following effects.

Magneto-Optical Effect

Historically the first experimental evidence of an interaction between light and an electromagnetic field, *Faraday rotation* is induced by a magnetic field.* Similar to optical activity, the field changes the index of refraction for right-circular polarization in a way different than for left-circular polarization.

But the two types of rotation are different. In optical activity, the sense of the rotation depends on the direction of the light, in Faraday rotation it depends on the direction of the field. Light passing through an optically active substance, and reflected back through it, shows no net rotation because the rotations cancel. In Faraday rotation, the total rotation is cumulative; it is twice the one-way rotation. The angle of rotation ϕ is proportional to the magnetic field strength **H** and to the distance L the light travels in the substance,

$$\phi = \mathbf{H}LV \qquad\qquad [17\text{-}6]$$

where V is called *Verdet's constant*.

Electro-Optical Effects

It is generally easier to generate an electric field than to produce a magnetic field. Hence, electro-optical materials are used more widely. Some of them make use of the *Kerr effect* where an electric field causes the substance to become birefrin-

---

* Named after Michael Faraday (1791–1867), British physicist. The son of a blacksmith, Faraday began his career as a bookbinder's apprentice. Reading some of the books that passed through his hands, he became interested in chemistry and worked for a while as Sir Humphry Davy's lab assistant. As a young man, Faraday proclaimed that women were nothing in his life and even published a poem ridiculing love. At age 30, he met, fell in love with, and married Sarah Barnhard, the 21-year-old daughter of a silversmith. Soon after Oerstedt's 1819 discovery that an electric current produced a magnetic field, Faraday, in 1821, placed a current-carrying wire near a magnetic pole and succeeded in making it rotate, opening the way to electric *motors*. Seeing this, he wondered whether the reverse might be true—whether a magnet could be made to produce a current. After casual experimentation over several years (during which time he liquefied gases, discovered benzene, and found a better way to make steel), he returned to his idea. For days he tried without success until, one day in 1831, in desperation he plunged a magnet down a coil: Suddenly there was an electric pulse. Why hadn't he seen that before? Clearly, it is the *change* of the magnetic field that did it. For this discovery, which opened the way to the electric *generator*, the whole scientific world honored him. In 1845, Faraday discovered the magnetic rotation of polarized light.

gent.* The magnitude of the induced index difference, Δn, is proportional to the square of the strength of the field, $\mathbf{E}$, and to the wavelength,

$$\Delta n = K\lambda \mathbf{E}^2 \qquad [17\text{-}7]$$

where K is *Kerr's constant*. Nitrobenzene has an unusually low Kerr constant, 2.4×10^{-10} cm/V^2.

A *Kerr cell* has two parallel metal plates immersed in a liquid such as nitrobenzene. When placed between crossed polarizers, no light is passing through, but when a voltage, in the kilovolt range, is applied to the plates, oriented at 45° to the direction of oscillation of the entering light, the light becomes elliptically polarized and the shutter *opens*. The Kerr effect is very fast, up to about 10^{10} Hz; it is limited mainly by the difficulty of producing electric pulses in the kilovolt range.

Another electro-optical effect is the *Pockels effect*.† In contrast to Kerr cells, which contain a liquid, Pockels cells contain a *crystal*. A material often used is potassium dihydrogen phosphate, KDP. Also in contrast to the Kerr effect, which is quadratic, the Pockels effect is a linear function of the electric field. The field may be applied either parallel to the optical axis (*longitudinal Pockels effect*) or normal to it (*transverse Pockels effect*). Most important, Pockels cells need much lower voltages, at least an order of magnitude less, than Kerr cells. For many applications such as optical communications networks and Q-switched lasers, this has made the Pockels cell the preferred type of a light modulator.

SUMMARY OF EQUATIONS

Brewster's law:

$$\tan I = \frac{n'}{n} \qquad [17\text{-}1]$$

Birefringence:

$$\varepsilon - \omega = \frac{\Gamma}{L} \qquad [17\text{-}2]$$

Malus' law:

$$I' = I_0 \cos^2 \theta \qquad [17\text{-}3]$$

---

* John Kerr (1824–1907), Scottish physicist. In 1875 Kerr used a block of glass (still on display at the Kelvin Museum of the University of Glasgow), with two wires from an induction coil attached to it and placed between crossed Nicol prisms. With the current on, the glass becomes positive uniaxial birefringent and light passes through.

† Named after German physicist Friedrich Carl Alvin Pockels (1865–1913).

PROBLEMS

17-1. Natural, unpolarized light is incident on a small volume of a scattering medium. How is the light polarized that is scattered at angles of 0°, 45°, and 135° from the forward direction?

17-2. Circularly polarized light passes through a colloidal suspension. How is the light polarized that is scattered 0°, 45°, 90°, and 180° from the forward direction?

17-3. What is the angle of incidence for complete polarization to occur on reflection at the boundary between water ($n = \frac{4}{3}$) and glass ($n = 1.589$) assuming the light comes from the side of:
   (a) The water?
   (b) The glass?

17-4. Light reflected by the polished surface of a block of glass has maximum polarization when the angle of *refraction* is 33.69°. What is the index of the glass?

17-5. A fisherman, bothered by glare reflected off the surface of a calm lake, uses Polaroid glasses to cut down on the glare. In which direction, to be most effective, should the crystals in the Polaroid material be oriented?

17-6. Polaroid filters can be made with the molecules forming concentric circles. If such an *axis finder* is placed in a beam of linearly polarized light, obtained by reflection at Brewster's angle, a dark band is seen extending across the filter in a direction normal to the plane of incidence at the Brewster surface. Therefore, are the molecules in the filter oriented *radial* or *tangential* with respect to the optical axis of the light passing through the filter?

17-7. Two polarizers are set for maximum transmission. If one of the polarizers is rotated through 30°, how much of the light is still going through?

17-8. Unpolarized light passes through two polarizers whose planes of transmission are parallel to each other. Through what angle must one of the polarizers be turned to reduce the intensity of the light to one-half of what it was before?

17-9. Two polarizers are crossed at 90° so that no light is going through. If a third polarizer, oriented at 45°, is placed between the two polarizers, what percentage of light will emerge from the system? (Try it; there *is* light going through.)

17-10. What is the least angle at which a π phase change occurs in linearly polarized light, the **E** vector oscillating parallel to the plane of incidence and the light internally reflected at a boundary between crown ($n = 1.52$) and carbon disulfide ($n = 1.63$)?

17-11. A beam of unpolarized light passes through a linear polarizer and a quarter-wave plate, is reflected by a plane mirror, and passes again through the quarter-wave plate and the polarizer. What happens to the light?

17-12. Ice is a positive uniaxial crystal with indices of refraction of 1.309 and 1.310. How thick must the ice be to act as a quarter-wave plate for light of 600 nm wavelength?

17-13. The specific rotation of quartz in light of 589 nm wavelength is 21.7°/mm. What thickness of quartz, cut perpendicularly to its axis and inserted between *parallel* polarizers, will cause no light to be transmitted?

17-14. A glucose solution of unknown concentration is contained in a 12-cm-long tube and seen to rotate linearly polarized light by 2.5°. Since the specific rotation of glucose is 52°, what is the concentration?

17-15. If the Verdet constant for monobromonaphthalene, in yellow D light, is 0.104 arc min/A and if a magnetic field of 2.6×10^5 A m^{-1} is applied and the light traverses a path 10 cm long, through what angle will the plane of oscillation turn?

SUGGESTIONS FOR FURTHER READING

BARRON, L. D. *Molecular Light Scattering and Optical Activity*. New York: Cambridge University Press, 1983.

COLLETT, E. *Polarized Light: Fundamentals and Applications*. New York: Marcel Dekker, 1993.

NESSE, W. *Introduction to Optical Mineralogy*. New York: Oxford University Press, Inc., 1986.

PINSON, L. J. *Electro-Optics*. New York: John Wiley and Sons, Inc., 1985.

SHURCLIFF, W. A. *Polarized Light, Production and Use*. Cambridge, Mass.: Harvard University Press, 1962.

Optical Data Processing

Even a simple lens is an optical data processor. It transforms a set of data, the object, into another set of data, the image, and it does so with the greatest of ease. Actually, the transformation takes place in two steps. In the first step, the object is transformed into a *diffraction pattern* or, as we call it today, a *Fourier transform*. In the second step, that pattern is transformed into the image. These two transformations occur immediately one after the other, at the speed of light; they are fundamental to both conventional image formation and to *optical data processing*.

ABBE'S THEORY OF IMAGE FORMATION

Ernst Abbe showed how the image is formed in a microscope. Light from a point source (at the bottom of Figure 18-1) enters a collimating lens (the condenser of the microscope) and is incident on the object. The object may be a grid with many closely spaced parallel lines.

Most of the light passes through the grid and forms the zeroth-order maximum. Other light is diffracted by the grid; that light forms the higher-order maxima. These maxima, the result of the first Fourier transformation, can easily be seen in the back focal plane of the objective. Another, subsequent transformation then converts the maxima into the (intermediate) image, which we observe through

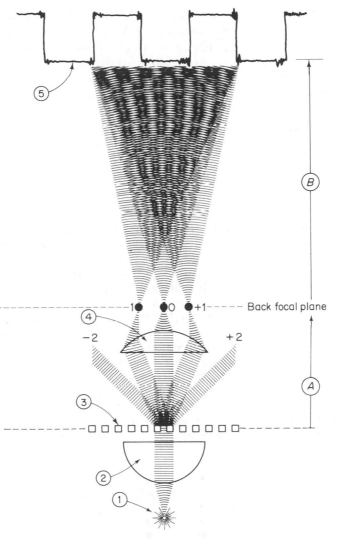

Figure 18-1 Image formation in the microscope: 1, light source; 2, condenser; 3, object; 4, objective lens; 5, image; *A* and *B*, Fourier transformations.

the eyepiece. The image, therefore, is the result of two successive Fourier transformations: this is *Abbe's theory of image formation.**

---

* Ernst Abbe (1840–1905), German physicist. Abbe attended first the University of Jena where he became acquainted with the chairman of the department of mathematics and physics, Karl Snell, so much so that he married Snell's daughter Elise. After four semesters he transferred to Göttingen, renowned for the physical sciences. In 1861 he received a Ph.D. with a thesis on the mechanical equivalent of heat. Two years later he returned to Jena to become an instructor in physics and

An image, in other words, is the result of two contributions: undiffracted light (the zeroth order), which provides the overall *illumination*, and diffracted light (the higher-order maxima), which provides the transfer of *information*. If more higher orders could be admitted and contribute to image formation, the image would contain more information, more detail. This is why a microscope objective of high numerical aperture yields a "better" image; "magnification" as such does not mean much.

At first, Abbe's theory was thought to apply only to periodic structures, to the microscope, and to coherent light. Today we know that it applies to any object, to any type of optical system, even to a simple lens or combination of lenses, and to all types of light.

Lab Experiment Illustrating Abbe's Theory

1. Use as the object a grid that has about 50 lines to the millimeter. Place the grid on an optical bench and illuminate it, from the left, with collimated white light. Orient the grid so that its lines are horizontal.

2. Set up a microscope, using a +10-diopter lens as the objective and a 10× eyepiece placed 30 cm to the right of the objective. Look through the eyepiece, and focus on the grid by slightly moving the objective back and forth.

3. Place a sheet of ground glass 10 cm to the right of the objective lens. Slowly move the ground glass back and forth. In one position the light forms a distinct pattern [Figure 18-2 (left)]. This is the Fourier transform; it appears in the back focal plane of the objective.

The bright spot in the center is the zeroth-order maximum. The less intense spots above and below are the higher-order maxima. The zeroth order is white. The higher orders are colored, red on the outside, blue on the inside.

4. Without moving anything else, replace the ground glass by an adjustable slit. Align the slit with the transform. Block out all maxima except the zeroth order [Figure 18-2 (center)]. A small piece of paper, held closely behind the slit, will help you do this. Look

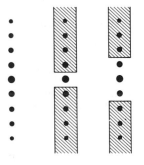

Figure 18-2 Complete Fourier transform (*left*). Same transform, but only the zeroth-order maximum is admitted to form the image (*center*). Zeroth- and two first-order maxima admitted (*right*).

mathematics. Soon he was sought out by Carl Zeiss, a highly skilled craftsman and instrument maker, who needed someone to help him with the design and construction of microscopes. Abbe recognized that it is the objective lens that is crucial. Objectives of low numerical aperture, however well they were made, gave images inferior to those produced by simpler lenses of high aperture, and on this basis formulated his theory of image formation: E. Abbe, "Beiträge zur Theorie des Mikroskops und der mikroskopischen Wahrnehmung," *Arch. mikrosk. Anat.* **9** (1873), 413–468.

through the eyepiece. Sufficient light passes through, but *no image is seen*; the zeroth-order maximum alone does not produce an image.

5. Open the upper and lower bracket of the slit so that the two first-order maxima are admitted also [Figure 18-2 (right)]. The lines in the grid become visible. Fully open the slit, admitting all maxima. The grid is seen even better; higher-order maxima are needed for seeing more detail.

FOURIER TRANSFORMATION

The crucial point in Abbe's theory is the role of *diffraction*. Diffraction (at the object) causes the higher-order maxima that are essential for a good image. Diffraction, in other words, as we had discussed it in Chapter 14, is not merely a nuisance, a byproduct that occurs whenever light passes by next to an edge or an aperture; instead, diffraction is a process that is most essential for image formation.

The proper mathematical tool to analyze this relationship is *Fourier transformation*. Assume that the light comes from a point at infinity (and somewhat below the optical axis) and that it produces plane wavefronts which pass through the aperture shown on the left in Figure 18-3. The fact that the light comes from infinity makes the diffraction process, in effect, a matter of *Fraunhofer diffraction*. A lens placed close to the aperture focuses the light into point P in the second focal plane, F_2 (on the right). Further assume that the height of the object (not shown) is x, the conjugate height of the image x', and that the diameter of the aperture is ξ.

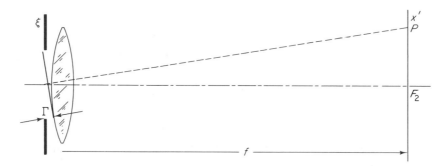

Figure 18-3 Light passing through aperture (*left*) is focused into the F_2 plane (*right*).

Since point P is off axis, each element of the wavefront of width $d\xi$ (inside the aperture) must be somewhat slanted, with a certain path difference, Γ, between its upper and lower ends. Such *path* difference corresponds to a *phase* difference

$$\delta = 2\pi\Gamma \qquad\qquad [18\text{-}1]$$

Each element (of width $d\xi$) then contributes part of the amplitude (at P) that may be represented by

$$d\mathbf{A} = d\xi \, \cos 2\pi\Gamma$$

or, since the cosine term represents the real part of a complex function,

$$d\mathbf{A} = d\xi e^{2\pi i\Gamma}$$

From similar triangles in Figure 18-3 we see that

$$\frac{\Gamma}{\xi} = \frac{x'}{f}$$

We solve this equation for Γ, substitute it in Equation [18-1] and for convenience set $f = 1$; that gives

$$\delta = 2\pi x'\xi$$

The total amplitude distribution in the image, in the $f(x')$ plane, is then found by integration:

$$f(x') = \int_{-\infty}^{+\infty} F(\xi)e^{2\pi ix'\xi} \, d\xi \qquad [18\text{-}2]$$

where $F(\xi)$ is the amplitude distribution inside the aperture, across the lens.

If there is no absorption, the propagation of the light is reversible. Consequently, we can reverse the process: knowing the amplitude in the image, we can determine the amplitude in the aperture. Therefore, the *Fourier transform*, $F(\xi)$, of a function $f(x)$ is

$$\boxed{F(\xi) = \int_{-\infty}^{+\infty} f(x)e^{-2\pi ix\xi} \, dx} \qquad [18\text{-}3]$$

and the *inverse transform*, $f(x)$, of a function $F(\xi)$ is

$$\boxed{f(x) = \frac{1}{2\pi} \int_{-\infty}^{+\infty} F(\xi)e^{2\pi ix\xi} \, d\xi} \qquad [18\text{-}4]$$

The two functions, $F(\xi)$ and $f(x)$, are called a *Fourier transform pair.**

Note the opposite signs of the exponents. It is immaterial whether the exponents are positive or negative. Equation [18-3] could be written with a positive

---

* Jean Baptiste Joseph Baron de Fourier (1768–1830), French physicist and mathematician. Because of his mathematical skills, Fourier became an artillery officer, served as Napoleon's science adviser to Egypt, later became prefect of Isère, near Grenoble, Secretary of the French Academy of Sciences, and head of the École Polytechnique. After his return from Egypt, Fourier wrote a 21-volume treatise on Egyptology. His *Analytical Theory of Heat* became the foundation of thermodynamics. In 1815 Fourier showed that any periodic oscillation, no matter how irregular, can be broken into a series of sine functions, an insight, first rejected, but today considered a classic of mathematical physics.

exponent but then the inverse transform, Equation [18-4], must have a negative exponent.

Signal-in-Noise Detection

Fourier transformation provides us with an efficient means of detecting certain signals, even in the presence of noise. The signal may have a given *spatial frequency*. This term, as we have seen earlier in Chap. 10, refers to the number of lines or dots that the object, or the image, has per unit of length or width.

Such spatial frequencies in the object or image are often easier to detect in the frequency domain than in the spatial domain. For example, a low contrast signal (the presence of many weak lines of constant spacing) may be masked by noise that is superimposed on the signal. That would make these lines hard to see. In the frequency domain, however, the signal is presumably much easier to see because all of its energy is then concentrated at one frequency, while the energy of the noise remains spread out over the whole bandwidth. Fourier transformation, in short, helps us to detect the *signal*, even in the presence of *noise*.

TWO-DIMENSIONAL TRANSFORMATIONS

The early experiments that led to the formulation of Abbe's theory were one-dimensional in nature. They made use of conventional one-dimensional gratings and showed the diffraction maxima, the Fourier transform, in the form of maxima arranged in a one-dimensional direction. This is entirely sufficient to present the theory and the principle of transformation as such.

Most images, though, are two-dimensional. Hence, we need to progress from one-dimensional to *two-dimensional Fourier transformations*. Now the grating lines, and the maxima, occur in two orthogonal directions, the same as the pixels of an image occur in two dimensions. This two-dimensionality is central to image formation, and to optical data processing in general. The difference between the two is whether or not there is any "intervention," that is, whether the first Fourier transform is acted on and in any way modified. If there is no intervention and the light is allowed to go on to the second transformation, it is *image formation*. But if the first transform is acted on and modified, it is *optical data processing*. The modification as such is called *spatial filtering*.

Assume that the object is a cross-grating, placed in the left-hand focal plane of a converging lens (as in Figure 18-4). The Fourier transform of the cross-grating is a two-dimensional array of maxima. Then another lens is placed to the right of the first lens and a screen is placed in the right-hand focus of the second lens. On that screen we see the transform of the transform, that is, we see an image of the original cross-grating.

To modify the image, we place a *mask* in the transform plane. A mask is a stop with some holes in it that will admit only certain parts of the light to the image

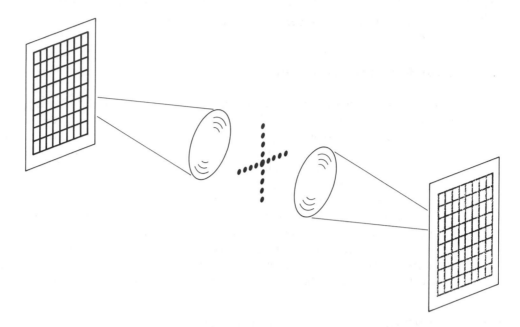

Figure 18-4   Image formation showing object (*left*), Fourier transform (*between the two lenses*), and image (*right*).

plane. First we use a mask that has merely one hole, in line with the axis and so small that only the zeroth order can go through. The higher orders are blocked. The result is that, while the screen receives enough light (from the zeroth order), *no image* will form [Figure 18-5 (top)].

Next we use a filter with a vertical narrow *slot* in it. Such a filter admits maxima that extend in the vertical direction. These maxima are caused by diffraction at the horizontal lines of the object; consequently, the image will show horizontal lines [Figure 18-5 (bottom)]. If the filter were wide open, or if we would use no filter, all of the grid, horizontal lines and vertical lines, would be visible and form the image.

But, we may also use a filter that selectively *blocks* certain maxima. Let us assume that this time we use as the object a set of flowers placed behind a grid (Figure 18-6). The grid consists of a regular cross-pattern of lines. Its Fourier transform is a regular, predictable array of maxima.

Now take a filter that has *holes* that coincide with the maxima. The result is that the grid (that causes these maxima) becomes visible. The flowers produce a more diffuse transform; this transform will for the most part be blocked by the filter and the flowers will not be visible (*top*). On the other hand, if the filter, instead of holes, has solid dots that *block the maxima* caused by the grid, the grid cannot be seen, but the flowers, with their more diffuse transform, can be seen (*bottom*).

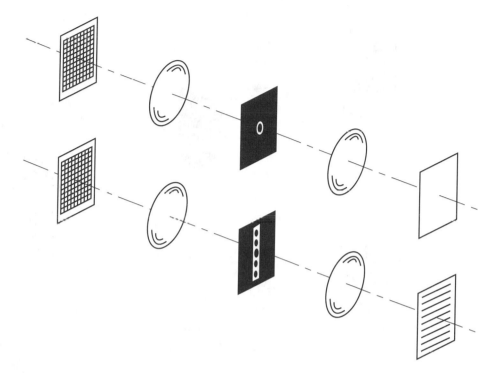

Figure 18-5 Spatial filtering using a pinhole (*top*) and a vertical slot (*bottom*).

Lab Experiment

A good example of spatial filtering is the process of *theta modulation*, applied perhaps to a scene that contains a house, a lawn, and sky (Figure 18-7). These elements are made out of small pieces of plastic diffraction grating oriented at different angles Θ, hence the term *theta modulation* (center).*

Project the scene by a lens onto a screen. Cover the lens with a sheet of aluminum foil, the dull side facing the light source. On the foil you will see three pairs of spectra that are produced by the three elements in the scene: the house, the lawn, and the sky, red always being diffracted most, blue least. Cut out part of the foil (such as hole *A* on the right in the illustration) and let some red pass through the foil, out of the light that belongs to the house. Then let some of the green for the lawn and blue for the sky pass through. The mask thus made transmits only part of the spectra and the image appears in these colors.

---

* J. D. Armitage and A. W. Lohmann, "Theta Modulation in Optics," *Appl. Opt.* **4** (1965), 399–403.

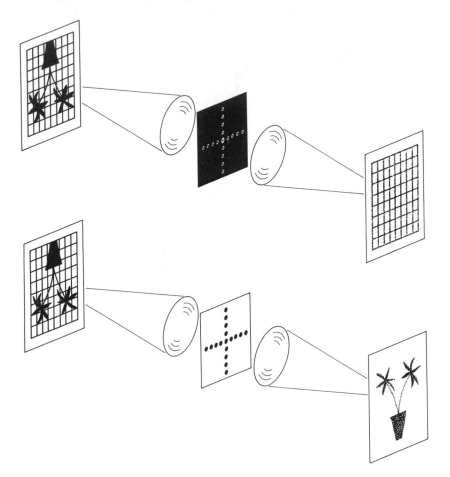

Figure 18-6   Spatial filtering using a filter that either transmits the maxima generated by a cross-grid (*top*) or that blocks them (*bottom*).

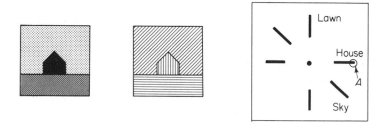

Figure 18-7   Theta modulation. Landscape scene in half-tone representation (*left*), theta-modulated (*center*), and Fourier transformed (*right*).

Optical Data Processors

Consider now in more detail the general form of an optical data processor. The object is placed on the left, in the *object plane* (Figure 18-8). The light emerging from that plane is made parallel by a first lens. A second lens focuses the light into the *image plane*, right. Since the light between the two lenses is parallel, their distance from one another is immaterial.

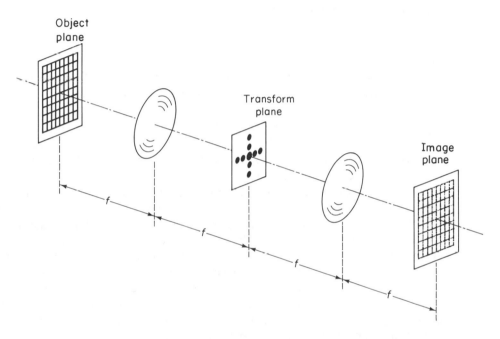

Figure 18-8   The $4f$ configuration of an optical data processor.

For reasons of simplicity, though, we use two lenses of the same focal length and separate the lenses by a distance equal to twice their focal length. The *transform plane* is then halfway between the lenses.

The system, in short, has three planes and two lenses, each of these separated from the next by one focal length ($1f$), with the total length from object to image equal to $4f$; that is called the *$4f$ configuration*.

But consider if the light, instead of coming from the object plane, were to come from infinity (on the left). Then the resulting (Fraunhofer) diffraction pattern would appear in the focal plane of the first lens, or again halfway between the two lenses. That plane, the Fraunhofer diffraction plane, is where the Fourier components of the input are displayed. This relationship shows once again the equivalence of Fraunhofer diffraction and Fourier transformation.

To describe the size and shape of the transform, we need to know the coordinates in the various planes. In the input plane, any point is defined by two variables, x and y. In the transform plane, we need other coordinates, following the notation of Fourier transformation. These are the coordinates ξ and η. In the output plane the coordinates are again x' and y'. The coordinates x and ξ, and y and η, are inversely proportional to each other: the closer the lines in a grating, for example, the wider apart the maxima.

From Equation [18-3] we have seen that the (one-dimensional) Fourier transform, $F(\xi)$, of a function $f(x)$ is

$$F(\xi) = \int_{-\infty}^{+\infty} f(x)e^{-2\pi ix\xi}\, dx$$

Now we go to two dimensions. As the input we use function $f(x, y)$; that gives the two-dimensional, double-integral Fourier transform

$$F(\xi, \eta) = \iint_{-\infty}^{+\infty} f(x, y)e^{-2\pi i(\xi x + \eta y)}\, dx\, dy \qquad [18\text{-}5]$$

Now we insert the filter in the transform plane. That will truncate the transform and change the image in various ways.

The simplest type of a spatial filter is a *binary filter*; they have a transmittance of either zero or one. Some binary filters block only the high space frequencies. Such a filter, in essence no more than a plate with a small hole in it, is called a *low-pass filter* [Figure 18-9(a)]. These filters are often used to eliminate a laser's granularity that may be due to inhomogeneities in the cavity windows or to dust.

A filter that eliminates the low space frequencies is called a *high-pass filter* (b). Such a filter enhances the details of an image (the high frequencies) and emphasizes its edges (*edge enhancement*). A *band-pass filter* is a ring-shaped, annular aperture (c); it enhances details of a certain, predetermined size.

For example, consider a cytological specimen such as a Pap smear. Without spatial filtering, we see large epithelial cells, medium-sized leukocytes, and small particles of undetermined origin, a wide range of microscopic detail. Using a band-pass filter, large structures and very small detail can be suppressed. The various cell nuclei, however, that in a Pap smear are the objects of interest, will stand out, a virtual necessity in any computer-aided screening of cytological samples.

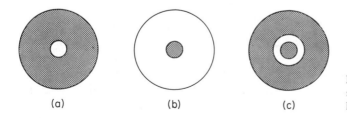

(a)   (b)   (c)

Figure 18-9   Various forms of simple spatial filters: (a) low-pass filter; (b) high-pass filter; (c) band-pass filter.

In addition to "yes-or-no" binary filters, we can also design *complex filters*. These filters act on both the amplitude of the light and on its phase. An example is *phase contrast*. Abbe had always assumed that the object is an *amplitude grating*. Its diffraction maxima all have the same phase. But the object may be a *phase grating* instead.

In Figure 18-10 we represent the various transforms by vectors. An amplitude grating, left, has an intense zeroth-order maximum and less intense higher orders, all of the same phase. A phase grating has an even more intense zeroth-order maximum (longer vector!), but the higher orders differ in phase by $\pi/2$ radian or 90° (center). The transform of a phase grating, therefore, is distinctly different from that of an amplitude grating.

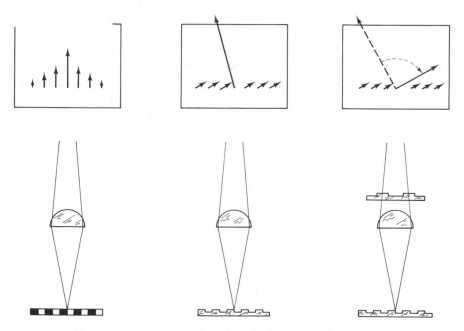

Figure 18-10 Vector representation of amplitude object (*left*), phase object (*center*), phase contrast (*right*).

It then occurred to Zernike* that if the transform were modified by attenuating the amplitude and delaying the phase of the zeroth order, a *phase object could be made to look like an amplitude object* [Figure 18-10 (right)]. That is of particu-

---

* Frits Zernike (1888–1966), Dutch physicist, professor of physics at the University of Groningen, received the Nobel Prize in physics in 1953 for the discovery of phase contrast. F. Zernike, "Beugungstheorie des Schneidenverfahrens und seiner verbesserten Form, der Phasenkontrastmethode," *Physica* **1** (1934), 689–704; and "How I Discovered Phase Contrast," *Science* **121** (1955), 345–49.

lar interest in the microscopic examination of living tissue where the details often differ from their surroundings only by their refractive index and, without staining, are hardly visible at all. Zernike called his process the "phase-strip method for observing phase objects in good contrast" or *phase contrast*, for short.

OPTICAL AND ELECTRONIC DATA PROCESSING

Optical and electronic data processing have much in common. Still, there are important differences. The first electronic computers had only one variable: time. Modern computers process information in parallel; in fact, modern supercomputers are massive parallel computers. But optical systems always process information in parallel. A well-corrected lens, for example, can easily form an image of a two-dimensional array of 1000×1000 pixels and it does so at the speed of light, and without scanning.

In addition, optical systems provide Fourier transformation. These transformations can be used either without or with spatial filtering, and often more conveniently than in an equivalent electronic system. The filtering operations as such make use of modulators or light switches based on polarization that we have discussed in the preceding chapter. These in effect are programmable filters. Systems of this kind play an important part in the theory and technology of *character recognition*, some examples of which we will discuss in the next chapter.

SUMMARY OF EQUATIONS

Fourier transformation,
   one-dimensional:

$$F(\xi) = \int_{-\infty}^{+\infty} f(x)e^{-2\pi i x \xi}\, dx \qquad\qquad [18\text{-}3]$$

   two-dimensional:

$$F(\xi, \eta) = \iint_{-\infty}^{+\infty} f(x, y)e^{-2\pi i(\xi x + \eta y)}\, dx\, dy \qquad\qquad [18\text{-}5]$$

PROBLEMS

18-1. If Abbe's theory can be reduced to the *grating equation*, what is the shortest distance, in terms of wavelengths, that can be resolved by a microscope?

18-2. Assume that the object is a cross-grating. If a *vertical* slot is placed in the transform plane so that it *blocks* the maxima spread out in the *horizontal* direction, what will the image look like?

18-3. The object illustrated in Figure 18-11 is known as an *Abbe diffraction plate*.
  (a) If such an object is placed in a microscope of high numerical aperture, what do you see in the Fourier transform plane?
  (b) How would the transform, and the image, change if the objective were of low numerical aperture?

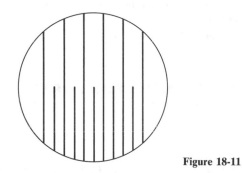

Figure 18-11

18-4. A grating is placed on the stage of a microscope such that the grating lines extend between the numbers 8 and 2 on a clock. If the aperture below the condenser, instead of being round, were an equilateral *triangle*, what would you see in the back focal plane (inside the tube of the microscope)?

18-5. The image sent back from space by a vehicle exploring Jupiter is marred by many closely spaced, slightly oblique but parallel scan lines. What kind of spatial filter will suppress these lines?

18-6. How can the grainy structure of a newspaper picture be smoothed out by a suitably shaped spatial filter?

18-7. An image contains a great many closely spaced concentric circles. If a long, narrow, horizontal slit is then placed in the transform plane, how will the image change?

18-8. A wild beast is kept in a cage made with many vertical bars. When in a photograph the bars are to disappear so that the beast seems to be out in the open, what kind of a spatial filter is needed to remove the bars while having the least effect on the rest of the picture?

18-9. The scene illustrated in Figure 18-12 is theta-encoded. How will the spectra, seen with white light, be oriented? What should a space filter look like that is to make the sky blue, the water green, the boat white, and the sail red?

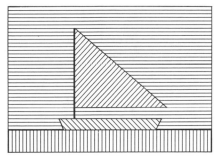

Figure 18-12

18-10. If the mask in theta modulation is made so that the barn is red, the sky blue, and the grass green, and if a red filter is placed in the path, how will it change the image?

18-11. A camera equipped with a spatial filter is used to take a picture of a group of high-rise office buildings. How much of the buildings, their contours and windows and other detail can be seen if the filter transmits:
   (a) Only the low space frequencies?
   (b) All frequencies?
   (c) Only the high space frequencies?

SUGGESTIONS FOR FURTHER READING

FEITELSON, D. G. *Optical Computing*. Boston: MIT Press, 1988.

REYNOLDS, G. O., J. B. DEVELIS, G. B. PARRENT, JR., and B. J. THOMPSON. *The New Physical Optics Notebook*: *Tutorials in Fourier Optics*. Bellingham, Wash.: SPIE Optical Engineering Press, 1989.

STEWARD, E. G. *Fourier Optics*: *An Introduction*, 2nd ed. New York: John Wiley and Sons, 1987.

Holography

Discoveries are often made because the time is right. Some discoveries have been made almost simultaneously yet independently. This is not so with *holography*. Holography is the discovery of one man, Dennis Gabor, at that time, 1948, working in a British industrial research laboratory. His discovery came at a time when the technology was not yet ready to take full advantage of it. It was several years later, with the invention of the laser, that Gabor's discovery reached its full potential.

Introductory Example

Holography, the same as image formation by Fourier transformations, produces images in two steps. In the first step the object is transformed into a photographic record, called the *hologram*, and in the second step, called *reconstruction*, the hologram is transformed further into the image. No lens is needed in either step.

Consider, as shown in Figure 19-1 (top left), that light is incident on a point object. Some of the light is diffracted by the point while other light reaches the screen directly. The two parts of the light interfere and, since the light comes from a point, form a series of concentric rings, like Newton's rings or like a zone plate. The rings are recorded photographically as a transparency, a *transmission-type hologram*. In the second step (top right), parallel light is made to fall on the hologram and, the same as with a zone plate, focused by it into a point.

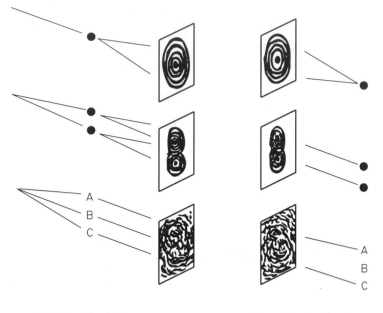

Producing the hologram   Reconstructing the image

Figure 19-1   The principle of holography. (*Top left*) With incident light coming from the left, a point object forms concentric rings; (*top right*) reconstruction of rings gives a point image. (*Center*) The same transformations with two points and (*bottom*) with a more complex object.

Next (center row), assume that the object consists of two points. The diffraction pattern now contains two sets of rings, in essence two overlapping zone plates. Each of these will focus the light; the image, thus, consists of two points. Finally (bottom row), the object may be an aggregate of many points, such as the letters ABC. These letters produce a multiplicity of lines and rings. But as before, each set of rings, each related to a given point in the object, focuses the light back into a point. The sum of these points is again an ABC.

PRODUCING THE HOLOGRAM

A hologram, as I said, is the result of interference. Interference occurs between two contributions, a *signal*, which is the light diffracted by, or scattered off, the object, and a "coherent background" or *reference*. If the signal alone would reach the photographic film, the film could only record the intensity, but it could not record the *phase* of the light. The phase information would be lost.

But, as Gabor argued when he invented holography or "wavefront recon-struction" as it was called then, the signal can be combined with a coherent background and then the phase of the diffracted light could be deduced from the position of the fringes on the film. What is more, if the coherent background is made sufficiently strong, the resultant phase of the combination would become similar to the phase of the background alone, the loss of phase would not matter much anymore and the hologram could be reconstructed again into a likeness of the original.*

To provide enough of the background radiation, Gabor used objects such as a wire mesh or slender letters with wide spaces between them. He then took light from a mercury arc (lasers had not been invented as yet), passed the light through a pinhole and a filter to isolate one of the Hg lines, and let it pass through the object. The signal beam and the reference, in other words, were coaxial, a virtual necessity with the inadequate coherence length (only about 0.1 mm) of mercury light.

A major advance was then made by Leith and Upatnieks who added the reference beam at an *off-set angle*, using either a prism or a series of mirrors. That made possible the recording of holograms of *solid, three-dimensional objects*.†

Today holograms are often recorded as shown in Figure 19-2.‡ Part of the incident light goes to the object and from there is reflected or scattered toward the photographic film. Another part is reflected by the mirror and then reaches the film. At the film the two contributions combine and, by interference, form fringes. These fringes are very closely spaced and cannot be seen by the unaided eye, hence the typical hologram appears to be uniformly gray. But under the microscope it is found to consist of a great many tiny domains, each containing fringes of various lengths and spacing.

---

* Dennis Gabor (1900–1979), Hungarian-born physicist, inventor, philosopher. At age 14 Gabor read advanced physics texts; at 17 he wondered what happens to light as it travels from object to image. Gabor studied at the Budapest Polytechnic and at the Technische Hochschule in Berlin, graduating in 1927 with a doctorate in electrical engineering. He became research engineer first in Germany and then at the British Thomson-Houston Co. in Rugby, Professor of Applied Electron Physics at the Imperial College of Science and Technology in London, and Staff Scientist at the CBS Laboratories in Stamford, Connecticut. His first publication on the subject was "A New Microscopic Principle," *Nature* (*London*) **161** (1948), 777–78. Shortly thereafter he wrote "Microscopy by Recon-structed Wave-fronts," *Proc. Roy. Soc. London* **A197** (1949), 454–87, and "Microscopy by Recon-structed Wave Fronts: II," *Proc. Physical Soc.* **B64** (1951), 449–69, which both became classics in holography. The term *holography* comes from the Greek, ὅλος = all, whole and γραφεῖν = to write. In 1971, Gabor received the Nobel Prize in physics.

† E. N. Leith and J. Upatnieks, "Wavefront Reconstruction with Diffused Illumination and Three-Dimensional Objects," *J. Opt. Soc. Am.* **54** (1964), 1295–1301.

‡ A laser is not really needed for holography; it is merely the use of solid, three-dimensional objects that calls for light whose coherence length exceeds the path differences due to the unevenness of such objects.

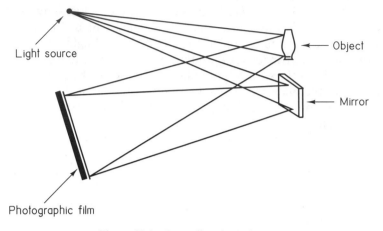

Figure 19-2   Recording the hologram.

RECONSTRUCTION

A hologram results from interference between two beams, the signal beam, $\mathbf{A}_1$, and the reference beam, $\mathbf{A}_2$. These letter symbols, $\mathbf{A}_1$ and $\mathbf{A}_2$, refer to amplitudes. These amplitudes can readily be added to one another and then be squared, but since photographic film responds to intensities, we find that for each point (in the hologram) the intensity is

$$I(x,\ y) = (\mathbf{A}_1 + \mathbf{A}_2)^2$$

$$= (\mathbf{A}_1 + \mathbf{A}_2)(\mathbf{A}_1 + \mathbf{A}_2)^*$$

$$= |\mathbf{A}_1|^2 + |\mathbf{A}_2|^2 + \mathbf{A}_1\mathbf{A}_2^* + \mathbf{A}_1^*\mathbf{A}_2 \qquad [19\text{-}1]$$

where the asterisk * indicates the complex conjugate.

In the reconstruction process, light of amplitude $\mathbf{A}_3$ is used to illuminate the hologram. The hologram has a transmittance function $T(x,\ y)$ and, since it had been produced by the amplitudes $\mathbf{A}_1$ and $\mathbf{A}_2$,

$$T(x,\ y) = \mathbf{A}_1\mathbf{A}_2^* + \mathbf{A}_1^*\mathbf{A}_2 \qquad [19\text{-}2]$$

The $\mathbf{A}_3$ light, therefore, is diffracted (*modulated*) by the hologram,

$$\mathbf{A}_4 = \mathbf{A}_3 T(x,\ y) \qquad [19\text{-}3]$$

Note that in Equation [19-1] it is only the two right-hand terms that carry the information because only *they* are the result of any interference between signal and reference. Hence, if we consider merely these terms, then

$$\mathbf{A}_4 \propto \mathbf{A}_3\mathbf{A}_1\mathbf{A}_2^* + \mathbf{A}_3\mathbf{A}_1^*\mathbf{A}_2 \qquad [19\text{-}4]$$

If the reconstruction amplitude $\mathbf{A}_3$ is equal, or at least proportional, to the reference amplitude $\mathbf{A}_2$, the amplitudes $\mathbf{A}_2$ and $\mathbf{A}_3$ in Equation [19-4] cancel and

$$\mathbf{A}_4 \propto \mathbf{A}_1 \qquad\qquad [19\text{-}5]$$

This means that the amplitude diffracted by the hologram is proportional to the amplitude initially diffracted by the object: *The image is a reconstruction of the object*.

At this point I emphasize again the fundamental difference between a hologram and a conventional photograph. In a photograph the information is stored with a one-to-one correspondence: each point in the object relates to a conjugate point in the image. In a hologram there is no such correspondence; light from every one point in the object goes to all of the hologram. This has two consequences. If a hologram were cut into small pieces, each fragment would still reconstruct the whole scene, not just part of the scene (although the resolution would be less).

In addition, a hologram often receives light not only from the side of the object facing the film but also from the adjoining sides (which could not be seen in conventional photography). This is the reason why frequently a small cube, or some other little toy, is used as the object. Taking a conventional photograph of such a cube would show just the side of the cube facing the camera, and nothing else. But in a hologram (whenever the holographic film is larger in size than the cube), the reconstructed image shows also the sides adjoining the front face and thus, as the observer's head moves sideways when viewing the hologram, the image is seen in three dimensions.

Holograms give both real and virtual images (Figure 19-3). A real image can be projected and focused on a screen. A virtual image can be seen but it cannot be projected. If both the $\mathbf{A}_2$ and the $\mathbf{A}_3$ waves are plane, the two images have the same magnification and are symmetric, with the hologram in the middle. This is an example of a *Fraunhofer hologram*. If the hologram has been recorded with divergent light, the real image is magnified and the virtual image is reduced in size. That is an example of a *Fresnel hologram*.

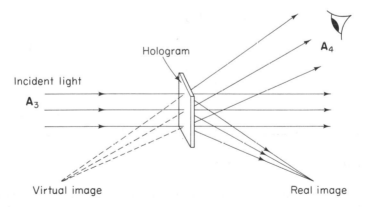

Figure 19-3 Viewing a hologram. The hologram diffracts the incident light $\mathbf{A}_3$, producing a real image and a virtual image.

In either case, the observer looking into the A_4 beam will find that beam hard to distinguish from the A_1 beam. Indeed, the image seen in holography looks startlingly similar to the object, including its three-dimensionality and the parallax typical of the real world. It is this realism that has made holography so intriguing a subject to both scientists and casual observers.

Practical Considerations

Holograms must be recorded on photographic film of high resolvance. Look again at Figure 19-2 and note that the reference, the light reflected by the mirror, and the signal, the light scattered by the object, subtend a certain angle at the film. If this angle is too large, the fringes formed between signal and reference are so close that even the best emulsions cannot resolve them. The experimental setup must also be stable and free from vibration to prevent the fringes from becoming obliterated during exposure.

Conventional silver-halide film is the most popular. Some emulsions particularly suited for holography are Agfa-Gevaert 8E75HD and Kodak 649F. For computer-generated holograms these films, or others, are used together with a laser printer, sometimes producing lines no wider than 1 μm.

Reconstruction always yields a photographic positive. If a contact print were made of a hologram and, hence, the blacks and whites reversed, the image would still look the same, just as the "negative" of a diffraction grating would show the same spectrum lines as the original "positive."

In an emulsion that is sufficiently *thick*, wavefronts traveling in one direction can be made to interfere with wavefronts traveling in the opposite direction. The wavefronts then form a three-dimensional standing-wave pattern, rather than the two-dimensional pattern of a conventional hologram. If viewed in white light, such *volume holograms* give reconstructions in full color.

APPLICATIONS OF HOLOGRAPHY

Holography has a great many fascinating applications. It extends to fields as diverse as precision measurements, pattern recognition, and even art. Most important, at least from the heuristic point of view, is that holography has provided us with *conclusive proof that any image formation is a two-step process*. Whereas in conventional image formation these two steps follow each other without delay, in holography they can be separated, conceptually as well as experimentally.

Holographic Interferometry

One of the practical applications of holography is to test for stresses and strains. For example, two exposures can be made of the object, one before applying a load and the other after. Both exposures are recorded on the same film. Then the hologram is reconstructed. Whenever, and wherever, the two recordings are different, they interfere, showing surface deformations in the form of fringes that are easy to see as well as convenient to measure.

Particle Size Determination

Particles such as aerosols and atmospheric pollutants in suspension are in constant motion. Thus, with conventional photography it is difficult to focus even on a single particle, and it is next to impossible to focus on all of them at the same time.

A hologram, though, can be recorded without focusing. Using a pulsed laser, for example, freezes the motion. Later, when viewing the image, the observer can examine at leisure any particles that were present at the instant the holographic recording was made.

Rainbow Holograms

Rainbow holograms can be viewed in white light. They give colorful images that are separate from each other, not superimposed as with other holograms. They are used on stick-on labels, buttons, and other commercial items.

Rainbow holograms are made in two steps. First a conventional "master" hologram is recorded. The hologram is then covered except for a slit about 1 cm wide. When illuminated, the hologram forms a real (and a virtual) image. The real image is combined with a reference beam to produce a "copy" hologram. In essence, a rainbow hologram is a hologram made of another hologram's reconstruction.

Computer-Generated Holograms

Assume that a hologram contains no more than a few concentric rings spaced like a zone plate. Such a hologram is easy to draw by computer. Reconstruction forms a point, the same as parallel light passing through a lens comes to a focus. But more complex *holographic optical elements* (HOEs) can be drawn as well such as mirrors, cylinder lenses, and aspherical lenses and mirrors. These HOEs can be used in place of conventional optical elements but, clearly, they work by diffraction rather than by refraction or reflection.

Holograms can also be produced of much more sophisticated objects. Some of such *computer-generated synthetic holograms* may show objects that perhaps do not even exist in reality but were derived from theoretical considerations. Indeed it has become possible to draw, first on a large scale and then to reduce photographically and to reconstruct, individual molecules and even single atoms, replete with nuclei and electrons rotating around them.

Synthetic landscapes, of cities seen from satellite photographs or of planets like Venus, with its wavy surface and clouds overhead, are known as examples of *virtual reality*.

Point-of-Sale Scanners

In recent years point-of-sale (POS) scanners have become a familiar sight at checkout counters in supermarkets and variety stores. They contain a light source, a fixed or rotating hologram, a series of mirrors, and a photodetector. The light scans across the bar-code pattern affixed to the merchandise and, even when the item is passed casually over the scanner, is reflected or scattered back and so identifies the item; the signal then goes to an in-store computer for further processing. In some POS scanners the hologram rotates; such a scanner is shown in Figure 19-4. In others, the hologram is stationary and a rotating pentagonal mirror provides the scan.

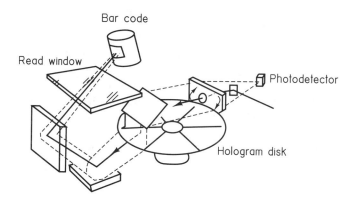

Figure 19-4   Point-of-sale scanner using rotating hologram. (Reproduced by permission, Fujitsu Laboratories Ltd., Atsugi, Japan.)

Pattern Recognition

Finally we come to one of the most exciting applications of holography, *pattern recognition*, also called *character recognition* in reference to alphanumeric characters. Early character recognition systems were based on geometrical optics. Consider, for instance, that we want to recognize the letter A (Figure 19-5). A set of characters, A, B, C, . . . , are printed as negatives on a strip of film which then is moved through the image plane. If the character sought matches the character on the film, the output from a photodetector is zero, triggering a printer. But in reality this doesn't work: the character and the negative must be aligned perfectly, both in position and in size, a requirement that virtually never can be met satisfactorily.

However, the two patterns can be made to interact with each other without perfect alignment and without even a lens between them. Such interaction is called *convolution*. Assume, for example, that two line drawings, each in the form

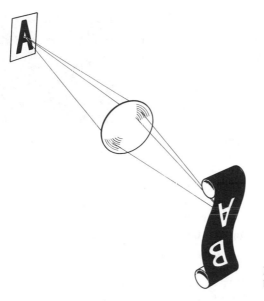
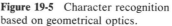

Figure 19-5 Character recognition based on geometrical optics.

of a transparency, are placed in front of a camera as shown in Figure 19-6. The pattern on the left is called the *target*. The pattern closer to the lens is the *mask*. The camera, as I said, is used without a lens; in fact, there are no lenses at all in the system, in contrast to the typical 4*f* configuration often used in optical data processing.

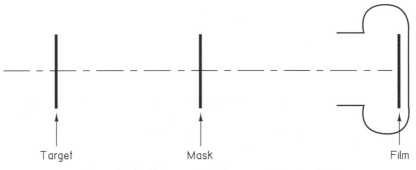

Target   Mask   Film

Figure 19-6 Placement of transparencies in front of camera.

First we use as the target a white line on a black background, a line drawing of a circle. As the mask we use a pinhole. The result is a (slightly blurred) image of the circle.

Next we use as the mask several pinholes arranged in a triangle. The system now acts as a multiple-pinhole camera that forms as many images of the circle as there are pinholes in the triangle; in short, it forms a series of circles arranged in a triangle.

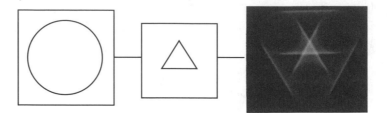

Figure 19-7 Cross-correlation between circle and triangle.

But then we use only line drawings, no pinholes. Again the target is a circle and the mask is a triangle. The result is shown in Figure 19-7.

The target may have a transmittance function $M(x, y)$ and the light an intensity I_0. Then the light passing through has an intensity distribution

$$I(x, y) = I_0 M(x, y) \qquad [19\text{-}6]$$

The mask, in turn, may have a transmittance function $N(x, y)$. After passing also through the mask, the light has an intensity distribution that is the product of both,

$$P(x, y) = I_0 M(x, y) N(x, y) \qquad [19\text{-}7]$$

The difficulty is that when one of the transparencies is moved in the x-y plane, either $M(x, y)$ or $N(x, y)$ changes while the pattern as such does not change. Hence, the function must be invariant under translation and the pattern have "internal coordinates" that move with it. The output, now called $P(\xi, \eta)$, is the integral of the product of $M(x, y)$ and $N(x - \xi, y - \eta)$ for all values of x and y over the extent of the pattern,

$$P(\xi, \eta) = \iint_{-\infty}^{+\infty} M(x, y) N(x - \xi, y - \eta) \, dx \, dy \qquad [19\text{-}8]$$

which is the *convolution integral*. The direction of translation, either $+$ or $-\xi$, and $+$ or $-\eta$, is immaterial. This is an example of a two-dimensional convolution between different functions, an example of *cross-correlation*.

Next assume that the pattern is convolved with another, identical pattern or, in other words, that it is convolved with itself. In that case, $\xi = \eta = 0$, and Equation [19-8] simplifies into

$$P(\xi, \eta) = \iint_{-\infty}^{+\infty} M(x, y) M(x, y) \, dx \, dy = \iint_{-\infty}^{+\infty} M^2(x, y) \, dx \, dy \qquad [19\text{-}9]$$

This is an example of *auto-correlation*. In that case the product in the integrand is as large as it can be for any given function so that in the center of the pattern detail the output reaches a maximum, forming a *bright spot of light*. The spot's intensity is a measure of the correlation between the two functions.

But now, instead of the mask containing a real image of the letter A, we use a hologram of an A. When such a hologram is illuminated with light that comes from another A, plane wavefronts result that can be focused, again, into *a bright spot* [Figure 19-8 (top)]. If the light were to come from a B, or from any other character different from that of the mask, it would merely form a diffuse patch of light (center). Hence, we can search through an array of letters and determine whether and where a given character is present (bottom).

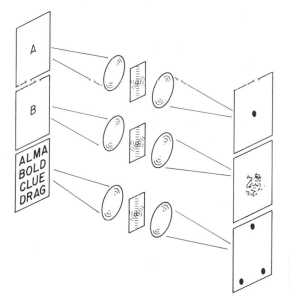

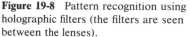

Figure 19-8 Pattern recognition using holographic filters (the filters are seen between the lenses).

The holograms used in Figure 19-8 seem to be amplitude filters. But because they were initially made by interference (between signal and reference), they represent in fact both the amplitude and the phase of the light; they are complex, matched *VanderLugt filters*. They are often made using a Mach–Zehnder interferometer; they play an important part in most any pattern recognition system.*

Plenty of difficulties still lie ahead before true *reading machines* can become a reality. Some letters and words are "inside" others. For example, F is inside E, P is inside R and B, T and L have the same horizontal and vertical lines, and *arc* is inside *search*. Clearly, the more alike the two characters, the less the power of discrimination. A major problem, also, will be to teach the machine to recognize the "meaning" of a letter set in different typeface; the letter A, for example, can

---

* The term *complex* is used, not because of the complexity of making them, but because of the analogy to complex numbers, which also carry both amplitude and phase information. The term *matched* comes from radar technology but now refers also to the phase of the filter that is conjugate, matched, to the phase of the Fourier transform of the signal.

be printed A, **A**, 𝔄, *ɑ̄*, a, **a**, α, *α*, and an almost infinite number of variations is possible when it comes to *handwriting*.

SUMMARY OF EQUATIONS

Convolution:

$$P(\xi, \eta) = \iint_{-\infty}^{+\infty} M(x, y)N(x - \xi, y - \eta) \, dx \, dy \qquad [19\text{-}8]$$

PROBLEMS

19-1. If, when making a hologram, the angle subtended at the hologram by the signal beam and the reference is 15°, if the wavelength is 518 nm, and if we consider the first order only, how far are the lines in the hologram apart from each other?

19-2. Holograms are taken with light of 632.8 nm wavelength. What is the limiting angle between the signal and the reference if the space frequency in the hologram is not to exceed 200 lines/mm?

19-3. Following an example by Gabor, assume that the amplitudes of the signal and the reference are related as 1 : 10. Since the two beams when they combine may be completely *in* phase, or completely *out of* phase, what is the maximum ratio of their intensities?

19-4. Take a hologram and, using high-resolution black-and-white film, make a contact *print* of it so that you now have in effect a "positive" and a "negative." View the new hologram in the same way as the original. How do the original and the print differ?

19-5. Reconstruction is made of a hologram that is part of a zone plate, *not* including its center. If the hologram is viewed in collimated light, show where the real and the virtual image will be formed.

19-6. A projectile is fired through a wind tunnel placed in one arm of a Mach–Zehnder interferometer. If the fringe pattern were "reconstructed," what would it show?

19-7. If a hologram is recorded with light of 400 nm wavelength and reconstructed with light of 800 nm, how does the image change compared with a reconstruction at 600 nm?

19-8. Consider a three-dimensional object and image and discuss the essential difference between *stereoscopic* and *holographic* image formation.

19-9. Determine the convolution of two transparencies, each containing three holes as shown in Figure 19-9.

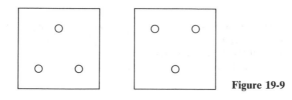

Figure 19-9

SUGGESTIONS FOR FURTHER READING

ABRAMSON, N. *The Making and Evaluation of Holograms*. London: Academic Press, 1981.

DENISYUK, YU. N. *Fundamentals of Holography*. Moscow: Mir Publishers, 1984.

HARIHARAN, P. *Optical Holography—Principles, Techniques and Applications*. New York: Cambridge University Press, 1986.

KASPER, J. E., and S. A. FELLER. *The Complete Book of Holograms*, *How They Work and How to Make Them*. New York: John Wiley and Sons, Inc., 1987.

Light Sources
and Detectors

Light is generated, transmitted, and received in the form of elemental units of energy, each of them called a *quantum*. The quantum aspect of light, the same as its ray aspect and its appearance as waves, is an integral part of optics. We begin our discussion of quantum optics with a description of phenomena seen with *light sources and detectors*.

LIGHT SOURCES

Many light sources, from the sun to candles to electric lights, are *thermal light sources*; they produce light because they are heated to *incandescence*.

The first electric light bulbs had carbon filaments.* Then came metal filaments, in particular those made of refractory metals that have a high melting point

---

* The first electric light bulbs were made in 1879 by Thomas Alva Edison (1847–1931), prolific American inventor. Out of a total of 1100 patents, in one four-year period alone he obtained 300, one patent every five days. Among them, besides the electric-lamp patent (No. 223,898, dated Jan. 27, 1880), were the phonograph (his own favorite) and the motion picture camera and projector. When Edison announced he would try producing light from electricity to compete with natural gas then used for illumination, gas stocks at the New York and London stock exchanges tumbled; so much faith had the public in his ability. Edison first tried various metals. It didn't work; so he turned to cotton thread, which, when "carbonized" (scorched), conducted electricity without melting. The second light bulb Edison made lasted for 40 hours. "I think we've got it," he exulted, "if it can burn 40 hours, I can make it last a hundred."

such as *tungsten*. Tungsten melts at 3410°C, but before that happens it slowly evaporates and forms deposits on the inner surface of the glass envelope, reducing the output. Some halogens, in particular iodine, retard this process. On the inside of the hot (quartz) envelope, the iodine combines with the tungsten deposits to form tungsten iodide. Near the even hotter filament the molecule dissociates again, depositing the tungsten back onto the filament and letting the iodine reenter the cycle. The filament can then be heated close to the melting point, which accounts for the high efficacy of such *halogen lamps*.

Fluorescent lamps, in contrast to incandescent lamps, do not produce much heat: fluorescence, together with phosphorescence, is part of a "cold" emission of light called *luminescence*. The difference between the two is that in fluorescence the emission lasts no longer than 10^{-8} s after excitation has ended while phosphorescence can persist up to several hours.

Light-emitting diodes (LEDs) are most often *semiconductors*. A semiconductor, as the name implies, conducts electricity better than an insulator but not as well as a conductor (that is, a metal). In an insulator, the electrons are tightly bound to their atoms. In a metal, the electrons can move freely; hence, even a small voltage applied to the conductor will cause a current to flow. A semiconductor lies somewhere in between.

The first LEDs were made from gallium arsenide, GaAs. By adding a small amount of a dopant, additional (**n**egatively charged) free electrons can be supplied to the conduction band; that results in an *n*-type material. Removing electrons from the valence band leaves (**p**ositively charged) "holes" in their place; that results in a *p*-type material. Both electrons and holes can be considered charge carriers, a hole moving from atom to atom the way a bubble moves through a liquid. With the *n*-side of a *p-n* junction connected to the negative terminal of a battery and the *p*-side to the positive terminal, electrons will flow from *n* to *p*, and holes from *p* to *n*. As the electrons combine with the holes, excess energy is released in the form of light as indicated in Figure 20-1.

Compared with other light sources, LEDs offer many advantages. They are small (about 1 mm³) and of light weight; they are used extensively in visual displays and fiber-optics communication systems.

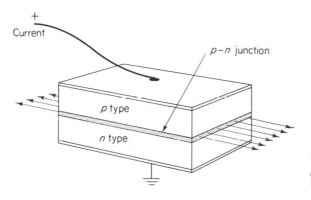

Figure 20-1 Schematic diagram of a *p-n* junction diode. Arrows show light being emitted.

Blackbody Radiator

A blackbody is a perfect source of radiation: no source, of a given temperature, can produce more radiation, at a given wavelength, than a blackbody. A blackbody can be approximated by a cavity such as a hollow block of cast iron with a small hole in it. When the cavity is heated, like an oven, it radiates energy out of the hole. The spectral distribution of that radiation is a function of temperature alone; the material as such plays no role.

The radiation is produced by oscillations of the molecules of the cavity's wall and, according to classical theory, is proportional to the number of standing waves generated by these oscillators. This assumption agrees well with experimental evidence in the infrared, but it does not agree in the ultraviolet. As the wavelength approaches zero, the number of waves theoretically would become infinite; but obviously this cannot be, and the amount of energy needs to become zero as well, a dilemma then called the "ultraviolet catastrophe."

The problem was solved by Max Planck.* Planck assumed that the oscillators emit energy, not in the form of a smooth continuous flow, but as discrete, elemental units of energy called *quanta* or *photons*. They have no mass or charge.

The energy E of a quantum of electromagnetic radiation is proportional to its frequency, ν,

$$\boxed{E = h\nu}$$ [20-1]

The factor of proportionality, h, is a fundamental constant of nature, called *Planck's constant*. It has the value

$$h = 6.6260755 \times 10^{-34} \text{ J s}$$

Planck's constant must be determined experimentally; it cannot be predicted from theory. It has been measured in a variety of ways, and all of these measurements agree.

The spectral distribution of the radiation emitted by a blackbody can be described by *Planck's radiation law*. This law is often written in terms of radiant power emitted per unit area, called *exitance*, M, as a function of wavelength, λ,

$$M(\lambda) = \frac{C_1}{\lambda^5} \frac{1}{e^{C_2/\lambda T} - 1}$$ [20-2]

---

* Max Karl Ernst Ludwig von Planck (1858–1947), German physicist, professor of theoretical physics at the University of Berlin, and president of the then Kaiser Wilhelm Gesellschaft, now Max Planck Gesellschaft. Planck introduced his concept of quantum units of light at a meeting of the Berlin Physical Society on 19 October 1900. Acceptance of his idea at first was slow; Planck himself resisted it for several years. "My futile attempts to fit the elementary quantum of action somehow into the classical theory continued for a number of years," he wrote, "and they cost me a great deal of effort." M. Planck, 'Ueber irreversible Strahlungsvorgänge,' *Ann. Phys.* (4) **1** (1900), 69–122. In 1918 Planck received the Nobel Prize for physics.

where T is the absolute temperature, in units of kelvin (K), and C_1 and C_2 are two *radiation constants*. The first radiation constant, C_1, is defined as, and has a numerical value of,

$$C_1 = 2\pi hc^2 = 3.741774 \times 10^{-16} \text{ W m}^2$$

and the second radiation constant, C_2, is defined as, and has a value of,

$$C_2 = \frac{hc}{k} = 0.01438769 \text{ m K}$$

where h is again Planck's constant, c the velocity of light, and k *Boltzmann's constant*, 1.380658×10^{-23} J K^{-1}.

Wien's Displacement Law

There are two sequels to Planck's radiation law. First, if we plot Planck's law for different temperatures, we find that with increasing temperature (a) *more energy is emitted* and (b) *the peak emission shifts toward the shorter wavelengths*. That is illustrated in Figure 20-2.

This shift agrees with common experience. For example, when a block of iron is gradually heated, it emits at first only infrared (which can be felt but not really *seen*). As the temperature is raised, the block begins to glow dark red. As the temperature is raised further, it becomes bright red ("red hot") and finally "white hot."

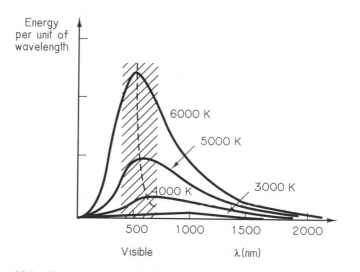

Figure 20-2 Distribution of radiant energy as a function of wavelength. At higher temperatures the peak emission shifts to the left.

The peak wavelength, λ_{max}, in nanometers, is given by

$$\lambda_{max} = \frac{2.8978 \times 10^6}{T} \qquad \text{[20-3]}$$

where T is again the absolute temperature ($0°C = 273.16$ K). This is *Wien's displacement law*.*

The *color temperature* of a light source refers to the temperature of a blackbody that emits radiation of the same color as the source. The color temperature of the sun, for example, is 5600 K, and that of a 200-W tungsten-filament lamp is about 3200 K. Accordingly, daylight color film is balanced for 5600 K and indoor color film for 3200 K.

Stefan–Boltzmann's Law

Second, the total radiation output of a blackbody is found by integration over all wavelengths,

$$M = \int_0^\infty M(\lambda)d\lambda = \sigma T^4 \qquad \text{[20-4]}$$

This is *Stefan–Boltzmann's law*.† It shows that the exitance increases as the fourth power of the absolute temperature. The constant of proportionality, σ, is *Stefan–Boltzmann's constant*; it has a numerical value of 5.67051×10^{-8} W m^{-2} K^{-4}.

All these radiation laws apply to blackbody radiators only. If the body is not black ("graybodies"), or if it is transparent or reflective, it is a poor radiator and absorber, with the amount of energy radiated less by a factor ε called *emissivity*. Stefan–Boltzmann's law, for example, then becomes

$$M = \varepsilon\sigma T^4 \qquad \text{[20-5]}$$

Only for a true blackbody will ε be unity. For other bodies, the emissivity, which also varies with wavelength, usually lies between 0.2 and 0.9, and for highly polished metals it may be as low as 0.01.

---

* Wilhelm Carl Werner Otto Fritz Franz Wien (1864–1928), German physicist. Wien studied under Helmholtz, became his assistant, later successor to Röntgen at both the universities of Würzburg and Munich. Wien contributed to thermodynamics and electric discharges but is best known for his displacement law. He received the Nobel Prize in physics in 1911. W. Wien, "Temperatur und Entropie der Strahlung," *Ann. Phys.* (Neue Folge) **52** (1894), 132–65.

† Josef Stefan (1835–1893), Austrian, professor of physics at the University of Vienna, found the relationship by experiment; Ludwig Edward Boltzmann (1844–1906), Stefan's one-time student and later successor as professor of physics at the same university up to his death by suicide, derived it independently from theory. Their discoveries were published in J. Stefan, "Über die Beziehung zwischen der Wärmestrahlung und der Temperatur," *Sitzungsber. Math.-Naturwiss. Classe kais. Akad. Wiss.*, Wien **79**, II. Abtheilung (1879), 391–428; L. Boltzmann, "Ableitung des Stefan'schen Gesetzes, betreffend die Abhängigkeit der Wärmestrahlung von der Temperatur aus der electromagnetischen Lichttheorie," *Ann. Phys.* (Neue Folge) **22** (1884), 291–94.

Example 1

If the peak emission from the sun is at 475 nm, what is the sun's surface temperature, in centigrades, or degrees Celsius?

Solution: From Wien's displacement law,

$$T = \frac{2.8978 \times 10^6}{475} = 6100 \text{ K}$$

This temperature, though, is given in units of kelvin; to change to degrees Celsius, we subtract 273°C:

$$6100 - 273 = \boxed{5827°C}$$

Example 2

What is hotter, heaven or hell?

Solution: Heaven, according to Isaiah 30:26, receives from the moon as much radiation as Earth does from the sun and in addition seven times seven (forty-nine) times as much as Earth from the sun, or fifty times as much radiation in all. If we assume that the average temperature of Earth is 27°C = 300 K, then from Stefan–Boltzmann's law,

$$\left(\frac{\text{heaven}}{300}\right)^4 = 50$$

which means that the temperature of heaven is

$$(300)(50^{1/4}) = 798 \text{ K} = \boxed{525°C}$$

Hell, according to Revelation 21:8, has as its main topographic feature a lake of molten brimstone (sulfur). Since the boiling point of sulfur is 444.7°C, the temperature of hell must be less than that: Above 444.7° the lake would evaporate, and below it, it would harden. Thus, *heaven is hotter than hell.* [From *Appl. Opt.* **11** (1972), 8, A14.]

LINE EMISSION SPECTRA

The spectra we have discussed so far were *continuous spectra*. They issue from solids and liquids and from gases under high pressure. Now we come to *line spectra*; they are produced by gases at low pressure where the atoms or molecules are far apart and do not significantly interact with one another.

Each chemical element has unique spectrum lines that are characteristic of the element (like a fingerprint is of a person) and therefore permit a positive identification. Some elements, such as hydrogen, have relatively few lines; others, such as iron, have thousands.

A group of lines as shown in Figure 20-3 is called a *series*. The most prominent, brightest line (of hydrogen) lies on the right; it has the longest wavelength

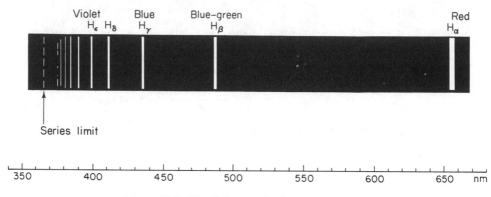

Figure 20-3   The Balmer series in the spectrum of hydrogen.

and is designated α. The next line to the left is called β, the third line γ, and so on. At the short-wavelength end, on the far left, the lines crowd together and form the *series limit*.

The exact position of the lines (in the spectrum of hydrogen) follows a simple relationship. It can be expressed in the form of *Rydberg's equation*,

$$\frac{1}{\lambda} = \mathcal{R} \left(\frac{1}{m^2} - \frac{1}{n^2} \right) \qquad \text{[20-6]}$$

where λ is the wavelength in nm, $\mathcal{R}$ is *Rydberg's constant*, ≈ 0.011 nm^{-1}, m is an integer, 1, 2, 3, . . . which is the same within any one series, and n is another integer, the "running figure," which runs through the series. The two integers are connected as $n = m + 1$.

If we set $m = 1$ and $n = m + 1 = 2$ and insert them in Rydberg's equation, we come to the first series (of hydrogen), the *Lyman series*. Its first line has a wavelength of

$$\frac{1}{\lambda} = (0.011 \text{ nm}^{-1}) \left(\frac{1}{1^2} - \frac{1}{2^2} \right)$$

$$\lambda = 121 \text{ nm}$$

It is the most powerful line emitted by the sun. The Lyman α line is completely absorbed by air; it is also absorbed by deoxyribonucleic acid, DNA, the building block of all living matter on Earth. On other planets, which have no protective atmosphere, therefore, life either cannot exist, or must be based on genetic material other than DNA.

The second series of hydrogen is the *Balmer series*. Here, $m = 2$ and $n = m + 1 = 3$. The Balmer α line, consequently, has a wavelength of

$$\frac{1}{\lambda} = (0.011 \text{ nm}^{-1}) \left(\frac{1}{2^2} - \frac{1}{3^2}\right)$$

$$\lambda = 655 \text{ nm}$$

which is near the red end of the visible spectrum.

The Balmer β line has

$$\frac{1}{\lambda} = (0.011) \left(\frac{1}{2^2} - \frac{1}{4^2}\right)$$

$$\lambda = 486 \text{ nm}$$

and the *series limit*

$$\frac{1}{\lambda} = (0.011) \left(\frac{1}{2^2} - \frac{1}{\infty}\right)$$

$$\lambda = 365 \text{ nm}$$

which is near the blue end of the visible spectrum.

The spectrum "lines" that we have described so far represent (the energies of) electrons that revolve around the nucleus of an atom. When the electron is in a certain orbit, however, the atom does not emit radiation. It is only when the electron makes a *transition* from one orbit to another that such radiation occurs. It is easier to represent these orbits, instead of in circles or ellipses, in the form of straight lines called *energy levels*. An example is shown in Figure 20-4.

Each horizontal line represents an allowed energy level. The vertical arrows between levels show various transitions. The lowest energy level is the level at which the electron revolves in the innermost orbit (where $n = 1$); that level is called the *ground state*.

If energy is supplied to the atom, the system is raised from a lower energy state, E_1, to a higher, *excited* state, E_2. Such a transition, $E_1 \rightarrow E_2$, is called *absorption*. The reverse process, the downward transition $E_2 \rightarrow E_1$, is called *emission*.

What does it all mean? First, the study of atomic emission has greatly clarified the structure of matter. Second, we now have a data base that gives computer access to some 100,000 spectra of just as many different substances. This helps us identify unknown chemicals by comparing their spectra to those of other, known substances.

For practical use, it has helped us to produce new and better light sources. For example, to sodium vapor, with its very yellow light, can be added a small amount of xenon. This causes the narrow sodium lines to widen into the green and red of the spectrum, providing a more pleasant and more efficient means of illumination. Finally, understanding atomic emission greatly helps us to understand *stimulated emission*. This is the basis of *laser* action, which we discuss in Chapter 23.

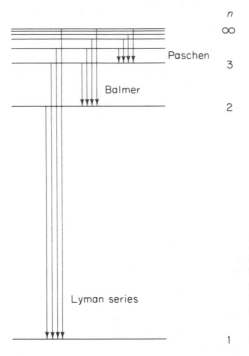

Figure 20-4 Energy levels of hydrogen showing some of the downward transitions in the Lyman, Balmer, and the next, the Paschen, series. The lowest level, 1, is the ground state. Radiation is emitted whenever the atom makes a transition from a higher energy state to a lower state.

LIGHT DETECTORS

Detectors of electromagnetic radiation, the same as light sources, can be divided into two major groups, *thermal detectors* and *quantum detectors*. Thermal detectors are based on absorption and heating; if the absorbing material is black, they are independent of wavelength. Quantum detectors are based on the *photoelectric effect*; some of them are so sensitive they respond to individual quanta.

Thermal Detectors

Thermal detectors are relatively slow to respond; it may take them almost a second to reach equilibrium. A good example of a thermal detector is the *Golay cell*, a thin black membrane placed over a small, gas-filled chamber. Heat absorbed by the membrane causes the gas to expand, which in turn can be measured, either optically (by a movable mirror) or electrically (by a change in capacitance). Golay cells are used in the infrared.

A *thermocouple*, in essence, is a junction between two dissimilar metals. As the junction is heated, the potential difference changes. In practice, two junctions are used in series: a *hot* junction exposed to the radiation and a *cold* junction shielded from it. The two voltages are opposite to each other; thus the detector, which without this precaution would show the absolute temperature, now measures the temperature differential. A *thermopile* contains several thermocouples and, therefore, is more sensitive. A *bolometer* contains a metal element whose electrical resistance changes as a function of temperature; if instead of the metal a semiconductor is used, it is called a *thermistor*. Unlike a thermocouple, a bolometer or thermistor does not generate a voltage; they must be connected to a voltage source.

Quantum Detectors

A beam of intense light, it seems, should cause more of an effect than a beam of dim light. With a quantum detector, though, that is not necessarily so. The wavelength of the light plays an important role; in fact, there is a certain upper limit of the wavelength above which there is no effect at all, no matter what the intensity.

Assume the radiation is incident on a plate M mounted in an evacuated tube as shown in Figure 20-5. The plate is made of a material that, when irradiated, releases electrons, then called *photoelectrons*. Opposite M is another plate, the collector plate C. If C is made *positive* with respect to M, the photoelectrons released by M are attracted by, and travel to, C. As the potential V, read on a high-impedance voltmeter, is increased, the current, i, read on an ammeter, increases too, but only up to a given *saturation level*, because then all of the electrons emitted by M are collected by C.

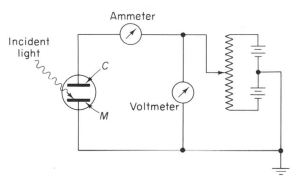

Figure 20-5 Photoelectric effect.

If more intense light falls on the photocathode, it will release more electrons but their energies, and their velocities, will remain the same. Instead, the energy of the photoelectrons depends on the *frequency* of the light: blue light produces more energetic photoelectrons than red light. That agrees with common experience: *blue light is simply more powerful than red light*. In biology, the same holds true: UV or blue may cause skin cancer while red or IR merely feels warm. In addition, the response of a quantum detector is all but instantaneous; there is *no time lag*, at least not more than 10^{-8} sec, between the receipt of the radiation and the resulting current.

On the other hand, if *C* is made *negative*, some photocurrent will still exist, provided the electrons ejected from *M* have enough kinetic energy to overcome the repulsive field at *C*. But as *C* is made more negative, a point is reached where *no* electrons reach *C* and the current drops to zero. This occurs at the *stopping potential*, V_0 (Figure 20-6). In short: A significant amount of photocurrent is present only if the collector, *C*, is made positive, that is, if it is the *anode*. The light-sensitive surface *M*, then, is the *photocathode*.

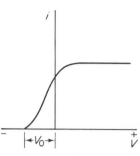

Figure 20-6 Dependence of photocurrent on potential applied to collector plate.

In earlier times these facts were hard to reconcile. Conceivably, the energy imparted to the electrons could come either from the light (but that would mean higher energies with more intense light and perhaps some time delay for very dim light), or it could come from heat energy stored in the material, the light merely acting as a trigger (but we do not find higher energies as the surface is heated).

The answer again follows from the fact that the light is received in the form of discrete quanta. But part of the energy contained in a quantum is needed to pry the electron loose from the surface; that part is called the *work function*, *W*. It is only the excess energy, beyond the work function, that appears as kinetic energy (of the electron). The maximum kinetic energy with which the electron can escape, therefore, is

$$\boxed{\mathrm{KE_{max}} = h\nu - W}$$ [20-7]

which is *Einstein's photoelectric-effect equation.**

It is often convenient to measure energies on an atomic scale not in joules but in *electron volts*, eV. By definition, 1 eV is the amount of energy acquired, or lost, by an electron moving through a potential difference of 1 volt:

$$1 \text{ eV} = (1e)(1V) = 1.60217733 \times 10^{-19} \text{ J}$$ [20-8]

---

* In 1905, Albert Einstein applied Planck's theory to the photoelectric effect, postulating that light is not only *emitted* but also *received* in the form of quanta. A. Einstein, "Über einen die Erzeugung und Verwandlung des Lichtes betreffenden heuristischen Gesichtspunkt," *Ann. Phys.* (4) **17** (1905), 132–48. This paper by Einstein on the photoelectric effect is much more radical than the one he wrote in the same year on the theory of relativity. In fact, it was this paper that earned him the Nobel Prize in physics.

If we then use $c = \lambda\nu$, solve for λ, and substitute $E = h\nu$, we find that

$$\lambda = \frac{hc}{E}$$

Inserting the actual values for h and c and substituting Equation [20-8] leads to

$$\lambda = \frac{(6.63 \times 10^{-34} \text{ J s})(3 \times 10^{8} \text{ m/s})}{1.6 \times 10^{-19} \text{ J/eV}} = \frac{1240 \text{ nm eV}}{E} \qquad [20\text{-}9]$$

which connects the wavelength, in nanometers, with the energy, in electron volts.

Example

The work function determines the longest wavelength to which a detector can respond: The lower the work function, the longer the wavelength. The lowest work functions are found among the alkali metals. Potassium, for example, as shown in Table 20-1, has a work function of 2.25 eV. At the threshold, *no* photoelectrons are released and Equation [20-7] becomes zero:

$$KE_{max} = h\nu - W = 0$$

so that

$$W = h\nu = E$$

Then, from Equation [20-9],

$$\lambda = \frac{1240 \text{ nm eV}}{W} = \frac{1240}{2.25} = \boxed{551 \text{ nm}}$$

which is the longest wavelength that can be detected using a potassium surface.

Table 20-1 Photoelectric properties of some alkali metals

Alkali	Work Function (eV)	Threshold (nm)
Sodium	2.28	543
Potassium	2.25	551
Rubidium	2.13	582
Cesium	1.94	639

Practical Quantum Detectors

The simplest type of a quantum detector is the *phototube*, illustrated in Figure 20-7. Light falling on the photocathode causes electrons to be released. If an electric potential is applied, a current is produced. This current may be amplified to produce a suitable signal.

A phototube is sensitive to all wavelengths at which the quantum energy of the light exceeds the work function of the photocathode. The tube is usually

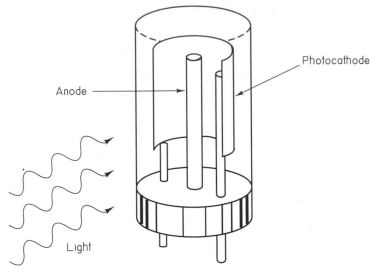

Figure 20-7   Phototube.

evacuated (vacuum phototube); sometimes it contains a small amount of a gas (gas phototube). Since electrons are released as a function of the light, or other radiation, that is incident on the cathode, phototubes are an example of *photo-emissive detectors*.

The ratio of the number of photoelectrons released to the number of photons received is called the *quantum efficiency*. Ordinarily, it is no better than a few percent. But if several *dynodes*, each at a higher voltage, are combined in series to form a *photomultiplier* as shown in Figure 20-8, the efficiency becomes much higher. A photomultiplier, therefore, is an amplifying phototube; it can provide a current of up to 10^6 times that of the first-step output.

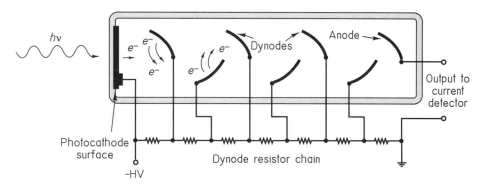

Figure 20-8   Photomultiplier. Electrons emitted by the photocathode are accelerated toward a series of curved dynodes, each producing additional electrons; these are collected by the anode.

A *photodiode* is the solid-state equivalent of the phototube. Most often it is a silicon semiconductor vacuum-deposited on glass. The deposited layer should be thick enough to absorb most of the light falling on it. With the light incident, the electrons in such a *p-n* junction start moving, but only in one direction, producing a current. Such *photovoltaic detectors* are used as solar cells and in exposure meters on cameras; they can have quantum efficiencies as high as 90 percent.

Photoconductive detectors change resistance when exposed to light. They are light-dependent resistors. In the dark they conduct electricity poorly; when exposed to light, they conduct very well. Materials often used are cadmium sulfide or selenide, CdS or CdSe, for use in the visible and near-infrared, and lead sulfide or selenide, PbS or PbSe, in the infrared.

An *image tube* is used not only to detect light but also to preserve the image. Some image tubes contain a photodetector inside a vacuum tube that can be read by an electron beam scanning across. In others, the photoelectrons emitted by the cathode are focused by an electron lens and made visible on a phosphor screen mounted in the same tube. Still other image detectors contain a great many capillaries fused into a wafer, each capillary corresponding to a single pixel. That is called a *microchannel image intensifier* (Figure 20-9).

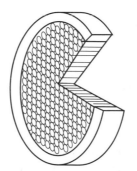

Figure 20-9   Microchannel image intensifier. (Reproduced by permission, Hamamatsu Corp.)

Today we can detect images at very low light levels. If the image is merely amplified, we have an *image intensifier*. If the image is formed in the X-ray range, the UV, or the IR and from there shifted into the visible, we have an *image converter*.

SUMMARY OF EQUATIONS

Quantum energy:

$$E = h\nu \tag{20-1}$$

Wien's displacement law:

$$\lambda_{\mathrm{max}} = \frac{2.8978 \times 10^6}{T} \tag{20-3}$$

Stefan–Boltzmann's law:

$$M = \sigma T^4 \qquad [20\text{-}4]$$

Rydberg equation:

$$\frac{1}{\lambda} = \mathcal{R}\left(\frac{1}{m^2} - \frac{1}{n^2}\right) \qquad [20\text{-}6]$$

Photoelectric-effect equation:

$$\mathrm{KE}_{\mathrm{max}} = h\nu - W \qquad [20\text{-}7]$$

Wavelength and electron energy:

$$\lambda = \frac{1240 \text{ nm eV}}{E} \qquad [20\text{-}9]$$

PROBLEMS

20-1. If the sun has a temperature of 5800 K, what is the peak wavelength of the light emitted by it?

20-2. Can daylight (peak wavelength 555 nm) be obtained from a tungsten filament lamp without using an additional filter?

20-3. Virtually everything emits electromagnetic radiation. For example, what is the peak wavelength sent out by the human body (normal temperature = 37°C)?

20-4. If a certain material is heated to 1000°C and then emits infrared energy of 2.3 μm wavelength, what must be its temperature to emit yellow light of 575 nm?

20-5. Assume that the light bulb in an ophthalmoscope is operated at a *higher* voltage than usual. How do blood vessels, which normally have a particular appearance and color, look now?

20-6. What is the temperature, in °C, of a surface 1 mm² in size that has an emissivity of 0.15 and an exitance of 2.27×10^5 W m^{-2}?

20-7. What is the wavelength of the tenth line in the Lyman series of hydrogen?

20-8. Using the Rydberg equation, find the wavelength and the color of the fifth line in the Balmer series of the spectrum of hydrogen.

20-9. When monochromatic light coming from a pinhole is collimated and focused into a small aperture in front of a photocell, as shown in Figure 20-10, the ammeter reads 10 mA. Now a glass plate A is inserted into one half of the beam (neglect plate B). Plate A is of a thickness such as to retard the light by one-half of a wavelength. How much light will reach the detector? How much photocurrent will result? What happens to the energy?

20-10. Continue with Problem 20-9. Leave plate A in place and insert in addition an opaque plate B, blocking one-half of the beam. What happens? How much current will result?

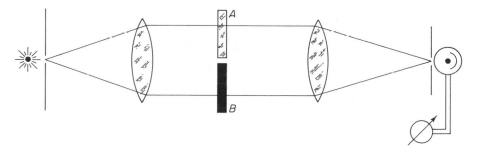

Figure 20-10

20-11. If a certain material has a work function of 3.1 eV, what is the longest wavelength (cutoff wavelength) that could produce any photoelectrons?

20-12. Light of 400 nm wavelength hits a surface that has a work function of 2.48 eV. What is:
   (a) The highest kinetic energy, in joules, of the electrons ejected from the surface?
   (b) The cutoff wavelength of the light?

20-13. Determine the wavelength of a quantum whose energy is 2.48 eV.

20-14. What is the energy, in eV, of a quantum of:
   (a) Infrared of 2.0 μm?
   (b) Ultraviolet of 250 nm wavelength?

SUGGESTIONS FOR FURTHER READING

KARIM, M. A. *Electro-Optical Devices and Systems*. Boston: PWS-Kent Publishing Company, 1990.

KNOLL, G. F. *Radiation Detection and Measurement*, 2nd ed. New York: Wiley-Interscience, 1989.

MEYSTRE, P., and M. SARGENT III. *Elements of Quantum Optics*. New York: Springer-Verlag, 1990.

TAYLOR, J. H. *Radiation Exchange, An Introduction*. Boston: Academic Press, Inc., 1990.

WILSON, J., and J. F. B. HAWKES. *Optoelectronics: An Introduction*, 2nd ed. Englewood Cliffs, N.J.: Prentice-Hall, Inc., 1989.

21

Radiometry/Photometry

Radiometry is the science of measuring radiant quantities; it applies throughout the electromagnetic spectrum. *Photometry* is part of radiometry; it applies only to that part of the spectrum that is perceived by the human eye as the sensation of *light*. Most radiometric quantities have the adjective *radiant* and all of their symbols carry the subscript e, for "electromagnetic." Most photometric quantities have the adjective *luminous* and all of their symbols carry the subscript v, for "visual."

Radiometry and photometry have long been confounded by ambiguity. Sometimes, different terms are used for the same quantities. Terms like *candlepower* are ill conceived. Still others, such as nox, phot, glim, skot and scot, bril and brill, helios, lumerg, pharos, stilb, and blondel, merely delight the historian.

In recent years much progress toward clarity has been made, especially since international agreement has been reached to adopt simple, logical, and easily convertible units, based on the International System of Units, SI. These terms and units are defined and used in this chapter.

RADIANT-TO-LUMINOUS CONVERSION

The fact that the human eye responds to only part of the electromagnetic spectrum requires that there be two systems of measurement. In *radiometry* we measure radiation in some uniform way; in *photometry* we take into account the spectral response of the eye. Moreover, we need to distinguish vision in daylight, called *photopic vision*, from vision at night, called *scotopic vision*. In photopic vision the peak sensitivity of the eye is in the bright yellow, at 555 nm; in scotopic vision, it shifts to 510 nm (Figure 21-1).

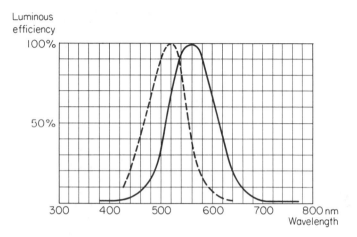

Figure 21-1 Sensitivity of the human eye. Photopic vision shown by solid line, scotopic vision by dashed line. Both curves are normalized to their own maxima.

But, how can we connect radiometric and photometric measurements, and convert readings taken in one system into quantities that are part of the other?

Many years ago, the "standard" of a luminous quantity was the flame of a candle. Then came the "new candle," a blackbody radiator heated to the temperature of solidification of platinum (1772°C). Today the connection is the SI unit of luminous intensity, the *candela* (cd). First we convert the wavelength of the photopic peak sensitivity, 555 nm, into frequency, using the general relationship

$$c = \lambda \nu$$

Solving for ν gives

$$\nu = \frac{c}{\lambda} = \frac{299792458 \text{ m/s}}{555 \times 10^{-9} \text{ m}} = 540 \times 10^{12} \text{ Hz} \qquad [21\text{-}1]$$

*The candela [then] is [defined as] the luminous intensity, in a given direction, of a source that emits monochromatic radiation of frequency 540 × 10^{12} hertz and that has a radiant intensity in that direction of (1/683) watt per steradian.**

---

* International agreement as adopted by the 16th General Conference on Weights and Measures, Paris, France, 1979.

At other wavelengths the eye's *luminous efficiency* is less (see Table 21-1). A blue lamp, for example, even if it emits the same amount of radiant power as a green lamp, appears dimmer because the eye is less sensitive to blue than to green. Luminous efficiency has no units; it is given as a percentage.

Table 21-1 Spectral distribution
of photopic luminous efficiency

Wavelength	Luminous Efficiency
420	0.004
430	0.012
440	0.023
450	0.038
460	0.060
470	0.091
480	0.139
490	0.208
500	0.323
510	0.503
520	0.710
530	0.862
540	0.954
550	0.995
560	0.995
570	0.952
580	0.870
590	0.757
600	0.631
610	0.503
620	0.381
630	0.265
640	0.175
650	0.107
660	0.061
670	0.032
680	0.017
690	0.008
700	0.004

Radiation Geometry

The term central to all radiometry and photometry is *power*, the former "flux." Depending on the geometry involved, we distinguish *exitance*, *intensity*, *radiance/luminance*, and *irradiance/illuminance*. These terms are shown in Figure 21-2 and, together with the associated symbols and units, are summarized in Table 21-2.

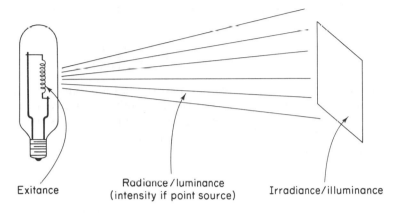

Exitance   Radiance / luminance   Irradiance/illuminance
              (intensity if point source)

Figure 21-2 Geometrical concepts used in radiometry and photometry.

Table 21-2 Radiometric and photometric terms, symbols, and units

Term	Symbol	Unit
Power	ϕ	
Radiant power	ϕ_e watt	W
Luminous power	ϕ_v lumen	lm
Exitance	M	
Radiant exitance	M_e watt/meter^2	W/m^2
Luminous exitance	M_v lumen/meter^2	lm/m^2
Intensity	I	
Radiant intensity	I_e watt/steradian	W/sr
Luminous intensity	I_v candela	cd
Radiance	L_e watt/(meter^2 steradian)	W/(m^2 sr)
Luminance	L_v candela/meter^2	cd/m^2
Irradiance	E_e watt/meter^2	W/m^2
Illuminance	E_v lux	lx

TERMS AND UNITS

Power

Power, ϕ, is the amount of energy, Q, that is generated, transmitted, or received per unit of time, t:

$$\boxed{\phi = \frac{Q}{t}}$$

[21-2]

The unit of *radiant power*, ϕ_e, is the same as that of power in general, *watt*, W:*

$$1 \text{ watt} = \frac{1 \text{ joule}}{1 \text{ second}}$$

Luminous power, ϕ_v, is that part of the radiant power that appears as light. Its unit is the *lumen*, lm. The ratio of light emitted to power consumed is called the *efficacy* of the light source; it is given in units of lumens/watts. With an electric light source, the electric power consumed is the product of voltage (in volts) and current (in amps).

Exitance

Exitance, M, is the amount of power, ϕ, that "exits" (emerges) from a source, per unit area, A:

$$M = \frac{\phi}{A} \qquad \text{[21-3]}$$

The unit of *radiant exitance*, M_e, is watts per square meter, W/m². The unit of *luminous exitance*, M_v, is lumens per square meter, lm/m².

Intensity

The term *intensity* means many things. In the theory of waves it means the square of the amplitude, in electrostatics it means force per unit charge, and in acoustics it means power per unit area. In optics, intensity, I, is the amount of power, ϕ, that is emitted by a *point* source and that proceeds within a cone of a given *solid angle*, ω:

$$I = \frac{\phi}{\omega} \qquad \text{[21-4]}$$

A solid angle is a three-dimensional entity. It is the (solid) angle subtended at the center of a sphere of a given radius R by an area A on the sphere's surface,

---

* The unit of radiant power is named after James Watt (1736–1819), Scottish instrument maker and civil engineer ("civil" in contrast to military engineers). When Watt repaired a Newcomen steam engine, used at that time to pump water out of coal mines, he hit upon the idea of keeping the cylinder with the reciprocating piston hot at all times, greatly increasing the engine's efficiency. By 1800 some 500 Watt engines were working throughout England, ushering in the Industrial Revolution.

The unit of energy is named after James Prescott Joule (1818–1889), British experimental physicist. The son of a wealthy brewer, Joule had the means to devote his life to research, mainly precision measurements of the heat produced by electric current, friction, compression, combustion, and irradiation. Even on his honeymoon, he took time out to measure the temperature of the water at the top of a waterfall and at its bottom, an idea that led to the concept of the mechanical equivalent of heat, now set at 1 cal = 4.184 J.

$$\omega = \frac{A}{R^2}$$

The *unit* of a solid angle is the ratio of the area on the sphere's surface to the distance, the radius of the sphere, squared. This unit is the *steradian*. It is defined as the solid angle subtended at the center of a sphere of 1 m radius by an area on its surface 1 m² in size, regardless of the shape of the area. Since a sphere has a (total) surface area of $4\pi R^2$, the total solid angle about a point is

$$\omega = \frac{4\pi R^2}{R^2} = 4\pi \ \text{sr}$$

Substituting in Equation [21-4] then shows that a point source, emitting uniformly into all space, has an intensity of

$$I = \frac{\phi}{4\pi} \qquad\qquad [21\text{-}5]$$

Radiant intensity, I_e, is radiant power per solid angle. Its unit is watts per steradian, W/sr. In *luminous intensity*, I_v, watts are replaced by lumens; its unit, therefore, is lumens per steradian, lm/sr, a *candela*, cd.

Example 1

A high-intensity lamp that, because of its small bulb, can be considered a "point light source" is rated at 25 W and has a luminous intensity of 60 cd. What is:
(a) Its luminous output?
(b) Its luminous efficacy?

Solution: (a) From Equation [21-5] we know that a 1-cd point source emits a total of 4π lumens. Thus, the luminous output is

$$\phi = (4\pi)(60) = \boxed{754 \ \text{lm}}$$

(b) The luminous efficacy is

$$\frac{e_m}{W} = \frac{754}{25} = \boxed{30 \ \text{lm/W}}$$

Example 2

If at 555 nm 1 lumen is equivalent to ⅟₆₈₃ watt, how many watts are equivalent to 1 lumen of red light of 630 nm?

Solution: From Table 21-1, the luminous efficiency at 630 nm is 0.265. Thus,

$$1 \ \text{W} = (683)(0.265) = 181 \ \text{lm}$$

and

$$1 \ \text{lm} = \frac{1}{(683)(0.265)} = \boxed{5.525 \ \text{mW}}$$

Radiance/Luminance

The terms *radiance* and *luminance* refer to radiation that comes from an *extended* source, rather than from a point source. It is the amount of power transmitted through a surface per unit area of that surface, per unit solid angle. A little shorter, radiance/luminance is intensity per unit area. Actually, it is not the area as such but the *projected* area, $A \cos \theta$,

$$L = \frac{\phi}{(A \cos \theta)(\omega)} = \frac{I}{A \cos \theta} \qquad [21\text{-}6]$$

where θ is the angle subtended by the surface normal and the direction of the radiation.

Consider this angle in more detail. In the direction of the surface normal, the angle is zero ($\theta = 0$) and its cosine is unity ($\cos \theta = 1$). In that direction, therefore, the amount of radiation is highest: it is $I_\perp$. In any other direction, away from the surface normal, the radiation is less:

$$I_\theta = I_\perp \cos \theta \qquad [21\text{-}7]$$

which is *Lambert's law*. Substituting Equation [21-7] in [21-6] gives

$$L = \frac{I_\perp \cos \theta}{A \cos \theta} = \text{constant} \qquad [21\text{-}8]$$

which shows that *the radiance/luminance of a Lambertian surface is the same in all directions*.

Example

Consider a flat metal disk with a rough surface, heated to incandescence. If the disk has an area of 1 cm^2 and a radiance of 1 W/(cm^2 sr), it radiates 1 W/sr in a direction normal to its surface. In a direction 45° to the normal, it radiates only (1 W/sr) (cos 45°) = 0.707 W/sr. However, when seen from the 45° direction, the disk appears compressed into an ellipse, and hence its projected area is less also. The two reductions cancel and the radiance remains the same.

The unit of *radiance*, L_e, is watts per square meter per steradian, W/(m^2 sr). The unit of *luminance*, L_v, is lumens per square meter per steradian. But since a lumen per steradian is a candela, a more convenient unit of luminance is candela per square meter, cd/m^2. Some examples are shown in Table 21-3.

Irradiance/Illuminance

Now let the radiation be *incident* on a surface. This means that the irradiance/illuminance, E, is equal to the power, ϕ, incident per unit area, A:

Table 21-3   Luminance of various objects

Object	Luminance (cd/m²)
Atomic bomb	2×10^{12}
Sun	2.3×10^9
Xenon arc	1×10^9
60-W light bulb	1.2×10^5
Snow in bright sunlight	4×10^4
Average landscape in summer	8000
Average landscape under overcast sky	2000
Luminance recommended for comfortable viewing	1400

$$E = \frac{\phi}{A}$$ [21-9]

the same relationship that defines the exitance of radiation from a surface.

Consider first light that comes from a *point* source. The total power emitted by a point source and proceeding into all space, following Equation [21-5], is $\phi = 4\pi I$. Therefore, since the surface area of a sphere is $A = 4\pi R^2$, a source of intensity I produces a total irradiance/illuminance of

$$E = \frac{4\pi I}{4\pi R^2}$$

Canceling 4π and changing R into d, for distance, gives

$$E = \frac{I}{d^2}$$ [21-10]

which is the *inverse-square law*. It shows that the irradiance/illuminance, E, at distance d, is directly proportional to the intensity, I, of the source and inversely proportional to the square of the distance (Figure 21-3).

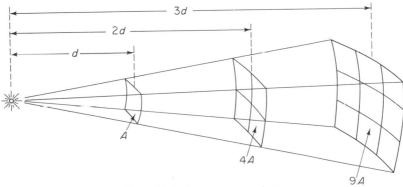

Figure 21-3   Inverse-square law.

There is an interesting similarity worth noting that exists between irradiance/illumi-nance and magnetic induction. Consider a wire such as that shown in Figure 16-2 (page 271). According to Biot–Savart's law, the induction, **B**, produced by a short section of the wire is given by

$$\mathbf{B} = \mu_0 \frac{1}{4\pi} i \int \frac{1}{R^2} d\mathbf{W} \times \hat{\mathbf{R}} \qquad [21\text{-}11]$$

that is, **B** is inversely proportional to the *square* of the distance.

But instead of integrating over the length of the wire, W, from $-\infty$ to $+\infty$, we could as well change variables and replace W by the angle θ subtended by the wire and the direction to the test point. That means that now the integration is performed from $\theta = 0$ (at infinity on one end) to $\theta = 180° = \pi$ rad (at infinity on the other end). Then

$$B = \mu_0 \frac{1}{4\pi} i \int_0^\pi \frac{1}{R^2} \sin \theta \, d\theta$$

$$= \mu_0 \frac{1}{4\pi} i \frac{1}{d} \int_0^\pi \cos \theta \, d\theta \qquad [21\text{-}12]$$

Solving the integral yields

$$B = \mu_0 \frac{1}{4\pi} i \frac{1}{d} (-\cos \pi + \cos 0) \qquad [21\text{-}13]$$

and thus

$$B = \mu_0 \frac{1}{2\pi} i \frac{1}{d} \qquad [21\text{-}14]$$

which shows that instead of the inverse-*square* relationship, we now have $1/d$.

This is exactly the same change that occurs when we proceed from a point source to an extended source. Of course, there are no true point sources, nor are there infinitely large sources. For a realistic light source, in short, the exponent in Equation [21-10] must lie somewhere between the limits 1 and 2, never reaching either one.

Example

Consider an illuminated slit 7 cm long and 25 cm away from a (small) photodetector. Determine the exponent that should replace the 2 in the inverse-square law for a point source.

Solution: If we examine Equations [21-13] and [21-14] we notice that they differ by a factor $(\cos \theta_1 - \cos \theta_2)/2$. To find to which power the distance d needs to be raised, we set

$$\frac{1}{d^x} = \frac{1}{d} \left(\frac{\cos \theta_1 - \cos \theta_2}{2} \right)$$

$$d^x = d \left(\frac{2}{\cos \theta_1 - \cos \theta_2} \right)$$

and, since by definition $\theta_2 = 180° - \theta_1$,

$$d^x = d \left(\frac{1}{\cos \theta}\right) \qquad \text{[21-15]}$$

Inserting the actual figures, $d = 25$ cm, $\theta_1 = \arctan [25/(7/2)] = 82°$ and $\theta_2 = 180° - 82° = 98°$, and using Equation [21-15] yields

$$25^x = 25 \left(\frac{1}{\cos 82°}\right) = 180$$

$$x = \frac{\log 180}{\log 25} = \boxed{1.613}$$

Indeed, with an extended light source, the exponent in the inverse-"square" law will always be *less than 2*.

So far we have assumed that the radiation reaches the surface at normal incidence. If it does not, as shown in Figure 21-4, the irradiance/illuminance is *less* by a cosine factor,

$$E' = \frac{I}{(d')^2} \cos \theta \qquad \text{[21-16]}$$

In addition, distance d' is *longer* than d:

$$d' = \frac{d}{\cos \theta}$$

Substituting in Equation [21-16] gives

$$E' = \frac{I}{(d/\cos \theta)^2} \cos \theta = \frac{I}{d^2} \cos^3 \theta \qquad \text{[21-17]}$$

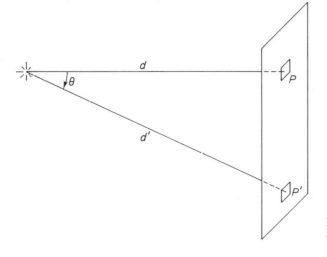

Figure 21-4 Light incident on an extended surface.

which shows that farther away from point P the irradiance/illuminance falls off as $\cos^3 \theta$.

Finally, assume that the light comes from an *extended* source. If the source has an area S, its radiance/luminance, following Equation [21-6], is $L = I/(S \cos \theta)$. Solving for I and substituting in Equation [21-17] gives

$$E' = \frac{LS}{d^2} \cos^4 \theta \qquad [21\text{-}18]$$

which shows that now the irradiance/illuminance falls off as $\cos^4 \theta$, a relationship known as the *cosine-fourth law*.

The ratio S/d^2 is the solid angle Ω subtended by the source at point P. Therefore, setting $\theta = 0$ gives

$$E = L\Omega \qquad [21\text{-}19]$$

This shows that the irradiance produced by an extended source depends only on the radiance of the source and on the solid angle it subtends; it does not depend on the distance between source and receiver.

The unit of *irradiance*, E_e, is watts per square meter, W/m^2, the same as that of radiant exitance. The unit of *illuminance*, E_v, is lumens per square meter, lm/m^2, or *lux*, lx, for short.

The relationship between candela and lux is illustrated in Figure 21-5. The source (A) has an intensity of 1 cd. Such a source, from Equation [21-5], has a *total* output, in all directions, of $\phi = 4\pi I = 12.57$ lm. But the candela is defined as lumens per steradian, and 1 steradian is the solid angle subtended by an area 1 m^2 in size at a distance of 1 m. That is just the area marked B in Figure 21-5; this area receives an illuminance $E = 12.57$ lm/12.57 m^2, which is 1 lux.

The unit of "*footcandle*," fc, is not a good choice. It seems to suggest that the luminous intensity, in units of "candles," is to be multiplied by a distance. Instead, it means lumens per square foot, lm/ft^2. If the area in Figure 21-5 were 1 ft^2 in size and 1 ft away from the source, it would have an illuminance of 1 footcandle (C).

Example

A 4-cd light source is suspended 80 cm above the center of a table 1 m in diameter. What is the illuminance:
  (a) In the center of the table?
  (b) Near the edge of the table?

Solution: (a) For a point light source, from Equation [21-10],

$$E = \frac{I}{d^2} = \frac{4}{(0.8)^2} = \boxed{6.25 \text{ lx}}$$

(b) With the dimensions given, the direction to the edge of the table, and therefore the angle of incidence at the edge, is

$$\theta = \arctan \frac{0.5}{0.8} = 32°$$

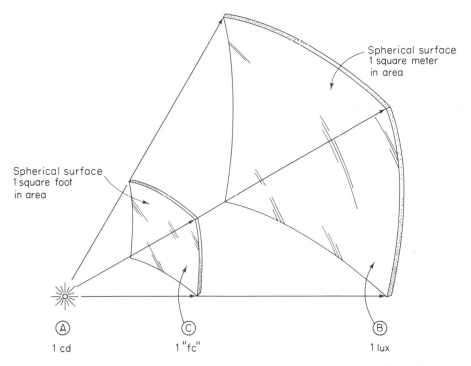

Spherical surface
1 square meter
in area

Spherical surface
1 square foot
in area

Ⓐ   Ⓒ   Ⓑ

1 cd   1 "fc"   1 lux

Figure 21-5 Diagram showing relationship between candela, footcandle, and lux.

The illuminance, therefore, is

$$E' = \frac{I}{d^2} \cos^3 \theta \ = \frac{4}{(0.8)^2} \cos^3 32° = \boxed{3.8 \text{ lx}}$$

Brightness is not the same as intensity or illuminance. Brightness is a function of illuminance and reflectivity. The term *illumination* should not be used either; illumination is the process of exposing an object to light, rather than a property of the object.

Throughput

In our discussion of magnification we saw that the optical invariant, nyU, is constant throughout an optical system. Now we extend this (two-dimensional) convention to three dimensions. Then the height of the object, y, becomes its area, A, and the plane angle U becomes a solid angle, ω. This leads to a new quantity, $n^2A\omega$, called *throughput*, T:

$$T = n^2A\omega \qquad\qquad [21\text{-}20]$$

Throughput is a geometric measure of how much light can be transmitted through a system of index n and cross section A and that accepts a solid angle ω.

The terms A and ω occur also in the definition of radiance, $L = \phi/(A\omega)$. Therefore, solving for $A\omega$, substituting in Equation [21-20], and setting $n = 1$ shows that throughput is also equal to the ratio of power to radiance, $T = \phi/L$. The unit of throughput is square meter steradian, $m^2 sr$.*

RADIOMETERS AND PHOTOMETERS

Comparison Photometers

Systems for measuring radiation in general are called *radiometers*. Systems for measuring light are called *photometers*. The classical example of a comparison photometer is the *Bunsen grease-spot photometer*.† It contains a sheet of white paper with a grease spot in the center that, by way of two mirrors, can be viewed from both sides at the same time. When the grease spot receives light from only one side, it appears, seen from that side, dark on a bright background; seen from the other side, it appears bright on a dark background. The distances, d_1 and d_2, from the two sources are adjusted so that the spot appears at least contrast. If we have two point sources, then, from the inverse-square law,

$$E_1 = \frac{I_1}{d_1^2} = E_2 = \frac{I_2}{d_2^2}$$

and thus

$$\frac{I_1}{I_2} = \left(\frac{d_1}{d_2}\right)^2 \qquad [21\text{-}21]$$

which could be called the *photometric comparison law*. With two extended sources, the exponent shrinks to unity and

$$\frac{L_1}{L_2} \approx \frac{d_1}{d_2} \qquad [21\text{-}22]$$

Photoelectric Radiometers

Basically, a photoelectric *radio*meter measures irradiance, although it may be calibrated instead in terms of power, intensity, or radiance. A photoelectric *pho-*

---

* There is as yet no agreement on whether *throughput* is the best term. W. H. Steel of the Australian National Standards Laboratory, for example, asks rhetorically: "If the input is doubled, what happens to the throughput?" *Answer*: Nothing. Input and output are actual quantities of light entering and leaving a system; throughput describes the *capacity* of the system to convey light, rather than the quantity being conveyed.

† Robert Wilhelm Bunsen (1811–1899), German chemist and professor of chemistry at the University of Breslau, later at Heidelberg. Bunsen invented the gas burner and the grease-spot photometer, both named after him, the carbon zinc battery, and the prism spectroscope, at that time containing a hollow prism filled with carbon disulfide.

*to*meter must have a detector whose spectral response, with or without additional filters, matches that of the eye. Various quantities are determined as follows.

1. *Power.* The total flux emitted by a light source is found by placing the source inside a hollow sphere with a diffusely reflecting inner surface, called an *integrating sphere.* The photodetector is placed near a small window in the sphere's surface.

2. *Irradiance.* Measuring irradiance is easy. All that is needed is a radiation detector. The surface of the detector must be completely filled with the radiation, and the surface must be oriented normal to the direction of the radiation.

3. *Intensity and radiance.* Measuring intensity and radiance is of much practical interest. Radiance, because it includes intensity, is the more general case. Measuring radiance requires certain precautions: (1) *all* of the radiation to be measured must reach the detector and (2) extraneous radiation must be excluded. With a weak source, it may not be enough to let the radiation simply fall on the detector. Instead, the source may need to be *imaged* on the detector's surface, using a lens, a suitably shaped stop, or even a telescope or a microscope.

The simplest luminance meters are the small light meters customary on photographic cameras. Aim the meter at a large, uniformly illuminated wall: You will get a certain reading. Then move back. The reading will be the same—as long as the solid angle subtended at the meter by the size of the wall does not become less than the meter's acceptance angle.

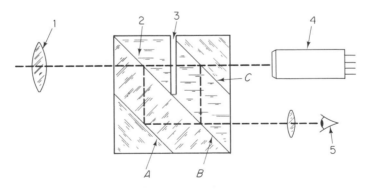

Figure 21-6 Schematic diagram of compact luminance meter. Light from the target proceeds to lens 1 and glass cube 2, which contains a slot 3 for a field-limiting mask, to photodetector 4. A series of partially coated interfaces *A*, *B*, and *C* direct light to observer 5, who can see the field both inside and outside the area limited by the mask. [Redrawn from D. A. Schreuder, "Optical System for a Universal Luminance Meter," *Appl. Opt.* **5** (1966), 1965–66, by permission.]

If the target is small and has an odd shape, the meter should look only at the sampling field. This is accomplished best by a see-through mask, such as that illustrated in Figure 21-6.

SUMMARY OF EQUATIONS

Radiant-to-luminous conversion (at 555 nm):

$$1 \text{ cd} = 1/683 \text{ W/sr} \qquad \text{[page 351]}$$

Intensity of point light source:

$$I = \frac{\phi}{4\pi} \qquad \text{[21-5]}$$

Radiance/luminance:

$$L = \frac{I}{A \cos \theta} \qquad \text{[21-6]}$$

Irradiance/illuminance:

$$E = \frac{I}{d^2} \qquad \text{[21-10]}$$

Photometric comparison law:

$$\frac{I_1}{I_2} = \left(\frac{d_1}{d_2}\right)^2 \qquad \text{[21-21]}$$

PROBLEMS

21-1. The total power emitted by the sun is estimated to be 3.9×10^{26} J/s. What is the power incident per square meter just outside Earth's atmosphere, assuming a distance of 150 million km?

21-2. How much energy does a 40-W light bulb consume in 15 min? How much is this in *kilocalories*?

21-3. What is the intensity of a light source that projects one-third of a watt into a hemisphere?

21-4. How much power is produced by a light source that emits 83 cd into all space?

21-5. A white screen receives 1940 lm of green light of 520 nm wavelength. How much radiant power is incident on the screen?

21-6. If a 60-cd light source consumes 0.68 A at 110 V and if it emits light uniformly in all directions, what is its efficacy?

21-7. If the luminance of a large, flat, diffusely reflecting surface, seen at a right angle and from a distance of 40 cm, is 3.6 cd m^{-2}, how does the luminance change when seen:
   (a) At an angle of 60°?
   (b) From a distance of 20 cm?

21-8. The luminance of an area 10 cm in diameter is measured from a distance of 40 cm, using a calibrated photometer with an acceptance angle of 20°. By what factor must the reading of the photometer be corrected?

21-9. A 4-cd high-intensity lamp is suspended 80 cm above the center of a table. What is the illuminance at the center of the table?

21-10. At what angle of incidence is the illuminance on a screen exactly one-half of what it is at normal incidence?

21-11. If a book is held first 1.2 m from a high-intensity lamp and then is moved to 30 cm from the lamp, by how much has the illuminance increased?

21-12. If a photographic print can be made in 10 seconds with the printing paper held 20 cm from a point light source, what is the correct time when the paper is held 30 cm away?

21-13. The top of a drafting table, tilted 30° from the horizontal, is illuminated by a single 20-cd lamp suspended 1.2 m above the center of the table. What is the illuminance?

21-14. An unknown lamp placed 60 cm from the screen of a Bunsen grease-spot photometer gives the same illumination as a 20-cd lamp placed 40 cm from the screen. What is the intensity of the unknown lamp?

21-15. Two high-intensity lamps of equal efficacy, one rated 25 W and the other 100 W, are mounted 90 cm apart from each other. How far from the 25-W lamp must the screen of a Bunsen grease-spot photometer be placed in order to have equal illuminance on both sides?

21-16. The flat center of a street receives light from two lamps of 355 cd each, mounted on poles 5 m high and 10 m apart from each other. What is the illuminance at that point?

SUGGESTIONS FOR FURTHER READING

BOYD, R. W. *Radiometry and the Detection of Optical Radiation*. New York: Wiley-Interscience, 1983.

SPIRO, I. J. *Selected Papers on Radiometry*. Bellingham, Wash.: SPIE Optical Engineering Press, 1990.

WYATT, C. L. *Radiometric System Design*. New York: Macmillan Publishing Company, 1987.

Absorption

When light passes through matter, some of the light energy changes into another form of energy, most often heat, which is molecular motion. That is called *absorption*.

FUNDAMENTAL PROCESS

In its most general definition, absorption is an irreversible thermodynamic process. Sometimes, the terms *absorption* and *extinction* are used synonymously. That is not correct. Extinction is the broader term; it includes attenuation by both absorption and scattering. But now consider a block of material that has a tint of neutral gray but otherwise is clear and of uniform refractive index. Light may be incident on the block from the left, and because some of the light changes into heat, a lesser amount emerges on the right.

Transmittance

We call ϕ_0 the amount of light incident on the block and ϕ' the amount of light that emerges from it [Figure 22-1 (left)]. The ratio of the amount of light transmitted to

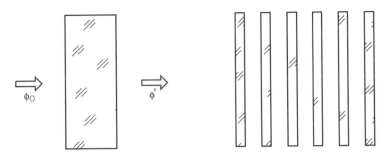

Figure 22-1 Block of absorbing material (*left*) and the same block thought to consist of several thin slices (*right*).

the amount of light incident, both measured at the same wavelength, is called *transmittance*, *T*,

$$T = \frac{\phi'}{\phi_0}$$ [22-1]

Next assume that the block is cut into a series of thin slices, each of thickness Δx. Each slice, because they all are alike, will absorb the same percentage of the light. For simplicity assume that this percentage, the loss per slice, is 50%. That means that, while 100% of the light is incident on the first slice, only 50% emerges from it and reaches the second. There again 50% is lost and only 25% (of the original total) reaches the third, 12.5% reaches the fourth, 6.25% the fifth, and so on. In short, while the *percentages* absorbed in each slice are the same, the actual *amounts* are progressively less.

That may have serious consequences. For example, when laser radiation is blocked by dark goggles, the first layers will absorb more radiation than the last layers. The material in front, therefore, will get hotter and expand more, and that expansion may cause the goggles to break.

In practice, such goggles should have two parts: a reflective coating applied to the front surface and absorptive material behind it. The coating will reject some of the radiation before it enters the absorptive layers, avoiding overheating and possible breakage. The absorptive material, in turn, gives protection even if the coating should become scratched and for that reason let some of the radiation pass through.

If we plot the amount of light that reaches a given depth, that is, if we plot the transmittance as a function of thickness, we obtain the curve shown in Figure 22-2.

Absorbance

To the casual reader, the terms *transmittance* and *absorbance* sound compatible, as if one were the opposite of the other. But this is not so. Transmittance is a

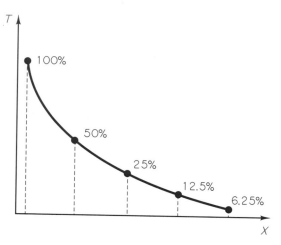

Figure 22-2 Transmittance as a function of thickness.

percentage but absorbance is not. *Absorbance, A*, sometimes called "optical density," *is defined as the logarithm to base 10 of the inverse of the transmittance,*

$$A = \log_{10}\left(\frac{1}{T}\right) \qquad [22\text{-}2]$$

Since transmittance, T, is the ratio of the amount of light transmitted, ϕ', to the amount of light incident, ϕ_0, absorbance can also be defined as

$$A = \log_{10}\left(\frac{\phi_0}{\phi'}\right) \qquad [22\text{-}3]$$

Consider again the block of material that we have discussed before. At the front surface of the block, no light is being absorbed as yet, $A = 0.00$. After passing through the first slice (where one-half of the light is absorbed and $T = 0.5$),

$$A = \log_{10}\left(\frac{1}{0.5}\right) = 0.30$$

After passing through the second slice (where again 50% is absorbed and $T = 0.25$),

$$A = \log_{10}\left(\frac{1}{0.25}\right) = 0.60$$

and after passing through the third slice,

$$A = \log_{10}\left(\frac{1}{0.125}\right) = 0.90$$

and so on. If we then plot the *absorbance* as a function of thickness, we arrive at the curve shown in Figure 22-3.

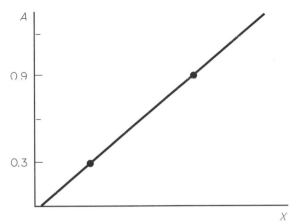

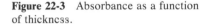

Figure 22-3   Absorbance as a function of thickness.

If we wish, we can always determine the amount of light transmitted through various thicknesses and plot the curve thus obtained. But, if we plot absorbances (rather than transmittances), the curve is *linear* and only two points are needed to establish a straight line.

The linearity means that absorbances are additive but transmittances are multiplicative. If we add several filters, the total absorbance of these filters is equal to the sum of the absorbances of the individual filters. But the total transmittance is equal to the *product* of the individual transmittances.

Exponential Law

If the block is of uniform matter, then in each slice a certain amount of light, $\Delta\phi$, is lost,

$$\phi_0 - \phi' = \Delta\phi$$

The magnitude of this loss is proportional to the amount of light incident, ϕ_0, to a constant of proportionality called *absorptivity*, α, and to the thickness of the slice, Δx:

$$\Delta\phi = -\phi_0\alpha\,\Delta x$$

the minus sign meaning that the loss is downward, rather than up. The absorptivity (the former "absorption coefficient") is characteristic of the material and also a function of wavelength.

Now we assume that the individual slices, instead of being of thickness Δx, are *infinitesimally thin*. The ratio $\Delta\phi/\phi_0$ then becomes

$$\frac{d\phi}{\phi_0} = -\alpha\,dx \qquad\qquad [22\text{-}4]$$

In order to find the total loss (within a block of thickness x), we integrate Equation [22-4] between the limits of ϕ and x:

$$\int_{\phi_0}^{\phi'} \frac{d\phi}{\phi} = -\alpha \int_0^x dx$$

Because

$$\int_a^b \frac{d}{dx} f(x) \, dx = f(b) - f(a) \qquad \text{and} \qquad \int \frac{dx}{x} = \ln x$$

we find that

$$\log_e \left(\frac{\phi'}{\phi_0} \right) = -\alpha x \qquad\qquad [22\text{-}5]$$

and therefore

$$\frac{\phi'}{\phi_0} = e^{-\alpha x}$$

and

$$\boxed{\phi' = \phi_0 e^{-\alpha x}} \qquad\qquad [22\text{-}6]$$

which is the *exponential law of absorption.**

Note that in deriving the exponential law, integration leads to *natural* logarithms (to base e), as in Equation [22-5]. But absorbance is based on *common* logarithms (to base 10).

To convert one into the other, we use the identities $\log_e x = 2.3026 \ldots$ $\log_{10} x$ or $\log_{10} x = 0.4343 \ldots \log_e x$, as needed. This gives us a set of equations that we will use frequently:

$$\boxed{\begin{aligned} A &= \log_{10} \left(\frac{\phi_0}{\phi'} \right) \\[2mm] A &= 0.4343 \alpha x \end{aligned}}$$

---

* The exponential law as presented has evolved through many efforts. Most notable among its originators were the following: Pierre Bouguer (1698–1758), French oceanographer, was the first to find a relationship similar to the exponential law. He published it in his book *Essai d'optique sur la gradation de la lumière* (Paris: Jombert, 1729). Johann Heinrich Lambert (1728–1777), a largely self-taught German physicist and mathematician, contributed to thermodynamics, photometry (cosine law), absorption, planetary motion, and cosmology, liked to reduce his thoughts on logic and philosophy to mathematical models. He described the law in *Lamberts Photometrie* (*Photometria, sive de mensura et gradibus luminis, colorum et umbrae*) (Augsburg, 1760), the same year in which a second, greatly enlarged edition of Bouguer's book, renamed *Traité d'optique sur la gradation de la lumière* (Paris: H. L. Guerin & L. F. Delatour, 1760) was published (posthumously) that contained virtually the same relationship. August Beer (1825–1863), German physicist, professor of mathematics at the University of Bonn, recognized the role of the concentration, as he reported in his book *Einleitung in die höhere Optik* (Braunschweig: F. Vieweg und Sohn, 1853).

$$T = \frac{1}{10^A}$$

$$\alpha = 2.3026 \frac{A}{x}$$

[22-7]

Comparison of Transmittances

From the second equation in [22-7] we see that the absorbance, A, of a given sample is directly proportional to its thickness, x:

$$A = (k)(x)$$

If we have two samples of the same material (same α), we divide one equation by the other:

$$\frac{A_1}{A_2} = \frac{(k)(x_1)}{(k)(x_2)}$$

cancel the constant, k, and rearrange:

$$A_2 = A_1 \frac{x_2}{x_1}$$

Multiplying by -1 and taking the logarithm of 10 to the Ath power gives

$$\log 10^{-A_2} = \log 10^{-A_1(x_2/x_1)}$$

and, substituting twice the third equation in [22-7],

$$T_2 = T_1^{x_2/x_1}$$

[22-8]

This is a very useful equation that we will need whenever we compare the transmittance of a sample of thickness x_1 with the transmittance of another sample, of the same material but of a different thickness, x_2.

Example 1

A neutral density filter of a given thickness absorbs 10% of the light incident on it. How much light will a filter absorb, made of the same material but 12 times as thick?

Solution: If 10% is being absorbed by the first filter, 90% is transmitted. Hence the transmittance is

$$T_1 = 90\% = 0.9$$

The transmittance of the second (thicker) filter, from Equation [22-8], is

$$T_2 = T_1^{x_2/x_1} = 0.9^{12} = 0.28$$

Therefore, if the second filter transmits 28% of the light, it absorbs

$$100 - 28 = \boxed{72\%}$$

Example 2

Another filter, 4 mm thick, absorbs 30% of the light. How thick a filter is needed to absorb 60%?

Solution: Converting these percentages into transmittances, we have $T_1 = 70\%$ and $T_2 = 40\%$. Also, $x_1 = 4$ mm, while x_2 is to be found. Looking once again at Figure 22-2, we see that if filter 2 were (only) twice as thick as filter 1, its transmittance, T_2, would be *less* than what we need and therefore filter 2 must be *more* than twice as thick. Using Equation [22-8],

$$T_2 = T_1^{x_2/x_1}$$

we find that

$$0.4 = 0.7^{x_2/4}$$

$$0.4^4 = 0.7^{x_2}$$

$$4 \log 0.4 = x_2 \log 0.7$$

(For this last step, either common or natural logarithms may be used.) The thickness of the second filter, therefore, must be

$$x_2 = \frac{4 \log 0.4}{\log 0.7} = \boxed{10.3 \text{ mm}}$$

SELECTIVE ABSORPTION

Some materials, such as metals, have high absorptivity from the ultraviolet through the infrared; even a thin metal membrane is virtually opaque to light. Other materials such as dyes block out only part of the spectrum. This is called *selective absorption*. In general, there are no substances that have no absorption at least somewhere in the electromagnetic spectrum: glass is opaque to UV, water absorbs below 190 nm and in the IR, and absorption by air extends from the region of soft X rays to and including the vacuum ultraviolet.

This makes somewhat obsolete the earlier distinction between *general absorption* and *selective absorption*. In general absorption, the absorptivity of the material is very nearly the same over a wide range of wavelengths. In selective absorption, attenuation occurs at certain wavelengths more than at others. Within the visible spectrum, this causes the sensation of *color*.* A sample of green glass,

---

* Early in 1672, Newton wrote to the Royal Society of London announcing that he wanted to report "a philosophical Discovery . . . in my judgment the oddest if not the most considerable detection wch hath hitherto beene made in the operations of Nature." He let sunlight fall on a "triangular glass-Prisme" and found that "Light it self is a Heterogeneous mixture of differently refrangible Rays." The colors obtained could not be divided further; they could be combined back into white. I. Newton, "A New Theory about Light and Colours," *Philos. Trans. Roy. Soc. London* **5**, No. 80 (Feb 19, 1672), 3075–87.

for instance, will absorb mainly red, and since red is complementary to green, the transmitted light will lack red and the sample will appear green.

Any color has three physical characteristics: *hue*, *lightness*, and *saturation*. Hue gives the color its name, such as blue, green, or red. Lightness refers to whether a color is pale (light) or dark (deep). Saturation is described by terms such as reddish white, moderate red, strong red, and vivid red. If more and more white is added to red paint, the dominant hue, red, stays the same but a series of tints will result, ending in a pale pink.

Absorption Spectra

Absorption spectra, the same as emission spectra, are either continuous spectra or line spectra. Continuous absorption spectra account for the colors of liquids and solids. Line absorption spectra result when (white) light passes through a gas at low pressure and at a temperature lower than that of the source. These lines have the same wavelengths as the lines which the gas would emit if it were hot; they identify an element just as emission spectra do. Absorption lines, seen in the continuous spectrum of the sun, are known as *Fraunhofer lines*.*

EXPERIMENTAL METHODS

In order to determine the absorption characteristics of a sample as a function of wavelength, we need a *spectrophotometer*, that is, a photometer or radiometer connected to a device for wavelength selection. If the sample is a solid such as glass or plastic, the sample is made into a plane-parallel, polished plate. If it is a liquid or a substance in solution, it is placed in a cuvette and a comparison made between the solution and the pure solvent. Modern spectrophotometers plot the transmittance, or the absorbance, or both, as a function of wavelength.

Types of Spectrophotometers

Spectrophotometers can be built as one-beam-indicator, one-beam-substitution, and two-beam instruments. The latter may have either one or two detectors. In the *one-beam one-detector method*, the light proceeds in a single path through a monochromator and through the sample to the photodetector (Figure 22-4). The

---

* In 1814, at his optics shop outside Munich, Germany, 27-year-old Joseph Fraunhofer looked at the spectrum of the sun, which to Newton 150 years earlier had seemed continuous. But Fraunhofer, by his own count, saw 574 dark lines crossing the spectrum, an observation he published in *Ann. Physik* (2) **26** (1817), 264–313 under the title "Bestimmung des Brechungs- und des Farbenzer-streuungs-Vermögens verschiedener Glasarten, in Bezug auf die Vervollkommnung achromatischer Fernröhre," Determination of the refraction and dispersion properties of various types of glass, relative to the optimization of achromatic telescopes.

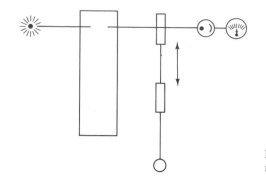

Figure 22-4 One-beam one-detector method.

sample and the reference (blank) are alternately brought into the path and the transmittance through each is measured at each wavelength desired.

In the *one-beam substitution* (zero) *method*, not shown, sample and reference again are alternately brought into the path. While measuring the reference, however, a neutral density filter or other attenuator is brought into the beam to make the two intensities equal. The advantage is that the response of the detector need not be linear.

In the *two-beam two-detector method*, the light coming from the monochromator is divided by a beamsplitter, passed through the sample and the blank, respectively, and is incident on two detectors (Figure 22-5). The potentiometer is adjusted so that the galvanometer reads zero. Two-beam instruments are independent of voltage fluctuations.

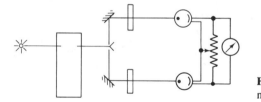

Figure 22-5 Two-beam two-detector method.

In the *two-beam one-detector method* (Figure 22-6), the light is split, passes through the sample and the blank, is recombined by another beamsplitter, and falls on *one* detector; thus there can be no mismatch between two photosurfaces.

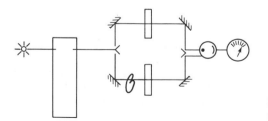

Figure 22-6 Two-beam one-detector method.

The reference beam is attenuated (chopped) and the degree of attenuation directly read in terms of transmittance or absorbance.

Photographic Emulsions

Absorption plays a key role in the process of *photography*. Photographic emulsions generally contain crystals of a silver halide such as silver bromide, AgBr, suspended in gelatin. When the emulsion is exposed to light, part of an AgBr crystal dissociates, the bromine ion releasing a free electron,

$$Ag^+Br^- \rightarrow Ag^+ + Br^0 + e^- \qquad [22\text{-}9]$$

The free electron becomes attached to a lattice defect in the silver ion, Ag^+, reducing the ion to a neutral, metallic silver atom,

$$e^- + Ag^+ \rightarrow Ag^0 \qquad [22\text{-}10]$$

That produces a *latent image*. The latent image as such is invisible, but it can be made visible by *development*. Developers often used are aminophenol derivatives and hydroquinone; they reduce the exposed crystal's other Ag^+ ions to grains of metallic silver. These grains render the exposed parts of the emulsion dark, forming a *negative*. To prevent the unexposed crystals from turning into silver, they are removed (dissolved) from the emulsion without affecting the developed image, using a "fixer" such as sodium thiosulfate, called "hypo," $Na_2S_2O_3$.

As more light is incident per unit area, the (developed) negative's absorbance is higher. A plot of *exposure* (which is the product of irradiance × time) *versus* absorbance gives a curve known as the *Hurter–Driffield (H–D) curve*.* The linear portion of the curve, *B–C* in Figure 22-7, has a slope called the *gamma* of the emulsion. The steeper the curve, the higher the gamma, and the higher the contrast in the emulsion. Parts *A–B* and *C–D* have less contrast. Very long

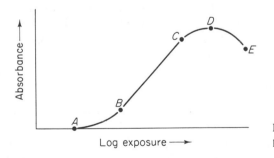

Figure 22-7 H–D curve of typical photographic film.

---

* F. Hurter and V. C. Driffield, "Photo-Chemical Investigations and a New Method of Determination of the Sensitiveness of Photographic Plates," *J. Soc. Chem. Ind.* (*London*), **9** (1890), 455–69.

exposures, or very bright light, however, may cause less of a response, which is called *reciprocity failure*, D–E.

The sensitivity, or *speed*, of the emulsion is expressed in units formulated by the International Standards Organization (ISO). These units have an arithmetic scale: twice a certain ISO number, as we see from Table 22-1, refers to a film twice as sensitive as another film.

Table 22-1   Speed and exposure time

Speed (ISO)	Exposure Time (in sec) Needed to Produce A Given Absorbance
400	1/250
200	1/125
100	1/60
50	1/30
25	1/15

Photochromic Glass

Materials are called *photochromic* if, under the influence of light, they become darker or change color. When the light is removed, they return to their initial state. The process of darkening, or *activation*, takes about a minute, until the absorbance reaches a certain saturation level. Light between 320 and 400 nm wavelength is most effective. The return process, or *clearing*, is a function of the incident IR radiation, that is, it depends on the ambient temperature (even at room temperature there is some *thermal fading*); clearing can be accelerated by light between 550 and 650 nm (*optical bleaching*). In the region between activation and bleaching, between 450 and 500 nm, light has no effect.

Photochromism is similar to photography: in both there is a dissociation of silver halides. But, in a photographic emulsion the halogen ion diffuses away while the silver remains in place. In photochromic glass, the glass matrix is so tight that the halogen cannot diffuse away; it remains close to the silver and recombines with it when activation ceases. Hence, the process of photochromism is *reversible*,

$$Ag\,Br \xrightleftharpoons{h\nu} Ag^+ + Br^- \qquad\qquad [22\text{-}11]$$

Samples have been exposed for half a million cycles of activation and bleaching, and have shown no fatigue. Typically, a small amount of copper, or other catalysts, is added to the glass; this increases the rate and depth of darkening by a factor of 100 or more.

Thermal fading, as the name implies, depends on the ambient temperature (on a hot day photochromic glass clears faster), but activation (darkening) does not. Thus the absorbance will reach a state of dynamic equilibrium that for a given irradiance decreases with increasing temperature. On a cold day, and with a given amount of light, photochromic glass becomes darker than on a hot day.

Photoelectric Coatings

The term *photoelectric* refers to materials, most often thin films, that under the influence of electric current darken or change color. Such films often come in the form of an electrochromic coating, usually a five-layer stack of vacuum-deposited thin films, an indium-tin-oxide (ITO) conducting layer, an electrochromic oxide layer, an electrolyte layer, an ion storage electrode layer, and another ITO layer. The whole coating is approximately 1.5 μm thick. Another two-layer coating of an inert polymer is applied to the top and bottom of the five active layers to provide protection against scratches and other external damage.

Whenever electrical power is applied to the two ITO layers, (positive) lithium ions flow from the ion storage layer through the electrolyte layer into the electrochromic layer, causing the coating to darken. The power needed is approximately 2 volts and 0.01 amps. After a time interval of 10 to 15 seconds, the coating changes from about 60% transmittance in the light state to about 15% transmittance in the dark. Tests have shown that at least 10,000 cycles can be achieved without fatigue.

The main purpose of such coatings is for sunglasses. But the same technology also seems useful for automobile windows and sunroofs, windows and skylights in commercial and residential buildings, and windows for boats, aircraft, and other vehicles.

SUMMARY OF EQUATIONS

Transmittance:

$$T = \frac{\phi'}{\phi_0}$$

[22-1]

Absorbance:

$$A = \log_{10}\left(\frac{1}{T}\right)$$

[22-2]

Exponential law of absorption:

$$\phi' = \phi_0 e^{-\alpha x}$$ [22-6]

Comparison of transmittances:

$$T_2 = T_1^{x_2/x_1}$$ [22-8]

PROBLEMS

22-1. If a dark filter blocks out most of the light falling on it, letting pass through only 5%, what is the absorbance of the filter?

22-2. If 37% of the incident light is absorbed in a sample, what is its absorbance?

22-3. If a neutral-density filter has an absorbance of 0.6, what is its transmittance?

22-4. A metallic coating absorbs 88% of the incident light. A thin film has one-half the absorbance of the coating. What is the transmittance of the film?

22-5. If a filter absorbs 20% of the light incident on it, how much light would the filter absorb if it were twice as thick?

22-6. A thin layer of a certain material absorbs 40% of the radiant energy passing through. How much energy will a layer absorb that is nine times as thick?

22-7. A 3-mm-thick color filter absorbs 74% of the light incident on it. What percentage of the light would pass through if the filter were 4.3 mm thick?

22-8. If 2.3-mm-thick sunglasses transmit 40%, how thick should the glasses be to transmit only 20%?

22-9. A certain type of protective glass has an absorptivity of $\alpha = 0.39$ mm^{-1}. What percentage of the light is transmitted through a plate 4 mm thick?

22-10. A pipe 5 m long contains a gas at normal atmospheric pressure. If the gas has an absorptivity of $\alpha = 0.08$ m^{-1}, how much light, in percent, is being absorbed?

22-11. A cylinder, 25 cm long and filled with a liquid, absorbs 62% of the light of a given wavelength. What is the absorptivity of the liquid?

22-12. The absorptivity of seawater at 570 nm is 0.22 m^{-1}. How long a path will reduce the transmittance to 80%?

22-13. Consider the Hurter–Driffield curve's section *D–E*, which represents severe *overexposure*. Therefore, what does the sun look like in the developed negative, compared to the surrounding sky?

22-14. If the memory unit in an optical computer were made of photochromic glass, which color of light should be used for:
   (a) Storing the information?
   (b) Reading the information?
   (c) Erasing the information?

SUGGESTIONS FOR FURTHER READING

BILLMEYER, F. W., JR., and M. SALTZMAN. *Principles of Color Technology*, 2nd ed. New York: John Wiley and Sons, Inc., 1981.

MEES, C. E. K., and T. H. JAMES. *The Theory of the Photographic Process*, 4th ed. New York: Macmillan Publishing Company, 1977.

OSTROFF, E., ed. *Pioneers of Photography*. Springfield: Society for Imaging Science and Technology, 1987.

23

Lasers

The term *laser* is an acronym for light amplification by stimulated emission of radiation. Compared with other sources of radiation, a laser stands out in several ways. Its output is highly directional and highly coherent. It can be generated in the form of very short pulses and, because of its high directionality, at very high power densities. Its beam can carry much more information than other light. These and other characteristics have made lasers a unique source of radiation in applications as diverse as data processing, materials processing, and communications, giving optics new impetus and wide publicity.

STIMULATED EMISSION OF RADIATION

Boltzmann Distribution

So far we have discussed the transitions between different atomic energy states, the upward transition from a lower energy state to a higher state, $E_1 \rightarrow E_2$, which occurs in absorption, and the downward transition, $E_2 \rightarrow E_1$, which occurs in emission. Now we consider the number of atoms, per unit volume, that exist in a given state. This number, N, called the *population*, is given by *Boltzmann's equation*,

$$N = e^{-E/kT} \qquad [23\text{-}1]$$

where E is the energy level of the system, k is Boltzmann's constant, and T is the absolute temperature.

In any absorption or emission, at least two energy states are involved. The ratio of the populations in these two states, N_2/N_1, is called *Boltzmann's ratio* or *relative population*,

$$\frac{N_2}{N_1} = \frac{e^{-E_2/kT}}{e^{-E_1/kt}}$$

from which it follows that

$$N_2 = N_1 e^{-(E_2-E_1)/kT} \qquad [23\text{-}2]$$

Then we plot the energy, E, versus the population, N; the result is an exponential curve known as a *Boltzmann distribution*. Under normal circumstances, when the system is in thermal equilibrium, it has more atoms in the lower state than in the higher state (as we will show later, in Figure 23-2). If we substitute Bohr's frequency condition in Equation [23-2], we find that

$$N_2 = N_1 e^{-h\nu/kT} \qquad [23\text{-}3]$$

Einstein's Prediction

Perhaps, it seems, the transition between two energy states can either be up (as in absorption) or down (as in emission). But Einstein, in a talk given in 1916 and published in writing a year later, postulated that there must be a third process.*

Assume first that an ensemble of atoms is in thermal equilibrium and *not subject to* an external radiation field. At higher temperatures, a certain number of atoms is in the excited state; on return to the lower state, these atoms will emit radiation, in the form of quanta $h\nu$. That is called *spontaneous emission*.

The *rate* of the transition is the number of atoms in the higher state that make the transition to the lower state, per second. (The reciprocal of the rate is the *lifetime* of the transition.) If, as before, N_2 is the number of atoms (per unit volume) in the higher state, the rate of the spontaneous transition, P_{21}, is

$$P_{21} = N_2 A_{21} \qquad [23\text{-}4]$$

where A_{21} is a constant of proportionality.

Assume next that the system *is subject to* some external radiation field. In that case, one of two processes may occur, depending on the direction (the *phase*) of the field with respect to the phase of the oscillator. If the two phases coincide, a quantum of the field may cause the emission of another quantum. This process, anticipated by Einstein, is now called *stimulated emission*. Its rate is

$$P_{21} = N_2 B_{21} u(\nu) \qquad [23\text{-}5]$$

---

* A. Einstein, "Zur Quantentheorie der Strahlung," *Mitt. Phys. Ges. Zürich*, **18** (1916), and *Phys. Z. XVIII* (1917), 121–28.

where B_{21} is another constant of proportionality and $u(\nu)$ is the energy density (in units of J m^{-3}), as a function of frequency, ν.

On the other hand, if the phase of the radiation field is opposite to that of the oscillator, the impulse transferred counteracts the oscillation, energy is consumed, and the system is raised to a higher state, as it occurs in *absorption*. Its rate is

$$P_{12} = N_1 B_{12} u(\nu) \qquad\qquad [23\text{-}6]$$

where B_{12} is still another constant of proportionality. These three constants, A_{21}, B_{21}, and B_{12}, are known as *Einstein's coefficients*. Before we proceed further, we summarize these findings in a schematic diagram (Figure 23-1).

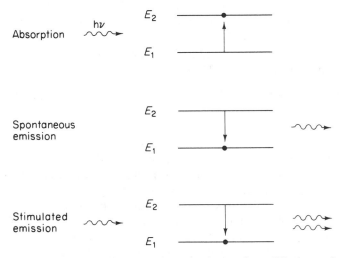

Figure 23-1 Transitions between energy states. If the atom rises from a lower to a higher energy state, it has absorbed a quantum of energy (*top*). If the atom goes from a higher to a lower state, it may release a quantum spontaneously (*center*) or it may be stimulated to release a quantum (*bottom*).

With the system in thermal equilibrium, the net rate of downward transitions must equal the net rate of upward transitions,

$$N_2 A_{21} + N_2 B_{21} u(\nu) = N_1 B_{12} u(\nu)$$

Dividing both sides by N_1 yields

$$\frac{N_2 A_{21}}{N_1} + \frac{N_2 B_{21} u(\nu)}{N_1} = B_{12} u(\nu)$$

$$\frac{N_2}{N_1} [A_{21} + B_{21} u(\nu)] = B_{12} u(\nu)$$

$$\frac{N_2}{N_1} = \frac{B_{12} u(\nu)}{A_{21} + B_{21} u(\nu)}$$

If we then substitute Equation [23-3], we obtain

$$\frac{B_{12} u(\nu)}{A_{21} + B_{21} u(\nu)} = e^{-h\nu/kT}$$

Solving for $u(\nu)$ gives

$$u(\nu) = \frac{A_{21}}{B_{12}} \frac{1}{e^{h\nu/kT} - B_{21}/B_{12}}$$ [23-7]

To maintain equilibrium, the system must release energy, probably in the form of electromagnetic radiation. The spectral distribution of this radiation follows Planck's radiation law, which we have discussed before,

$$M(\lambda) = \frac{C_1}{\lambda^5} \frac{1}{e^{C_2/\lambda T} - 1}$$

If Planck's law is written in terms of energy density, rather than exitance, $M(\lambda)$ changes to $u(\lambda)$ and, since u is related to M as

$$u = \frac{4}{c} M$$

the first radiation constant becomes

$$C_1 = 8\pi hc$$

and Planck's law

$$u(\lambda) = \frac{8\pi hc}{\lambda^5} \frac{1}{e^{-hc/\lambda kT} - 1}$$ [23-8]

The energy density must be consistent with Planck's law for any value of T. This is possible only when

$$B_{21} = B_{12}$$ [23-9]

and

$$\frac{A_{21}}{B_{21}} = \frac{8\pi h\nu^3}{c^3}$$ [23-10]

These two equations are called *Einstein's relations*. They show that (1) the coefficients for both stimulated emission and for absorption are numerically equal and (2) the ratio of the coefficients of spontaneous *versus* stimulated emission is proportional to the third power of the frequency of the transition radiation. This explains why it is so difficult to achieve laser emission in the X-ray range, where ν is rather high.

Population Inversion

Assume that an atomic system has three energy states, a *three-level system*. These levels may be designated E_1, E_2, and E_3. With the system in thermal equilibrium, the uppermost level, E_3, is populated least, and the lowest level, E_1, is populated most (Figure 23-2).

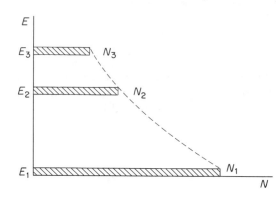

Figure 23-2 Population of a three-level system in equilibrium.

Ordinarily, when the system is in thermal equilibrium, absorption and spontaneous emission take place side by side. But, because $N_2 < N_1$, absorption dominates: an incident quantum is more likely to be absorbed than to cause emission.

But, if we can find a material that could be *induced* to have a majority of atoms in the higher state so that either N_3 or N_2, or both, exceed N_1, the system becomes top-heavy and reaches a condition called *population inversion*, a precondition to the operation of a laser (Figure 23-3).

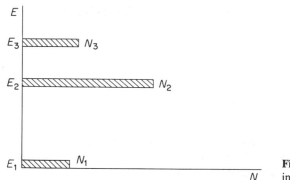

Figure 23-3 Three-level system showing population inversion.

GENERAL CONSTRUCTION

First, a laser requires a *power supply* to furnish the energy needed for raising the system to the excited state. That is called *pumping*, in analogy to pumping water up to a higher level of potential energy. It also requires an *active medium* that, when excited, reaches population inversion and lases. The medium may be a solid, liquid, or gas, and it may be one of thousands of materials that have been found to lase.

A *cavity* is optional. Some systems are built as laser *amplifiers*; they have no cavity. Most systems, however, are laser *oscillators*; here the medium is enclosed in a cavity that provides feedback and additional amplification. Generally, the cavity is formed by two mirrors facing each other. One of the mirrors is coated to full reflectance; the other mirror is partially transparent to let some of the radiation pass through to be used (Figure 23-4).

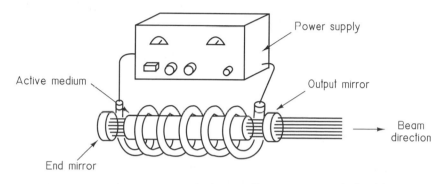

Figure 23-4 Basic components of a laser oscillator: Power supply furnishes energy to a medium contained between two mirrors. Beam emerges through right-hand mirror.

When the system is raised to the excited state, the first few photons will be emitted spontaneously. These photons trigger the release of more photons and become part of a continually growing wave, an *avalanche*. As the wave is reflected back and forth between the mirrors, the chance for triggering more transitions continually increases, thus a laser oscillator is much more efficient than a laser amplifier. These processes, excitation, population inversion, and avalanche action, are fundamental to stimulated emission; they have made lasers a reality.*

---

* The laser is an example of the same idea having come to different people, in different parts of the world, at about the same time. The idea first occurred to Charles Hard Townes (1915–), then professor of physics at Columbia University, while, one day in 1951, sitting on a bench in Franklin Park in Washington, D.C. Independently, Alexandr Mikhailovich Prokhorov (1916–), head of the Oscillation Laboratory of the P. N. Lebedev Institute of Physics at the University of Moscow, and his assistant Nicolai Gennadievich Basov (1922–), a 1943 graduate of the Kiev School of Military Medicine, now director of the Lebedev Institute, in 1952 presented a paper at an All-Union (SSSR) Conference on radio spectroscopy in which they discussed the possibility of building a "molecular generator." In 1958, Arthur L. Schawlow (1921–) and Townes showed that the same principle, first developed for microwaves and hence called *maser*, could be extended into the visible. Subsequently, in 1960, Theodore H. Maiman built the first *laser* using ruby as the active medium. The pertinent publications are C. H. Townes, *Production of Electromagnetic Energy*, U.S. Patent 2,879,439, Mar. 24, 1959; N. G. Basov and A. M. Prokhorov, *Zh. Eksp. Teor. Fiz.* **28** (1955), 249–50, Engl. trans. "Possible Methods of Obtaining Active Molecules for a Molecular Generator," *Sov. Phy. JETP* **1** (1955), 184–85; and A. L. Schawlow and C. H. Townes, "Infrared and Optical Masers," *Phys. Rev.* **112** (1958), 1940–49. In 1964, Townes, Prokhorov, and Basov, and in 1981 Schawlow, were awarded the Nobel Prize in Physics.

Excitation

There are several ways of pumping a laser and producing the population inversion required. A *direct conversion* of electrical energy into radiation, for example, occurs in light-emitting diodes, LEDs, and in semiconductor lasers that derive from them. In *optical* pumping, a light source supplies the energy. Short, intense flashes of light were already used in Maiman's ruby laser and still are customary today with solid-state lasers.

Pumping can be accomplished also using one or several other lasers, often a series of semiconductor lasers, which are reliable, produce little heat, and are highly efficient. Excitation by *atom-atom collisions* occurs in the helium-neon laser. *Thermal excitation* is typical of the gasdynamic laser where a hot gas is forced through a nozzle.

Cavity Configurations

The simplest type of a resonant cavity is a combination of two *plane mirrors* [Figure 23-5 (top)]. Such mirrors require precise alignment. This configuration is very efficient because of its good *filling* but, because of difficult alignment and low stability, is rarely used today.

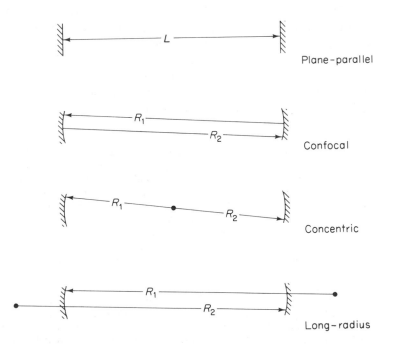

Figure 23-5   Cavity configurations. *L*, distance between mirrors; *R*, radii of curvature.

A *confocal cavity* has two concave mirrors of the same radius of curvature, R, separated by a distance L that is equal to R, $L = R$. The focal length of the mirrors, therefore, is one-half of L and the focal points coincide in the center, hence the term "confocal." A confocal cavity is much easier to align than plane-parallel mirrors but the filling is poor; halfway between the mirrors, for instance, only a small fraction of the medium is utilized.

A *concentric cavity* (also called *spherical cavity*) has two concave mirrors of the same radius, one-half of the distance between them, $L = 2R$. Again, the filling is poor and alignment not easy. A *long-radius cavity* has two concave mirrors, their radii of curvature significantly longer than the distance between them, $R_1 = R_2 > L$. This is a good compromise between the plane-parallel and the confocal variety; it is the type of cavity used most often in today's commercial lasers.

Mode Structure

Assume for simplicity that the cavity is limited by two plane-parallel mirrors. Since the mirrors form a *closed* cavity, the standing-wave pattern inside the cavity has nodes, rather than loops, at both ends. If, as shown earlier in Figure 23-5 (top), L is the length of the cavity, the longest wavelength possible is $\lambda_1 = 2L$. The next shorter wavelength is $\lambda_2 = L$, the next shorter wavelength $\lambda_3 = \frac{2}{3}L$, and so on; thus,

$$\lambda = \frac{2}{q} L \qquad [23\text{-}11]$$

where q is the number of half-wavelengths, or *axial modes*, that fit into the cavity.

It is often better to write Equation [23-11] in terms of frequency, ν. We take the relationship $v = \lambda \nu$, replace v by the velocity of light c, solve for ν, and substitute Equation [23-11]. That gives

$$\nu = q \frac{c}{2L} \qquad [23\text{-}12]$$

But a laser cavity must, by necessity, contain a medium (of index n), rather than free space. Hence, we replace the actual length of the cavity, L, by the optical path length, $S = Ln$, so that, instead of Equation [23-12], we have

$$\nu = q \frac{c}{2S} \qquad [23\text{-}13]$$

which is the *resonance condition for axial modes*. A wave of frequency ν that travels along the axis of the cavity, therefore, forms within the cavity a series of standing waves, called *stationary axial modes*.

Now consider two consecutive modes (which differ by $q = 1$). These modes have frequencies that, following Equation [23-13], are separated by a frequency difference $\Delta\nu$,

$$\Delta\nu = \frac{c}{2S} \qquad \text{[23-14]}$$

Slightly different frequencies are closely, and evenly, spaced. Often several of them lie within the width of a single emission line. That means that the output of the laser then consists of a number of lines separated by $c/2S$, as shown in Figure 23-6.

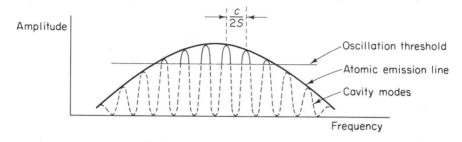

Figure 23-6 Output of laser containing several resonance frequencies.

In addition to the axial modes we also have *transverse modes*, called TEM, for transverse electromagnetic, modes. The TEM modes are generally few in number, and they are easy to see. Aim the laser at a distant screen and spread the beam out by a negative lens. Most often the light forms several bright patches, separated from one another by intervals called "nodal lines." Within each patch, the phase of the light is the same, but between patches the phase is reversed. Figure 23-7 shows several examples.

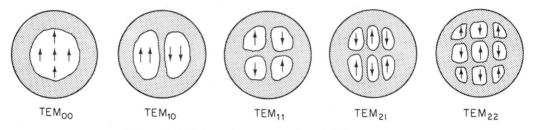

TEM_{00} TEM_{10} TEM_{11} TEM_{21} TEM_{22}

Figure 23-7 Schematic representation of different TEM modes.

In the lowest possible transverse mode, TEM_{00}, there is no phase reversal across the beam (the beam is "uniphase"), the spatial coherence is the highest possible, and the beam can be focused to the smallest spot size and reach the highest power density. If, in addition, the laser also oscillates in the lowest possible axial mode, the cavity will select, and amplify, out of several resonant frequencies only one frequency. This results in the highest possible temporal coherence, that is, the light will be as monochromatic as light can be.

Gain

The gain of a laser depends on several factors. Foremost among them is the separation of the energy levels that provide laser transition. If the two levels are farther apart, the gain is higher because then the laser transition contains a larger fraction of the energy compared to the energy in the pump transition. If the two levels are closer, the gain is less.

In a way, gain is the opposite of *absorption*. Consider once more the exponential law,

$$\phi' = \phi_0 e^{-\alpha x} \qquad [23\text{-}15]$$

and note that the exponent $(-\alpha x)$ is negative and, therefore, the absorptivity α positive. This is true for thermal equilibrium (normal Boltzmann distribution), where $N_2 < N_1$. Conversely, in population inversion where $N_2 > N_1$ the exponent is positive and α negative. Indeed, laser emission could be considered *negative absorption* and Equation [23-15] be written

$$\phi' = \phi_0 e^{\beta x}$$

where β is the *gain coefficient*, the negative of the "absorption coefficient," $\beta = -\alpha$.

As the wave is reflected back and forth between the mirrors, it will lose some of its energy, mainly because of the limited reflectivity of one of the mirrors. The limited reflectivity, of course, is needed to let radiation pass through the mirror to be utilized.

Beam Shape

Look at the light beam produced by a laser. You will be impressed by how slender the beam is and how far it proceeds, seemingly without spreading apart. No other source of light, with or without the help of lenses or mirrors, produces a beam like that.

There are two characteristics that contribute: the beam *profile* and the beam's *divergence*. The term profile refers to the energy distribution across the beam's diameter. With the laser operating in the TEM_{00} mode, the energy has a Gaussian distribution: at a given distance r from the axis, the intensity I falls off exponentially,

$$I(r) = I_0 e^{-(2r/w)^2} \qquad [23\text{-}16]$$

The width of the beam is twice its radius w, measured from the axis where the intensity is I_0 to where it has dropped to $1/e^2$ of I_0.

A beam retains its profile both inside and outside the cavity; the profile merely contracts or expands. For example, with a confocal cavity (but also with a concentric and a long-radius cavity) the beam has its least diameter, or *beam waist*, halfway between the mirrors. There, the radius is

$$w_0 = \sqrt{\frac{L\lambda}{2\pi}} \qquad [23\text{-}17]$$

where L, as before, is the distance between the mirrors.

Twice that radius is the diameter of the beam at the waist. Its size accounts for (part of) the *divergence* of the beam, simply because the shape of the beam outside the cavity is an extension of its shape inside, a relationship illustrated in Figure 23-8.

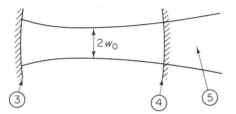

Figure 23-8   Confocal cavity causes beam to contract to minimum beam diameter $2w_0$. (3) Fully reflective mirror, (4) semitransparent mirror, (5) laser output.

Farther away from the laser, in the *far field* where the beam's shape can be considered a linear function of the distance, the beam's contour subtends with the axis an angle θ,

$$\theta = \frac{\lambda}{\pi w_0} \qquad [23\text{-}18]$$

Twice that angle is the *full-angle divergence*,

$$2\theta = \frac{4\lambda}{\pi d_0} \qquad [23\text{-}19]$$

an expression where I have replaced the radius at the waist by the diameter, d_0.

Note that the divergence of the beam is inversely proportional to d_0. With a small waist, the divergence is large. Conversely, for a well collimated beam (of small divergence), the waist must be large—as we have already seen from Figure 23-8.

Now we place a converging lens in the path of the light, causing the beam to contract to a "focus." That focus, though, is merely another waist where the beam's wavefronts are essentially, if temporarily, plane. Beyond the waist, the beam expands again. Assume the beam has a radius r and the lens a focal length f. The half-angle vergence of the beam is then

$$\theta = \frac{r}{f}$$

Taking twice that angle, combining it with Equation [23-19] by eliminating 2θ, and solving for $2r$ gives

$$2r = \frac{4f\lambda}{\pi d_0}$$ [23-20]

where $2r$ is the beam's diameter at the focus.

With a typical He-Ne laser, the beam divergence is about 1 milliradian. Such a beam, in other words, increases in diameter about 1 mm for every meter that the beam travels.

Another part of the beam's divergence is due to diffraction. Its magnitude is found from Rayleigh's criterion,

$$\theta \approx 1.22 \frac{\lambda}{D}$$ [23-21]

where D is the diameter of the laser's aperture, or of a lens if it expands, and then limits, the beam. With most gas lasers, the diffraction divergence is about twice as large as the beam-waist divergence.

Equation [23-21] tells us that the diffraction divergence can be reduced if the beam is passed through a *beam expander* (which makes D larger). A beam expander is typically an inverted telescope, either of the astronomical or the Galilean type. The astronomical type is used when a spatial filter is needed (which then is placed in the focal plane common to the two lenses). The inverted Galilean type, shown in Figure 23-9, is preferred with high-power lasers because the radiation is not brought to a focus (which might cause ionization and breakdown of the air).

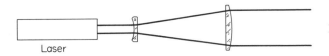

Laser

Figure 23-9 Beam expander. Type shown is inverted Galilean telescope.

TYPES OF LASERS

Solid-State Lasers

The classical example of a solid-state laser is the *ruby laser*.* Ruby is synthetic aluminum oxide, Al_2O_3, with 0.03 to 0.05% of chromium oxide, Cr_2O_3, added to it. The Cr^{3+} ions are the active ingredient; the aluminum and oxygen atoms are inert. The ruby crystal is made into a cylindrical rod, several centimeters long and several millimeters in diameter, with the ends polished flat to act as cavity mirrors. Pumping is by light from a xenon flash tube.

---

* First described by T. H. Maiman, "Stimulated Optical Radiation in Ruby," *Nature (London)* **187** (1960), 493–94, and "Stimulated Optical Emission in Fluorescent Solids: I. Theoretical Considerations," *Phys. Rev.* **123** (1961), 1145–50.

As shown in Figure 23-10, chromium-doped ruby has three energy levels, E_1, E_2, and E_3. The uppermost level, E_3, is fairly wide (it will accept a wide range of wavelengths) and has a short lifetime; the excited Cr^{3+} ions rapidly relax and drop to the next lower state, E_2. This transition is nonradiative.

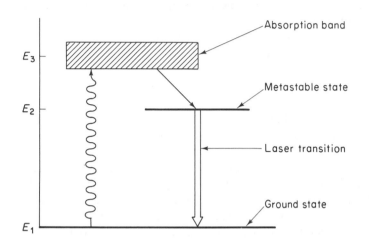

Figure 23-10 Three-level energy diagram typical of ruby. Pumping transition is shown by wavy line, nonradiative transition by simple arrow, laser transition by hollow arrow.

The E_2 state is *metastable* (that is, nearly stable); it has a lifetime of about 10^{-3} s, considerably longer than that of E_3, and the Cr^{3+} ions remain that much longer in E_2 before they drop to the ground state, E_1. The $E_2 \rightarrow E_1$ transition is radiative; it produces the spontaneous, incoherent red fluorescence typical of ruby, with a peak near 694 nm. But as the pumping energy is increased above a critical threshold, population inversion occurs in E_2 with respect to E_1 and the system lases, with a sharp peak at 694.3 nm.

The *neodymium : YAG laser* has four energy levels. As shown in Figure 23-11, the laser transition begins at the metastable state and ends at an additional level somewhat above the ground state.

The active ingredient is trivalent neodymium, Nd^{3+}, added to an yttrium aluminum garnet, YAG, $Y_3Al_5O_{12}$. As before, excitation raises the system from the ground state, E_1, to the highest of the four levels, E_4. From there the system returns to the metastable state, E_3. Since both E_4 and E_2 drain rapidly, laser emission will commence as soon as E_3 has reached population inversion with respect to E_2. This inversion can be maintained even with moderate pumping. That allows such lasers to operate at high repetition rates as well as in the *continuous wave* (c.w.) mode, in contrast to most three-level lasers, which emit only pulses. The Nd : YAG laser is fairly efficient; its emission is at 1.064 μm.

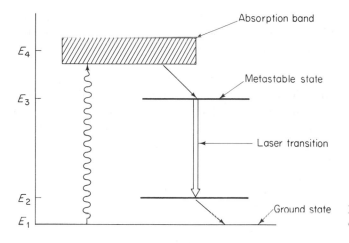

Absorption band

Metastable state

Laser transition

Ground state

E_4

E_3

E_2

E_1

Figure 23-11 Four-level system typical of neodymium.

Gas Lasers

The widely used *helium-neon laser* is an example of an atomic gas laser.* Typically, it consists of a tube about 30 cm long and 2 mm in diameter, with two electrodes on the side and fused silica windows at both ends. The tube contains a mixture of 5 parts helium and 1 part neon, kept at a pressure of approximately 1 mm Hg. The mirrors are placed outside the tube, as shown in Figure 23-12.

The end windows are set at Brewster's angle. This causes light, returning from the outside mirrors back to the cavity and oscillating *normal* to the plane of

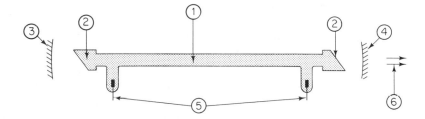

Figure 23-12 Helium-neon laser with (1) gas discharge tube, (2) Brewster windows, (3) fully reflective mirror, (4) partially transparent mirror, (5) electrodes connected to power supply, (6) output.

---

* First described by A. Javan, W. R. Bennett, Jr., and D. R. Herriott in "Population Inversion and Continuous Optical Maser Oscillation in a Gas Discharge Containing a He-Ne Mixture," *Phys. Rev. Lett.* **6** (1961), 106–10, who discovered emission at 1.15 μm. The familiar red radiation at 632.8 nm was obtained a year later by A. D. White and J. D. Rigden, "Continuous Gas Maser Operation in the Visible," *Proc. IRE* **50** (1962), 1697.

incidence, to be reflected away from the cavity. Consequently, the component oscillating *parallel* to the plane of incidence becomes dominant and will sustain laser emission; the radiant energy that emerges from the laser, therefore, is linearly polarized.

Pumping is by electric current. Initially the excitation affects only the helium. But, two of the excited helium states have about the same energies as two of the higher neon states, and thus the excited helium atoms will, by collision, transfer their energy to the neon, raising the neon while the helium returns to the ground state. The neon E_3 states are metastable and quickly reach population inversion relative to E_2. The neon returns to the lower states by various transitions, most notably by laser emission at 632.8 nm. The essence is that the helium atoms, which are fairly light, can easily be excited; the neon atoms, which are much heavier, could not be pumped up efficiently without them.

The *carbon dioxide laser*, a four-level laser, is virtually in a class by itself because of its high power. The first CO_2 lasers had a continuous output of a few milliwatts. Today we have powers of some 200 kW, more than enough to cut through steel plates several centimeters thick in a matter of seconds. In fact, CO_2 lasers may be the most practical lasers of all. Their efficiency in converting electrical energy into radiation is better (more than 10%) than that of any other laser. Their construction and operation are relatively simple. Excitation is by way of nitrogen, which takes the place of the helium in the He-Ne laser, and subsequent collision with CO_2. Often other gases, such as He, Xe, and H_2O, are added because they enhance the homogeneity of the output and the yield. Emission is at 10.6 μm.

Dye Lasers

Most dye lasers are liquid lasers. They contain an excitable fluorescent dye, often rhodamine 6G. Their major attraction is that they can be *tuned*, often over a wide range of wavelengths, from the ultraviolet to the near infrared. That has completely revolutionized the field of absorption spectroscopy. The tuning elements can be prisms, diffraction gratings, etalons, or birefringent filters. Often the laser is combined with a wavelength-measuring device such as a Michelson or a Fabry–Perot interferometer. Another advantage is that liquids do not crack or shatter and hence dye lasers can be made in almost unlimited sizes. In fact, some of them have powers as high as solid-state lasers, at a fraction of their cost.

A *chemical laser* does not require an external source of power; instead, the energy comes from a chemical reaction. The first chemical lasers used a mixture of hydrogen and chlorine. Today we often use hydrogen + fluorine, H_2 + F, or, even better, deuterium + fluorine. These lasers produce very high c.w. (continuous-wave) power, mainly in the 3–5 μm wavelength range. Since Earth's atmosphere is transparent in this range, such lasers are of some military interest.

Semiconductor Lasers

Semiconductor lasers, also known as *diode lasers*, are derived from the light-emitting diode, LED, that we have discussed before. Ordinarily, when a *p-n* junction is connected to a battery, electrons flow from *n* to *p*, and holes from *p* to *n*, the combination of both producing light. But, as the current is increased, hole-electron pairs that have not had time to recombine spontaneously are forced together causing *stimulated emission*. Such emission was first observed in the IR, and with an efficiency much higher than with optical pumping (40% versus 3%).

Diode lasers are particularly well suited for data processing and fiber-optics communication, mainly because of their small size (no more than a cubic millimeter), their high efficiency, the speed by which they can be turned on and off, and because of their compatibility with fiber and integrated optics and with semiconductor electronics. No external mirrors are needed. Gallium arsenide lasers emit in the red or infrared, which is fine for fibers. But blue semiconductor lasers provide better data storage (because they allow higher storage densities); they also are part of large-screen displays (where blues and greens are needed to cover the full range of the spectrum).

Since Maiman first used ruby, numerous other materials have been found to lase, emitting radiation from the extreme ultraviolet (XUV), with wavelengths as short as a few nanometers, to millimeter waves. All gaseous elements and many molecules are known to lase, and the same is true of most metals. Any dye that fluoresces will lase if only enough power is supplied. Today an estimated 5000 laser transitions are known. The types of lasers most commonly used are listed in Table 23-1.

Table 23-1 Summary of typical lasers

Ion	*Matrix*	*Principal Wavelength*
Cadmium	Helium	441.6 nm
Neon	Helium	632.8 nm
Chromium	Aluminum oxide (ruby)	694.3 nm
Gallium arsenide		Near IR
Neodymium	YAG	1.064 μm
Carbon dioxide	Nitrogen	10.6 μm

APPLICATIONS

In the James Bond movie *Goldfinger*, the villain forces his way into the gold bullion depository at Fort Knox using a laser mounted on top of an armored personnel carrier. In the mind of the public, this episode has come to represent the wondrous capabilities of the laser. Today some industrial lasers may indeed be able to cut through a heavily fortified door but their principal uses lie somewhere else. Still, it is the fact that a laser can deliver high power to a small area that has made it such an exciting and unique tool.

Energy, Power, and Power Density

A typical laser produces perhaps 10 J of energy. That does not seem to be very much. But if this energy is delivered within a pulse only 0.5 ms long, its power is 20 kW, a respectable amount. The power of a *pulsed* laser, in short, is considerably higher than that of a *continuous-wave* laser, and if the pulse can be made shorter, the power is even higher. Figure 23-13 shows a comparison.

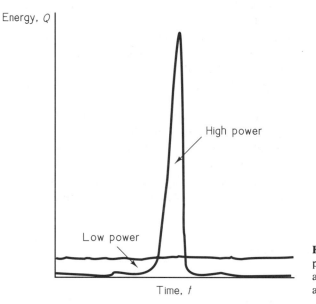

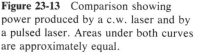

Figure 23-13 Comparison showing power produced by a c.w. laser and by a pulsed laser. Areas under both curves are approximately equal.

Some lasers generate 300 J in 5 ns, which translates into 60 GW (gigawatt, 10^9 W). But even pulses as short as 8 femtoseconds (8×10^{-15} s) have been reached, and powers as high as 2500 TW (terawatt, 10^{12} W). That is an incredible amount, about as much as the output of all power plants in the world combined. These plants, of course, generate power continually, whereas a laser pulse may last for perhaps only a fraction of a nanosecond.

Such high powers are generally produced by *Q switching*, which means compressing the energy into a very short period of time. One possibility of doing this is by using a rotating mirror at one end of the laser cavity (Figure 23-14). Then, after the system has been pumped, oscillation cannot occur *except* when one of the facets of the rotating mirror is exactly parallel to the stationary mirror; at that instant all of the accumulated energy is dumped into one single, very short "giant" pulse.

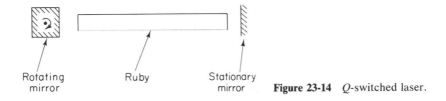

Figure 23-14 *Q*-switched laser.

In addition to the high power produced by a laser, it is also its *high power density* that accounts for many of its useful, and often its deleterious, effects. In the lowest transverse mode, TEM_{00}, the beam can be focused down to the smallest spot size and the laser reach its highest possible power density.

Nonlinear Effects

Ordinarily, the refractive index and the absorptivity of a material are properties of the material, and independent of the intensity of the light that passes through. With very intense light, that can be different and one or both of these characteristics may well be a linear or a *nonlinear* function of the intensity. This causes a number of phenomena, long predicted by theory but observed only recently since the advent of high-powered lasers. Light from such lasers can, in fact, change the refractive index; heat and thermal expansion, as they occur with absorbing materials, are not involved.

Some lasers produce power densities so high that the radiation strips electrons from atoms and molecules creating a *plasma* (a mixture of ions and free electrons rarely found in nature except in lightning and in the atmosphere of the sun and the stars). The electric field strength can be as high as 10^9 V/m, more than enough to cause the breakdown of air (3×10^6 V/m).

Nonlinear effects worth noting are *self-focusing*, where a beam of light contracts into thin, short-lived, powerful threads of light that quickly shatter the material through which they pass. *Optical bistability* arises in saturable systems (whose absorptivity increases as the material becomes more excited). The resulting increase in absorption makes the material yet more excited, causing even more absorption, and so on.

The term *phase conjugation* refers to a reflection of light by a gas acting as a flexible mirror. Assume that the light passes first through a random medium such as turbulent air. That causes some of the light to advance farther while other parts are lagging behind. The light then enters a tube containing methane or CO_2 under high pressure. The gas will reflect the light just like a mirror although a mirror unlike any known from geometrical optics: the different parts of the light are returned in precisely the same direction as that from which they came. The distortions, in other words, are canceled as the light retraces its path through the random medium once again.

Frequency doubling is the generation of second harmonics. An example is shown on the cover of the July 1963 issue of *Scientific American*: A beam of light from a ruby laser, a vivid red, enters a colorless crystal of ammonium dihydrogen phosphate. As the light comes out of the crystal, it is doubled in frequency, to one-half the wavelength of the incident light: it is blue, a good example of nonlinear optics. Similarly, the 1.064-μm emission from an Nd : YAG laser can by successive frequency doubling, first in a crystal of KDP, potassium dihydrogen phosphate, and then in potassium dideuterium phosphate, be converted to its fourth harmonic, 266 nm.

Industrial Applications

The high powers available from lasers have led to a great many industrial and medical applications, from more mundane tasks such as the *machining* of materials to the exotic, such as trying to release energy from nuclear fusion.

Cutting materials by laser is commonplace. Its advantage is that lasers can be easily aligned, which makes them well suited for computer-controlled material processing (robotics); they can be used with almost any material, from cardboard to cast iron. They are clean, reliable, and economical to run and never grow dull.

Drilling by laser is another common application. Diamond, about the hardest material known, has been drilled before, but the process is tedious and time consuming. A laser does it quickly, producing traces of (black) graphite, which facilitates absorption. Holes have also been drilled into teeth, paper clips, even into single human hairs.

Earlier, *welding* by laser meant only welding on a microscopic scale, such as in integrated circuits. But in recent years, as lasers have become more powerful, even heavy steel plates have been welded. For example, steel plates 5 cm thick can be fused using a 90-kW laser, at speeds higher than 2.5 m per minute.

Lasers are used extensively for *surveying*. Some distance measurements are indeed very precise; the distance to the moon, for example, has been determined to an accuracy *better than 15 cm*.

Medical Applications

Laser effects on biological tissue are either thermal or nonlinear. Thermal effects depend on absorption, mainly in pigments such as melanin (as it occurs in the skin and the iris and choroid of the eye) and hemoglobin (blood). Absorption causes a conversion of radiant energy into heat, increasing the temperature of the tissue and resulting in a denaturation of proteins called *coagulation*. With the process produced by light, it is called *photocoagulation*.*

---

* Photocoagulation is of interest especially in ophthalmic surgery, much of it pioneered by Gerd Meyer-Schwickerath (1920–92), German ophthalmologist and longtime chairman of the Department of Ophthalmology at the University of Essen. Meyer-Schwickerath's most dramatic result came when he was asked to treat a one-eyed ophthalmologist who had a completely updrawn iris in an aphakic eye, leaving him blind. Because of his darkly pigmented iris, laser irradiation caused a sudden, almost

Nonlinear effects are entirely different. The rapid expansion of a plasma that is characteristic of a nonlinear effect produces a shock wave that mechanically (rather than thermally) causes a disruption of the tissue called *photodisruption*. The difference between thermal and nonlinear effects can be shown by aiming a c.w. laser at the tip of a match and seeing it burst into flame. Firing a *Q*-switched laser (which because of its short pulse length has much higher power) at a match makes the tip crumble, without ignition.

Medical applications include the treatment of *retinal detachment*, where the focused beam causes small burns that on healing keep the retina back in place by the formation of scar tissue. Similar is the treatment of *diabetic retinopathy*, where local distentions of small blood vessels (*microaneurysms*) and newly formed blood vessels (*neovascularization*) are obliterated by coagulation. A cloudy posterior capsule, often left in place after removal of a cataract, can be opened quickly and without discomfort by *posterior capsulotomy*. Lasers are used also in the treatment of *melanoblastoma*, a heavily pigmented tumor of the skin and the choroid of the eye that sometimes erupts into highly malignant growth.

Laser Speckle

Laser light, as we have seen, is very coherent and very intense. So it comes as a disappointment when we notice that this light, when projected on a screen, is rather mottled and uneven. Laser light has a definitely grainy structure called *speckle*.

The theory of laser speckle is very complex. With some simplification we can say that any surface has a certain roughness, with a vast number of individual facets and scatterers. The light reflected from such a surface consists of contributions from all of these facets and, since the light has high spatial coherence, these contributions form loci of interference, distributed at random in three-dimensional space in front of the surface. For example, look at a screen diffusely illuminated by laser light. While looking, slowly move your head from side to side as if saying "no." The path lengths traversed by these contributions will change and the speckle pattern will move, an effect aptly called a *"red snowstorm."*

The same effect is seen when, instead of moving the head, the screen is moved slowly transverse to the line of sight. If the head, or the screen, is moved slowly in one direction and the speckle grains are seen to move in the *opposite* direction, the observer is myopic (nearsighted). If the grains move

---

explosive reaction, producing a hole 2 mm in diameter and instantly opening the line of sight. The patient jumped up from the table, stared at Meyer-Schwickerath, and said: "Now I can see you! Thank you very much! You look very young."

in the *same* direction, the observer is hyperopic (farsighted). This can be explained by the position of the far point of the eye, which, as shown on page 166, in myopia is located in front of the eye, and in hyperopia behind it. The apparent motion of the speckle pattern, therefore, can be used to test for visual refractive deficiencies.

LASER SAFETY

Any light, no matter where it comes from, can cause damage. But *laser radiation is particularly dangerous*, mainly because of its high spatial coherence (which means that it can be focused down to very high power densities). In addition, many lasers emit in a region of the spectrum (red) where the light just does not seem to be as powerful as if it were white.

Of all the parts of the human body, clearly the *eye* is the most vulnerable. Which part of the eye is subject to injury is a matter of wavelength. In the IR, above 10 μm, much of the energy is absorbed by water. Since water is the main constituent of most any biological tissue, it is the *cornea* that is severely damaged first.

Light between 400 nm and 1.4 μm, on the other hand, will penetrate through the eye and be absorbed in the pigment epithelium next to the *retina*. The pigment epithelium then virtually explodes, ejecting black granules into the vitreous, forming vapor bubbles and causing hemorrhages all around the destroyed tissue.

Example

Compare the irradiances at the retina that result when looking:
(a) Directly at the sun.
(b) Into a 1-mW He-Ne laser.

Solution: (a) The sun subtends an angle of 0.5° = 0.0087 rad. Outside Earth's atmosphere, the sun's irradiance is 1.4 kW/m². At Earth's surface, it is approximately 1 kW/m² or, easier to comprehend, 1 mW/mm². Assume that the pupil of the bright-adapted eye is 2 mm in diameter. Its area, therefore, is

$$A = \tfrac{1}{4}\pi D^2 = \tfrac{1}{4}(\pi)(2 \text{ mm})^2 \approx 3 \text{ mm}^2$$

With an irradiance of 1 mW/mm², a power

$$\phi = (1 \text{ mW/mm}^2)(3 \text{ mm}^2) = 3 \text{ mW}$$

will pass through the pupil. The focal length of the eye, from Example 2, page 33, is 22.5 mm. The image of the sun on the retina, therefore, has a diameter

$$D = (22.5 \text{ mm})(0.0087 \text{ rad}) \approx 0.2 \text{ mm}$$

and an area

$$A = \tfrac{1}{4}(\pi)(0.2 \text{ mm})^2 \approx 0.03 \text{ mm}^2$$

The irradiance at the retina then is the power (3 mW) divided by the area of the sun's image,

$$E = \frac{\phi}{A} = \frac{3 \text{ mW}}{0.03 \text{ mm}^2} = \boxed{100 \text{ mW/mm}^2}$$

(b) The laser may have a beam 2 mm in diameter. Using Equation [23-20] and setting $\lambda = 633$ nm, $f = 22.5$ mm, and $d_0 = 2$ mm, the focal spot has a diameter

$$2r = \frac{4f\lambda}{\pi d_0} = \frac{(4)(22.5)(633 \times 10^{-6})}{(\pi)(2)} = 9 \text{ } \mu\text{m}$$

which corresponds to an area

$$A = \tfrac{1}{4}(\pi)(9 \times 10^{-3} \text{ mm})^2 = 64 \times 10^{-6} \text{ mm}^2$$

and an average irradiance

$$E = \frac{1 \text{ mW}}{64 \times 10^{-6} \text{ mm}^2} \approx 0.016 \times 10^6 \text{ mW/mm}^2 = \boxed{16 \text{ W/mm}^2}$$

That is *160 times the irradiance brought about by the sun*! Clearly, **EXTREME CAUTION MUST BE USED IN ANY WORK INVOLVING ANY TYPE OF A LASER!**

SUMMARY OF EQUATIONS

Resonance condition for axial modes:

$$\nu = q \frac{c}{2S} \qquad\qquad [23\text{-}13]$$

Full-angle divergence:

$$2\theta = \frac{4\lambda}{\pi d_0} \qquad\qquad [23\text{-}19]$$

Beam's diameter at the focus:

$$2r = \frac{4f\lambda}{\pi d_0} \qquad\qquad [23\text{-}20]$$

Wavelength of emission of:

He-Ne laser	632.8 nm	
Ruby laser	694.3 nm	
Nd : YAG laser	1.064 μm	[Table 23-1]
CO_2 laser	10.6 μm	

PROBLEMS

23-1. If laser action occurs by the transition from an excited state to the ground state, $E_1 = 0$, and if it produces light of 693 nm wavelength, what is the energy level of the excited state?

23-2. Transition occurs between a metastable state E_3 and an energy state E_2, just above the ground state. If emission is at 1.1 μm and if $E_2 = 0.4 \times 10^{-19}$ J, how much energy is contained in the E_3 state?

23-3. Rate, in order of decreasing efficiency, the filling provided by different cavity configurations.

23-4. Ruby has a refractive index of $n = 1.765$. If the crystal is 4 cm long, the wavelength 694.3 nm, and the laser operating in the TEM_{00} mode, what is the least number of standing waves that fit into the cavity?

23-5. Emission from a semiconductor laser occurs in the narrow junction between two types of material. If the wavelength is 905 nm and the width of the junction, the "slit width," 5.2 μm, what is the full-angle divergence of the output?

23-6. What is the divergence of a beam that at first is 2 mm wide and, after a distance of 10 m, has spread to a diameter of 16 mm?

23-7. What is the full-angle divergence of a He-Ne laser, oscillating in the TEM_{00} mode, whose beam is 1 mm in diameter?

23-8. If the beam produced by a CO_2 laser is 4 mm wide, what should be the focal length of a lens that will focus the beam into a spot 0.25 mm in diameter?

23-9. If the light from a He-Ne laser is 1.93 mm in diameter and limited only by diffraction, what is the divergence of the beam as it leaves the aperture?

23-10. A laser beam of a certain divergence is projected, eyepiece first, through an afocal telescope built with an objective of 24 cm focal length and an eyepiece of $f = 16$ mm. How will the divergence change?

23-11. If a laser delivers 1.48 mW in a beam 3.6 mm in diameter, what is the power density in a spot 1 μm in diameter, assuming a loss of 20% in the focusing system?

23-12. Show, in the form of a simple schematic drawing, what the wavefront of a laser beam may look like:
(a) Before the beam enters a sheet of ground glass.
(b) After the beam emerges from the ground glass.
(c) After phase conjugation.

23-13. While looking at the speckle pattern produced on a diffuse screen, an uncorrected *myopic* observer turns his/her eyes slowly from left to right. That will cause the speckle grains to move.
(a) How do the loci of interference, generated in front of the screen, move relative to the screen?
(b) How do their conjugate images move on the retina?
(c) Therefore, what does the observer see?

23-14. A He-Ne laser of only 1 mW power projects a beam of 2 mm diameter and 6 mrad divergence into the eye. Using 22.5 mm for the focal length of the reduced eye, how large a focal spot will form on the retina?

SUGGESTIONS FOR FURTHER READING

BERTOLOTTI, M. *Masers and Lasers: An Historical Approach*. Bristol, England: Adam Hilger Ltd., 1983.

BUTCHER, P. N., and D. COTTER. *The Elements of Nonlinear Optics*. New York: Cambridge University Press, 1990.

DAINTY, J. C., ed. *Laser Speckle and Related Phenomena*, 2nd ed. New York: Springer-Verlag New York, Inc., 1984.

LUXON, J. T., and D. E. PARKER. *Industrial Lasers and Their Applications*. Englewood Cliffs, N.J.: Prentice-Hall, Inc., 1985.

MILLONI, P. W., and J. H. EBERLY. *Lasers*. New York: Wiley-Interscience, 1988.

SCHÄFER, F. P. *Dye Lasers*, 3rd ed. New York: Springer-Verlag New York, Inc., 1990.

TARASOV, L. V. *Laser Physics*. Moscow: Mir Publishers, 1983.

WINBURN, D. C. *Practical Laser Safety*. New York: Marcel Dekker, Inc., 1985.

WITTEMAN, W. J. *The CO_2 Laser*. New York: Springer-Verlag New York, Inc., 1987.

Relativistic Optics

We have almost reached the end of our *Introduction to Classical and Modern Optics*. Now comes a major constraint. Whenever the light source or the observer or any part of the system is moving, virtually all the laws of optics change. The example of relativistic optics cited most often is the Michelson–Morley experiment. Actually, the Michelson–Morley experiment, while of historic interest, has not materially contributed to the development of the theory of special relativity and has little bearing on our thinking today. Accordingly, we go directly to a discussion of the facts of relativity, both Galileo's and Einstein's, and then continue with a series of examples, from space flight to gyroscopes, that illustrate how relativistic optics replaces the laws of conventional optics.

TRANSFORMATIONS

Galileo's Transformation

Assume that somewhere in three-dimensional space there occurs a single physical *event*, such as a collision of two particles. To describe the event fully we need four coordinates, three space coordinates, x, y, z, and a time coordinate, t. It is understood that the coordinate system is that of the laboratory including the observer and that it is an *inertial system*, with a frame of reference in which Newton's law of inertia holds true. Such a system may either be stationary or it may be in uniform motion, but it may not contain any element subject to acceleration: a space vehicle merely drifting along, without spinning and with its engines cut off, is a good example of an inertial system.

Now consider a given inertial frame, S, and another inertial frame, S', that is in uniform motion relative to S. For simplicity assume that the three sets of axes are parallel to each other and that the frame S' moves, at velocity v, along the x-x' axis, in the direction of $+x$. The observer attached to S ascribes to the event the coordinates x, y, z, t. But to an observer moving along with S' the same event occurs at x', y', z', t'. Since the two frames did coincide at time $t = t' = 0$ and since $x = vt$,

$$\boxed{\begin{aligned} x' &= x - vt \\ y' &= y \\ z' &= z \end{aligned}}$$

[24-1]

which is *Galileo's space coordinate transformation*. The time at which the event is seen to occur, to both the observers in S and S', is the same; thus we add the statement that

$$\boxed{t' = t}$$

[24-2]

Now assume that instead of a point we have an extended (one-dimensional) object such as a meter stick and that we wish to determine its length. The endpoints of the stick are called A and B and these are at rest in the S frame. If the stick is parallel to the x axis, the observer in S assigns to these points the coordinates x_A and x_B and the observer in S' the coordinates x'_A and x'_B. Using Galileo's transformation with respect to x, we find that

$$x'_A = x_A - vt_A \quad \text{and} \quad x'_B = x_B - vt_B$$

Subtracting the first of these equations from the second gives

$$x'_B - x'_A = x_B - x_A - v(t_B - t_A)$$

Since the two endpoints, A and B, are measured at the same time,

$$t_A = t_B$$

the last term in the preceding equation drops out and

$$x'_B - x'_A = x_B - x_A$$

[24-3]

The meter stick, therefore, has the same length in both the S frame and the S' frame.

Similar arguments can be made for the velocity, the acceleration, momentum, angular momentum, kinetic energy—in short, for Newton's laws and all other laws of mechanics that follow from them. These laws are the same in all inertial frames. The laws of electrodynamics, however, are not. This dilemma could be resolved by *Einstein's postulates* (assumptions).

Einstein's Postulates

The two Einstein postulates lie at the heart of the special theory of relativity.* They state that:

1. The laws of physics, including electrodynamics, are the same in all uniformly moving coordinate systems (inertial frames). There is no preferred frame. It is not possible to detect any absolute motion of bodies in space but only relative motions of one body with respect to another. This is the *principle of special relativity*.

2. The velocity of light in a vacuum is the same for all observers, independent of the motion of the source emitting the light and independent of the velocity of the observer. This is the *principle of the constancy of the speed of light*.† The unique aspect of Einstein's work is that it was not based on prior experimental evidence. Rather, it led to a *prediction* of experiments whose results, over the years, have conclusively shown that Einstein was right.

Lorentz' Transformation

We now try to measure a simple parameter, the *length* of an object. If we measure the length of the object *at rest*, in its own frame of reference, S, we measure its *proper length* ("rest length"). But the object may by moving, in a frame S', relative to S. This requires the two observers, in S and S', to determine the coordinates of the endpoints of the object simultaneously. If the endpoints are not determined at the same time, then, because of the finite speed of light, one point

---

* Albert Einstein (1879–1955), German-born physicist, mathematician, philosopher, and humanitarian. From 1902 to 1909, Einstein worked as a patent examiner in the Swiss Federal Patent Office in Bern, then became professor at the universities of Zürich and Prague, the Swiss Polytechnic Institute, and the Prussian Academy of Sciences in Berlin. In 1933, Einstein came to the United States and joined the then newly organized Institute for Advanced Studies in Princeton, New Jersey, where he remained for the rest of his life.

Despite his eminence as a scientist, or because of it, Einstein remained a humble man. When new in Princeton, his neighbor's little daughter came by every afternoon, so regularly that her mother felt obliged to apologize for it. "Don't worry," replied Einstein, "I like Rita's visits. Each time she brings me some candy, and in return I do her homework in math."

The principal paper that laid the foundation for the special theory of relativity is A. Einstein, "Zur Elektrodynamik bewegter Körper," *Ann. Phys.* (4) **17** (1905), 891–921, available in an English translation in A. Einstein et al., *The Principle of Relativity* (New York: Dover Publications, Inc., 1981). By his own account, Einstein began his work at the age of 16 and continued working on it, off and on, for 10 years.

† At first sight, this is hard to comprehend. If a bullet is fired from a moving truck, the velocity of the bullet is different depending on whether the gun is pointed forward or backward with respect to the motion of the truck. This is not so with light. Although experiments have been tried to prove the "ballistic theory," all of them failed, as shown, for example, by G. C. Babcock and T. G. Bergman, "Determination of the Constancy of the Speed of Light," *J. Opt. Soc. Am.* **54** (1964), 147–51.

may be seen to have moved more, or less, than the other point, and the length measured may not be the proper length.

Simultaneity is the crucial word. But how can we tell that two events are simultaneous or, in a more general sense, how can we *synchronize* two clocks? Within the same frame this is easy. But we have two frames, moving relative to each other. If it were possible to transmit a signal with infinite speed, there would be no problem. But we do not know any signals that could reach all points of the universe in zero time. The fastest signals known are those transmitted by light or other electromagnetic radiation, such as radio waves. No faster method of sending a signal has ever been found.

It is actually classical physics which makes the fictitious assumption of zero time (science fiction). Relativistic physics requires a finite, limiting speed. Nature itself shows that relativistic physics is *real* and not a philosophical concept.

Will the synchronized clocks of the observer in S also be in synchrony with the synchronized clocks of the observer in S'? Because of the finite speed of light, they will not. This means that simultaneity is not independent of the frame of reference. Time and space are both relative; to find a connection between the unprimed and the primed coordinates, we somehow have to tie the concept of time into the concept of space.

To do this we use a hypothetical *light clock*, illustrated in Figure 24-1. A source A emits a pulse of light that travels through distance L to a mirror, M, and back again to A. To an observer in the same frame as the clock, the time interval for the pulse to go from A to M and back is

$$\Delta t = \frac{2L}{c}$$

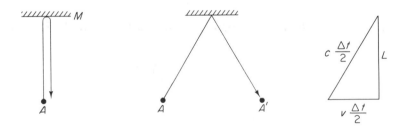

Figure 24-1 Light clock at rest (*left*) and moving at velocity v, left to right (*center*). (*Right*) graphical construction.

Now let the clock be moving, in frame S', left to right. To the observer in S the path from A to the mirror and back to A' is *longer*. The new time interval is found from the triangle in Figure 24-1 (right), and Pythagoras' theorem:

$$\left(c\,\frac{\Delta t}{2}\right)^2 = L^2 + \left(v\,\frac{\Delta t}{2}\right)^2$$

$$c^2\,\Delta t^2 - v^2\,\Delta t^2 = 4L^2$$

$$\Delta t^2 = \frac{4L^2}{c^2 - v^2}$$

and thus

$$\Delta t = \frac{2L}{\sqrt{c^2 - v^2}} = \frac{2L}{c}\,\frac{1}{\sqrt{1 - (v/c)^2}}$$

Substituting $2L/c = \Delta t$ yields

$$\Delta t' = \Delta t_0 \sqrt{1 - (v/c)^2} \qquad\qquad [24\text{-}4]$$

which means that to the observer in S the (moving) clock in S' *runs slow*, a phenomenon called *time dilation*.

Consider again the two endpoints, A and B, of the object, and their coordinates, x_A and x_B. In the S frame, these coordinates determine the proper length, L_0, of the object:

$$L_0 = x_B - x_A$$

To the observer in S, it takes time interval Δt for the object to move past a given point:

$$\Delta t = \frac{L_0}{v}$$

But in frame S', it takes

$$\Delta t' = \frac{L'}{v} \qquad\qquad [24\text{-}5]$$

We solve Equation [24-5] for L', substitute Equation [24-4], and set $v\Delta t_0 = L_0$. This gives

$$\boxed{L' = L_0 \sqrt{1 - \left(\frac{v}{c}\right)^2}} \qquad\qquad [24\text{-}6]$$

which is the *FitzGerald–Lorentz length contraction.**

---

* Named after George Francis FitzGerald (1851–1901), Irish physicist and professor of natural philosophy at Trinity College in Dublin, and Hendrik Antoon Lorentz (1853–1928), Dutch physicist and professor of mathematical physics at the University of Leiden, winner of the Nobel Prize in physics in 1902. FitzGerald and Lorentz, working at the time of intense search for the aether, used their contraction to explain the Michelson-Morley experiment. H. A. Lorentz, "De relatieve beweging van de aarde en den aether," *Versl. Zitt. Wis. Natuurkundige Afdeeling Koninklijke Akad. Wetenschappen, Amsterdam* **1** (1892), 74–79.

Evidently, an object has its greatest length when seen in its rest frame (where its velocity is zero). When seen from another frame, its length, measured in a direction parallel to the motion, becomes less by a factor $\sqrt{1 - (v/c)^2}$.

Does the object *really* become shorter? No, it retains its proper length for an observer flying along with it in the S' frame. But to an observer in S the object not only *appears* to be shorter, it really *is* shorter. The length contraction is not an illusion; it is real for an observer not moving along with the object.

If we apply the contraction factor to the x dimension only, and leave y and z unchanged, we obtain a new set of equations:

$$
\begin{aligned}
x' &= \frac{x - vt}{\sqrt{1 - (v/c)^2}} \\
y' &= y \\
z' &= z
\end{aligned}
\qquad \text{[24-7]}
$$

the *Lorentz' space coordinate transformation equations*.

OPTICS AND THE SPECIAL THEORY OF RELATIVITY

Headlight Effect

Now we come to the specific applications of the theory of relativity to optics. When a firework explodes in the sky, it scatters sparks uniformly in all directions. But when the firework explodes while in rapid motion, it scatters most of its sparks in the forward direction.* It is much the same with a source of light.

Consider a point source that emits light uniformly in all directions. A screen with a hole in it is placed over the source so that half of the light goes to the left of the screen, and half to the right [Figure 24-2 (left)]. With the source standing still, the screen is flat and the angle subtended by the screen and the $+x$-axis, θ, is 90°.

But now let the source and the hypothetical screen travel to the right at a velocity v. To an observer looking at the source from another frame of reference, the screen folds into a cone (with the angle θ becoming less than 90°) but it still divides the light into equal parts, one-half emitted outside the cone and the other half inside; it is merely the *distribution* of the light that changes. (To an observer moving along with the source, nothing happens: it is the relative motion between two reference frames that matters.)

---

* As quoted from W. Rindler, *Essential Relativity: Special, General, and Cosmological*, 2nd ed., p. 260 (New York: Springer-Verlag New York, Inc., 1977).

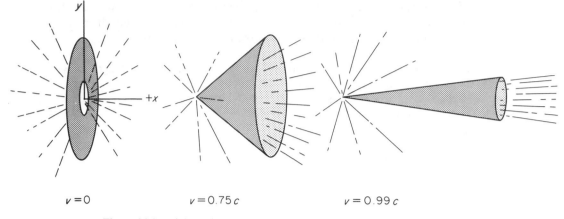

$v = 0$ $v = 0.75\,c$ $v = 0.99\,c$

Figure 24-2   Light emitted from moving source becomes concentrated in the forward direction.

During time interval Δt, light traveling along the cone's surface covers a distance $c\Delta t$. The projection of $c\Delta t$ on the $+x$-axis is Δx. But Δx is also the distance, $v\Delta t$, that the source travels. The ratio of these two distances, therefore, is the cosine of θ,

$$\cos\theta = \frac{v\Delta t}{c\Delta t} = \frac{v}{c} \qquad [24\text{-}8]$$

As the source moves faster, the cone becomes tighter and the light more concentrated in the forward direction, a result that gave the phenomenon its name, *headlight effect*.

Reflection of Light by a Moving Mirror

Let a plane mirror move in a direction normal to its surface and assume that the observer is moving along with the mirror. Neglecting the sign, light incident on the mirror at a given angle, I_1, will be reflected at the same angle, I_2.

But then let the mirror move in a frame of reference different from that of the observer. Again the light is incident on the mirror at an angle I_1 (Figure 24-3). The light may be represented by a series of plane wavefronts that are distances $c\Delta t$ apart from each other (not shown).

During the time interval Δt between any two wavefronts, the mirror advances by a distance $v\Delta t$. Therefore, with the mirror moving in the direction shown, the lower edges of the wavefronts reach the mirror earlier than they would if the mirror were standing still. This makes the mirror appear as if it were tilted

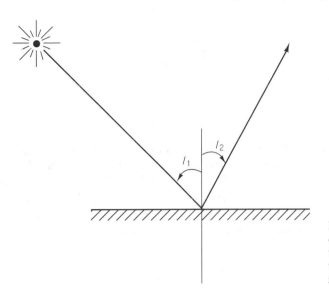

Figure 24-3 Seen from another frame of reference, if the mirror is moving toward the source, the angle of reflection becomes *less* than the angle of incidence.

with respect to the real mirror and reflection returns the light at an angle different from the angle of incidence, as shown. Numerically, if v is the velocity of motion of the mirror, then*

$$\boxed{\frac{\tan \frac{1}{2}I_1}{\tan \frac{1}{2}I_2} = \frac{c + v}{c - v}}$$

[24-9]

Example

What velocity is needed to observe the effect, assuming the angle of incidence is 45°?

Solution: The most sensitive goniometer, that is, angle-measuring instrument, is the Michelson stellar interferometer. It can resolve angles as small as 0.022 arc sec, or $0.022/3600 \approx 0.000\,006°$. Then, setting $I_2 = 45° - 0.000\,006°$ and using Equation [24-9], we have

$$\frac{0.414\,213\,562}{0.414\,213\,501} = 1.000\,000\,147 = \frac{300\,000\,000 + v}{300\,000\,000 - v}$$

$$300\,000\,000 + v = 300\,000\,044 - 1.000\,000\,147\,v$$

---

* Following W. Rindler, *Introduction to Special Relativity*, p. 54 (New York: Oxford University Press, 1982).

so that

$$v \approx \frac{44}{2} = \boxed{22 \text{ m/s}}$$

a surprisingly low velocity.

Doppler Effect

The light reflected by a moving mirror also changes in *wavelength*. That is easy to understand. As the mirror is moving toward the source and against the flow of photons incident on it, the momentum of the mirror raises the energy $h\nu$ of the reflected photons, thus increasing the frequency and reducing the wavelength [Figure 24-4 (left)]. As the mirror moves away and with the flow of the photons, the opposite happens and the wavelength increases (right).

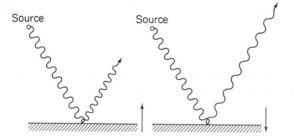

Figure 24-4 With the mirror moving toward the source, the wavelength of the reflected light becomes shorter (*left*); with the mirror moving away, it becomes longer (*right*).

Light reflected by a mirror can be considered as coming from a virtual source located behind the mirror. Thus, as the mirror moves, the virtual source moves as well (at twice the velocity of the mirror). We call x_0 the distance through which the light travels during time interval Δt,

$$x_0 = c\Delta t$$

With the source at rest, that distance contains N waves of length λ_0,

$$x_0 = N\lambda_0$$

and thus

$$\lambda_0 = \frac{x_0}{N} = \frac{c\Delta t}{N} \qquad\qquad [24\text{-}10]$$

But then let the source be moving at velocity v,

$$v = \frac{\Delta x}{\Delta t}$$

The new distance x' contains the same number of waves but now of length λ':

$$x' = x_0 - \Delta x = N\lambda'$$

$$\lambda' = \frac{x_0 - \Delta x}{N} \qquad\qquad [24\text{-}11]$$

Subtract Equation [24-11] from [24-10]; that gives

$$\lambda_0 - \lambda' = \frac{x_0}{N} - \frac{x_0 - \Delta x}{N} = \frac{\Delta x}{N}$$

Substituting $\Delta x = v\Delta t$ and $N = c\Delta t/\lambda_0$ and setting $\lambda_0 - \lambda' = \Delta\lambda$ yields

$$\boxed{\Delta\lambda = \lambda_0 \frac{v}{c}} \qquad\qquad [24\text{-}12]$$

which is the *classical, first-order Doppler effect.** With the source approaching, we find that $\lambda' < \lambda_0$, which gives a "blue shift." With the source receding, we have $\lambda' > \lambda_0$, which gives a "red shift."

Relativity has added a correction to the first-order Doppler effect. The reason is that the distance $\Delta x = v\Delta t$ changes according to Lorentz' transformation and the wavelength of the light (of a receding source) becomes

$$\lambda' = \lambda_0 \sqrt{\frac{1 + v/c}{1 - v/c}} \qquad\qquad [24\text{-}13]$$

which is the longitudinal *relativistic Doppler effect.*

Example

> One of the largest red shifts on record is shown by the galaxy 3C123. There, lines normally found in the UV are shifted into the red, the actual figures indicating that this galaxy is moving away from Earth at a velocity of $0.637\,c$, more than one-half the speed of light. This ties in well with the concept of the "Big Bang," the assumption that the evolution of the Universe started with a cataclysmic explosion and subsequent expansion of matter.

---

*Named after Johann Christian Doppler (1803–1853), Austrian high school mathematics teacher, later professor of experimental physics at the University of Vienna. In his publication, "Ueber das farbige Licht der Doppelsterne und einiger anderer Gestirne des Himmels," *Abh. Königl. böhm. Ges.* (Prag: Borrosch und André, 1842), p. 465, Doppler attributed the different colors of certain stars to their motion toward or away from Earth. He was wrong; the speed of light is so high that in order to change color, the stars would have to move at velocities too high even on an astronomical scale.

The Michelson–Morley Experiment

Of all the interference experiments connected with relativistic optics the *Michelson–Morley experiment* is the most famous.* A Michelson interferometer is set up so that one arm is parallel to the orbital motion of Earth and the other arm normal to it (Figure 24-5). The two arms are adjusted to equal lengths.

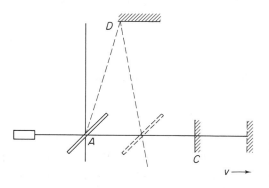

Figure 24-5 Michelson–Morley experiment. Notice similarity to the "light clock" shown earlier in Figure 24-1.

But, due to Earth's motion, the beamsplitter and the two mirrors actually are at positions shown by the dashed lines. The time needed for the light to make a round trip in the upper (perpendicular) arm is

$$t_\perp = \frac{2L}{\sqrt{c^2 - v^2}} = \frac{2L}{c} \frac{1}{\sqrt{1 - (v/c)^2}} \qquad [24\text{-}14]$$

In the parallel arm it is

$$t_\parallel = \frac{L}{c + v} + \frac{L}{c - v} = \frac{2cL}{c^2 - v^2} = \frac{2L}{c} \frac{1}{1 - (v/c)^2} \qquad [24\text{-}15]$$

If we expand both Equations [24-14] and [24-15] by the binomial theorem and drop terms higher than $(v/c)^2$,

$$t_\perp = \frac{2L}{c} \left[1 + \frac{1}{2} \left(\frac{v}{c} \right)^2 \right]$$

---

* Albert A. Michelson, at the time of the experiment, was professor of physics at the Case School of Applied Science in Cleveland; Edward Williams Morley (1838–1923) was professor of chemistry at Western Reserve University, also in Cleveland. The pertinent publication is A. A. Michelson and E. W. Morley, "On the Relative Motion of the Earth and the Luminiferous Aether," *Philos. Mag.* (5) **24** (1887), 449–63. The Michelson–Morley experiment was later repeated with extraordinary care by Joos, with the same result. G. Joos, "Die Jenaer Wiederholung des Michelsonversuchs," *Ann. Phys.* **7** (1930), 385–407. For a detailed account of the experiment see R. S. Shankland, "Michelson–Morley Experiment," *Amer. J. Physics* **32** (1964), 16–35.

and

$$t_{\parallel} = \frac{2L}{c}\left[1 + \left(\frac{v}{c}\right)^2\right]$$

The interferometer is now turned through 90°, which means the time difference between the two positions is

$$\Delta t = t_{\parallel} - t_{\perp} \approx \frac{2L}{c}\left(\frac{v}{c}\right)^2$$

The quantity of interest is the optical path difference, Γ:

$$\Gamma = c\,\Delta t = 2L\left(\frac{v}{c}\right)^2$$

and since the number of fringes, m, is related to Γ as $m = \Gamma/\lambda$,

$$m = \frac{2L}{\lambda}\left(\frac{v}{c}\right)^2 \qquad\qquad [24\text{-}16]$$

In the actual experiment, L was 11 m, $\lambda = 590$ nm, and $v = 30$ km/s. This should have given a fringe shift of

$$m = \frac{(2)(11)}{590 \times 10^{-9}}\left(\frac{3 \times 10^4}{3 \times 10^8}\right)^2 = 0.37 \text{ fringe}$$

which would have been easy to see.

But *there was no fringe shift*. This result of the Michelson–Morley experiment, at that time, came as a complete surprise. Today we know that, according to the second postulate, the velocity of light is the same in all inertial frames and independent of the motion of the source.

Velocity of Light in Moving Matter

As a medium moves, its refractive index changes. That is easy to understand: As the medium moves toward the source, the light encounters more molecules per unit of time and the refractive index appears to be *higher*; as the medium moves away from the source, the refractive index appears to be *lower*.

In *Fizeau's experiment*, shown in Figure 24-6, the light is divided by a beamsplitter B. One beam is traveling clockwise through the path, the other counterclockwise. The beams recombine at B and form interference fringes that are observed through a telescope. Normally, with the water inside the pipe flowing, one of the beams is traveling *with* the direction of flow and the other *against* it. If then the flow of the water is reversed, the fringes are seen to shift.

As before, c is the speed of light in free space and v the velocity of light in a medium of refractive index n (where $v = c/n$). We call w the velocity of the

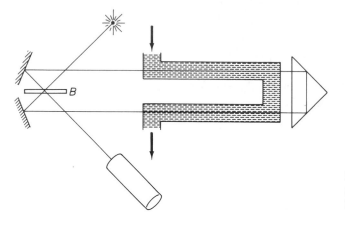

Figure 24-6 Fizeau's experiment for measuring the velocity of light in moving water.

medium and v' the new velocity of the light (in the moving medium). If we merely add the two velocities,

$$v' = \frac{c}{n} + w$$

the new velocity would be too high. Instead, we add the two velocities relativistically,

$$v' = \frac{v_1 + v_2}{1 + v_1 v_2/c^2} \qquad [24\text{-}17]$$

Expansion by the binomial theorem, neglecting second-order terms because $w \ll c/n$, and substituting $v_1 = c/n$ and $v_2 = w$ gives

$$v' = \frac{c}{n} + w \left(1 - \frac{1}{n^2}\right) \qquad [24\text{-}18]$$

The new velocity of the light, therefore, does not change by the full velocity of the moving medium but only by a fraction thereof, with the last term in parentheses, $(1 - 1/n^2)$, called *Fresnel's drag coefficient.**

Rotation Sensors

Closely related to Fizeau's experiment is the *gyroscope*. Modern gyroscopes contain a coil of fibers as shown in Figure 24-7. The light is split into two, one

---

* Fresnel was the first to derive Equation [24-18]: A. Fresnel, "Sur l'influence du mouvement terrestre dans quelques phénomènes d'optique," *Ann. Chim. Phys.* (2) **9** (1818), 57–66. Fizeau then confirmed it, performing the experiment with flowing water: H. Fizeau, "Sur les hypothèses relatives à l'éther lumineux, et sur une expérience qui parait démontrer que le mouvement des corps change la vitesse avec laquelle la lumière se propage dans leur intérieur," *Compt. rendu* **33** (1851), 349–55.

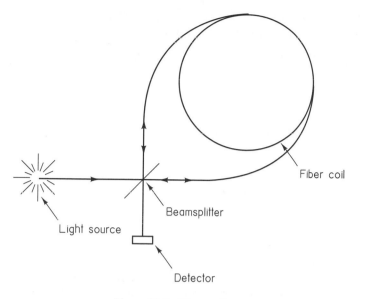

Fiber coil

Beamsplitter

Light source

Detector

Figure 24-7   Fiber-optics gyroscope.

beam traveling clockwise through the fiber, the other counterclockwise. The two beams combine again at the beamsplitter and go to a photodetector. But, since there is one π phase change (on reflection at the front surface of the beamsplitter), the two contributions cancel and the detector registers no light.

But now let the whole apparatus (light source, beamsplitter, coil, and detector) rotate about a vertical axis. We call v the (tangential) velocity of the fiber, n its index, and L the length of one loop. Then the time it takes the light to travel through L in direction of rotation is

$$t_1 = \frac{nL}{c - v}$$

and in the opposite direction it is

$$t_2 = \frac{nL}{c + v}$$

The difference between the two is

$$\Delta t = t_1 - t_2 = \frac{2vnL}{c^2 - v^2} \approx \frac{2vnL}{c^2}$$

Note that tangential velocity and angular velocity are related as $v = \omega R$ and that $L = 2\pi R$, where R is the radius of the loop. Then the time difference is

$$\Delta t \approx \frac{4\pi \omega nR^2}{c^2} = \frac{4A\omega n}{c^2} \qquad [24\text{-}19]$$

where A is the area enclosed by the loop. Within a time interval Δt, light of velocity c proceeds through a path difference Γ,

$$\Gamma = c \, \Delta t$$

Since

$$\Gamma = m\lambda$$

a fringe shift m and the time difference Δt are related as

$$m = \frac{c \, \Delta t}{\lambda} \qquad [24\text{-}20]$$

Inserting Equation [24-19] in [24-20] and assuming that the coil has N loops of fiber gives

$$m \approx \frac{4A\omega nN}{c\lambda} \qquad [24\text{-}21]$$

which is *Sagnac's formula*.* It describes the fringe shift that occurs on rotation. Using such a type of a gyroscope we can monitor any rotation with high precision and provide for correct navigation and stabilization.

De Broglie Waves

The equation

$$E = mc^2 \qquad [24\text{-}22]$$

is Einstein's much quoted *mass-energy relation*. If we combine this expression with Planck's

$$E = h\nu$$

we obtain

$$h\nu = mc^2$$

Replacing ν by v/λ and c by v leads to

$$h\frac{v}{\lambda} = mv^2$$

---

* Georges Marc Marie Sagnac (1869–1928), French physicist, professor of physics at the University of Lille, later at Paris. In one of his experiments, Sagnac used a square path about 1 m on each side. It took a velocity of 120 rpm to see the fringes shift but Sagnac, who retained a lifelong dislike for relativity, attributed the result to the "aether wind." G. Sagnac, "L'éther lumineux démontré par l'effet du vent relatif d'éther dans un interféromètre en rotation uniforme," *Compt. rendu* **157** (1913), 708–10.

and thus

$$\lambda = \frac{h}{mv} \qquad [24\text{-}23]$$

This is an important concept. It assigns a wavelength to particles (moving with a momentum mv), an idea that originated with Louis de Broglie, who thought that if light is acting at some times as waves and at others as particles, perhaps "real" particles would show some wave behavior too. The waves themselves are called *de Broglie waves* and the wavelength *de Broglie wavelength*.*

De Broglie's hypothesis has been extensively tested and completely verified. It applies to all particles, of any size, even to a 10-ton truck, although *its* diffraction pattern can probably never be observed. But electrons, neutrons, and similarly small particles, when incident on a crystal of the proper interplanar spacing, are diffracted much the same as X rays are diffracted—which shows that such "real" particles have wave properties as well.

The Principle of Complementarity

The "dualistic nature of light" is a term that has intrigued many writers and implies some kind of dichotomy. I take a more *unitary* view and think of light as something *unique*. Light is neither a true wave (its oscillations come in finite wavetrains, like no other wave) nor is light made up of conventional particles (quanta can only move at the speed of light, while "real" particles can have any speed less than c). Light's wave nature and quantum nature are not mutually exclusive; they are, as Bohr put it, *complementary*.

Light is both, quanta and waves. The quanta contain the energy. The waves guide them. At shorter wavelengths, and in the emission and absorption of light, the quantum aspect becomes predominant. At longer wavelengths, and in diffraction and interference, the wave aspect becomes predominant. But wavelength is not the divisive factor: Light of the same wavelength can act as waves in one experiment and as quanta in another.

OPTICS AND THE GENERAL THEORY OF RELATIVITY

Newton's law of gravitation tells us that the force F causing acceleration between two masses m_1 and m_2 is given by

$$F = G\,\frac{m_1 m_2}{R^2} \qquad [24\text{-}24]$$

---

* Louis Victor Pierre Raymond Duc de Broglie (1892–1987), French physicist. After graduating from the Sorbonne with a degree in medieval history, de Broglie, pronounced "d' Bro'lie," turned to science. He wrote a thesis on quantum theory, became professor at the Sorbonne, and in 1929 won the Nobel Prize in physics. L. de Broglie, "A Tentative Theory of Light Quanta," *Philos. Mag.* (6) **47** (1924), 446–458.

where G is the universal constant of gravitation, 6.672×10^{-11} N m^2 kg^{-2}, and R is the distance between the two masses. But does a force between masses act instantaneously? Probably not, because that would violate one of the basic tenets of relativity that no signal can go faster than light.

According to the *general theory of relativity*, a gravitational field changes the propagation of the light.* Just as a projectile moves through a gravitational field in the form of a parabola, so does light follow a *curved path* rather than a straight line (Figure 24-8). For light passing by near the edge of the sun, Einstein predicted a deflection of 1.745 arc sec. Indeed, stars close to the solar disk are found to be displaced 1.70 ± 0.10 arc sec away from the sun, a brilliant confirmation of Einstein's theory.

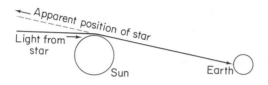

Figure 24-8   Bending of light in a gravitational field.

Sir Arthur Eddington, British astronomer, was once asked: "Is it true, Sir Arthur, that you are one of three men in the world who understands Einstein's theory of relativity?" The astronomer appeared reluctant to answer. "Forgive me," said the questioner, "I should have realized that a man of your modesty would find such a question embarrassing." "Not at all," said Eddington, "I was just trying to think who the third could be."

SUMMARY OF EQUATIONS

FitzGerald–Lorentz length contraction:

$$L' = L_0 \sqrt{1 - \left(\frac{v}{c}\right)^2} \qquad [24\text{-}6]$$

Reflection on moving mirror:

$$\frac{\tan \frac{1}{2}I_1}{\tan \frac{1}{2}I_2} = \frac{c + v}{c - v} \qquad [24\text{-}9]$$

---

* A. Einstein, "Über den Einfluss der Schwerkraft auf die Ausbreitung des Lichtes," *Ann. Phys.* (4) **35** (1911), 898–908.

First-order Doppler effect:

$$\Delta\lambda = \lambda_0 \frac{v}{c} \tag{24-12}$$

De Broglie wavelength of a particle:

$$\lambda = \frac{h}{mv} \tag{24-23}$$

PROBLEMS

24-1. A passenger walks forward in the aisle of a train at a brisk pace, 1.5 m/s. If the train moves along a straight track at 80 km/h, how fast is the passenger moving with respect to ground?

24-2. A gun with a muzzle velocity of 20 m/s, mounted on a vehicle, subtends an angle of 45° with the forward direction. If the vehicle moves at 40 km/h, what is the velocity of the projectile as it leaves the barrel?

24-3. If a 15.3-m-long space vehicle passes a planet at a velocity of 6×10^7 m/s, how long does the vehicle appear to be to an observer on that planet?

24-4. How fast must a meter stick travel for an observer to conclude that it is only one-half as long?

24-5. What is the (total) apex angle of a hypothetical cone equally dividing the light emitted by a point source traveling at 96% of the speed of light?

24-6. What velocity is needed to concentrate one-half of the light into a cone no wider than 5°?

24-7. Light is incident at an angle of 40° on a plane mirror moving toward the source at a velocity of $0.15c$. Find the angle of reflection.

24-8. A plane mirror is receding in a direction normal to its surface at a velocity of one-half the speed of light. Show how the light is reflected if it is incident on the mirror at an angle of:
(a) 30°.
(b) 60°.

24-9. A certain star shows a first-order Doppler shift of 1.4 Å involving the red (656 nm) hydrogen line. How fast does the star move relative to Earth?

24-10. If a distant star is receding from Earth at a velocity of 2.4×10^7 m/s, by how much does the 656 nm hydrogen line shift? Consider the first-order Doppler effect only.

24-11. The spectrum of a distant nebula shows the 434-nm hydrogen line to have shifted to 752 nm. How fast is the nebula moving away from Earth?

24-12. How fast do you have to go through a red traffic light ($\lambda = 600$ nm) so that it appears green (500 nm) to you? Consider the first-order effect only.

24-13. A simple way of introducing the Michelson–Morley experiment is by the *rowboat analogy*. A man is rowing through a distance of 100 m, at a velocity of $c = 0.2$ m/s, across a river flowing at $v = 0.1$ m/s (Figure 24-9).

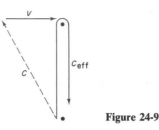

Figure 24-9

(a) What is the effective velocity of the boat?

(b) What is the time needed to complete a roundtrip?

24-14. Continue with Problem 24-13 and now let the man row *parallel* to the river, downstream and upstream, again covering each way a distance of 100 m. How much longer does that take him?

24-15. When Fizeau performed his experiment with water flowing at a rate of 7 m/s:

(a) What drag coefficient did he use?

(b) By how much did the velocity of the light (in water) change?

24-16. An optical system containing a medium with a refractive index of $n = 1.8$ moves at a velocity of 50 km/s toward the sun. What is the new refractive index?

24-17. Sagnac performed his experiment with light of 530 nm wavelength. How much of a fringe shift did he see?

24-18. What is the de Broglie wavelength of an electron ($m = 9.1 \times 10^{-31}$ kg) traveling at 1% of the speed of light?

SUGGESTIONS FOR FURTHER READING

RESNICK, R. *Introduction to Special Relativity*. New York: John Wiley and Sons, Inc., 1968.

RINDLER, W. *Introduction to Special Relativity*, 2nd ed. New York: Oxford University Press, Inc., 1991.

TAYLOR, E. F., and J. A. WHEELER. *Spacetime Physics*: *Introduction to Special Relativity*, 2nd ed. New York: W. H. Freeman, 1992.

Well, dear Reader, we have come to the end of the fourth edition of my *Introduction to Classical and Modern Optics*. Do you have any comments or criticism? If you do, I would appreciate it if you would send them either to the publisher or to me. Many thanks!

Jurgen R. Meyer-Arendt

Answers
to Odd-Numbered
Problems

Chapter 1

1-1. 6.25 Hz
1-3. 4.8×10^{14} Hz
1-5. 4 m
1-7. 6.25 m
1-9. 11.25 mm
1-11. 48.6°
1-13. 1.31
1-15. 1.696
1-17. 60°
1-19. 3 cm
1-23. 9.74°; 0.1437°

Chapter 2

2-1. −1 diopter
2-3. +6 cm
2-5. 128 mm
2-7. −40 cm

2-9. +50 mm
2-11. +4 cm
2-13. 50 cm to the left of the lens
2-15. +20 cm
2-17. +5 diopters
2-19. −0.5×
2-21. 12 diopters

Chapter 3

3-3. +10 diopters
3-5. 25 cm
3-7. 10 diopters; 2 diopters
3-9. at the center of curvature
3-11. 8.1 diopters
3-13. 1.8 diopters
3-15. $\begin{bmatrix} 6 \\ 1 \end{bmatrix}$
3-17. $\begin{bmatrix} 8 \\ 1 \end{bmatrix}$
3-19. +10 cm

Chapter 4

4-1. 91 cm; no change
4-3. −6 cm
4-5. 8 cm
4-7. −45 cm
4-9. real, inverted, and magnified
4-11. −2.75 diopters
4-13. 30 cm
4-15. −20 diopters

Chapter 5

5-1. 0.4 mm
5-3. −1/3; −2/3
5-7. +1.50 diopters
5-9. 16.1 mm
5-11. +2.5 diopters
5-13. 1.6448
5-15. +11.8 diopters; −5.8 diopters

Chapter 6

6-1. *f*/8
6-3. 6.25 m; from 3.125 m to infinity
6-5. 53.13°
6-9. −7.5 cm to the left of the lens
6-11. +12.5 mm to the right of the lens;
15 mm

Chapter 7

7-1. 61 mm
7-3. 2.29°
7-5. 2.8; immaterial
7-7. 72°
7-9. 7.6 mm

Chapter 8

8-1. −18.21°
8-3. 32 mm
8-5. 19.8°
8-7. 24 mm

Chapter 9

9-1. +1.5 diopters; +7.5 diopters
9-3. 16 ×; 2 cm
9-7. 6 cm
9-9. +1.6 ×; 18.75 mm
9-11. 2 mm
9-13. −2.4 cm; 62.5 ×
9-15. −2.5 diopters
9-17. 20 cm in front of the eye; 14.3 cm
9-19. 180 mm

Chapter 10

10-1. 0.0625 lines/mm
10-3. 6.1 diopters
10-5. 5 units
10-7. not much; severe degradation; not
much

Chapter 11

11-1. 480 nm; blue
11-3. 9 mm
11-5. 2.75 μm
11-7. 0.13 sec
11-9. 776 638
11-11. 1.0004

Chapter 12

12-1. 1; 608 nm
12-3. 1; 0; 1; $2\pi = 0$
12-5. 0.02 mm
12-7. 55 cm
12-9. 1.8 mm
12-11. 6.25%
12-13. 0.47%
12-15. 1.70

Chapter 13

13-1. 0.8 mm
13-3. 13.2 μm; 4.4×10^{-14} sec

13-5. 5 units
13-7. 2 m
13-9. 88 m

Chapter 14

14-1. 2λ
14-3. blue inside, red outside
14-5. they look the same
14-7. 8 km
14-9. 4 mm
14-11. 3.6 mm; any wavelength

Chapter 15

15-1. 625
15-3. 500 lines/mm
15-5. prism: blue spread out more; grating: linear
15-7. 3
15-9. 0.25 nm
15-11. 1.735 Å
15-13. 29°

Chapter 16

16-1. 1.5×10^{-7} *T*
16-3. 14.51
16-5. 10%
16-7. 20%
16-9. no better penetration

Chapter 17

17-1. 0° unpolarized; 45° and 135° partial linear
17-3. 50°; 40°
17-5. horizontal
17-7. 75%
17-9. 25%
17-13. 4.15 mm
17-15. 45°

Chapter 18

18-1. 1 wavelength
18-3. **(b)** bottom pattern not resolved
18-5. slot parallel to scan lines
18-7. only the circles' vertical segments are visible

Chapter 19

19-1. 2 μm
19-3. 1 : 1.49
19-5. real near observer, virtual opposite
19-7. with 800 nm more magnification, with 600 nm less

Chapter 20

20-1. 500 nm
20-3. 9.3 μm
20-5. light more blue; vessels darker
20-7. 92 nm
20-9. none; no current; diverted to other maxima
20-11. 400 nm
20-13. 500 nm

Chapter 21

21-1. 1.38 kW
21-3. 53 mW/sr
21-5. 4 W
21-7. no change; no change
21-9. 6.25 lx
21-11. 16 $\times$
21-13. 12.0 lx
21-15. 30 cm

Chapter 22

22-1. 1.3
22-3. 25%
22-5. 36%

22-7. 14.5%

22-9. 21%

22-11. 3.87 m^{-1}

22-13. bright, on a dark background

Chapter 23

23-1. 2.87×10^{-19} J

23-3. plane-parallel; long radius; confocal; concentric

23-5. 20°

23-7. 0.8 mrad

23-9. 0.4 mrad

23-11. 1.5 kW/mm²

Chapter 24

24-1. 85.4 km/h

24-3. 15 m

24-5. 32.52°

24-7. 30°

24-9. 64 km/sec

24-11. $v = \frac{1}{2}c$

24-13. 0.1732 m/s; 19.245 min

24-15. 0.4375; by $\frac{1}{2}$ of the velocity of the water

24-17. $\frac{1}{15}$ of a fringe

Index

Note: Entries in italics refer to biographical and historical information in footnotes.